HEAVILY DOPED SEMICONDUCTORS

MONOGRAPHS IN SEMICONDUCTOR PHYSICS

HEAVILY DOPED SEMICONDUCTORS

Victor I. Fistul'

Institute for Fine Chemical Technology
Academy of Sciences of the USSR, Moscow

Translated from Russian by
Albin Tybulewicz
Editor, *Soviet Physics — Semiconductors*

ℚ PLENUM PRESS • NEW YORK • 1969

Viktor Il'ich Fistul' was born in 1927. A physicist, he completed his higher educa-
tion in 1949, under Academician A. F. Ioffe, in the Physico-Mechanical Department
at the Leningrad Polytechnical Institute. Initially, he was concerned with investigations of physical phenomena in electronic vacuum devices. He was awarded
the degree of Candidate of Technical Sciences in 1957 for a dissertation on the
process of gas evolution from electrical vacuum materials, which he had investigated by mass-spectroscopic methods. Since 1958, Fistul' has been investigating
physical phenomena in semiconductors and attained the degree of Doctor of
Physico-Mathematical Sciences in 1965 for his investigations of heavily doped
semiconductors. At present, in addition to his researches, Professor Fistul'
lectures on the physics of semiconductors and on experimental methods for
investigating these materials at the M. V. Lomonosov Institute for Fine Chemical
Technology in Moscow.

Library of Congress Catalog Card Number 68-28095

The original Russian text, published by Nauka Press in Moscow in 1967 as part
of a series on "Physics of Semiconductors and Semiconducting Devices," has
been revised by the author for this English edition.

Виктор Ильич Фистуль

Сильно легированные полупроводники

SIL'NO LEGIROVANNYE POLUPROVODNIKI

ISBN 978-1-4684-8823-4   ISBN 978-1-4684-8821-0 (eBook)

DOI 10.1007/978-1-4684-8821-0

Preface to the American Edition

Recently, there has been a considerable upsurge of interest in heavily doped semiconductors. This interest is due primarily to the expanding range of applications of such materials. Moreover, the heavy doping of semiconductors produces new effects (the formation of impurity aggregates, the appearance of allowed states in the forbidden band, etc.), which are of great interest in solid-state physics.

The rapid growth in the number of papers on heavily doped semiconductors makes it difficult to review the results obtained so far. Therefore, many investigations carried out in 1966-7, particularly those on $A^{III}B^{V}$ semiconductors, are not discussed in the present monograph, which represents the state of the knowledge in 1965. Nevertheless, the author hopes that, in spite of this, the book will be useful. An attempt is made, first, to review investigations of heavily doped semiconductors from a certain viewpoint and, secondly, to suggest some ideas (Chap. 5) which may be controversial but which are intended to stimulate further studies of heavily doped semiconductors which can be regarded as a special case of disordered systems.

The work of American scientists investigating heavily doped semiconductors, in particular the efforts of E. O. Kane, J. I. Pankove, R. N. Hall, R. A. Logan, W. G. Spitzer, F. A. Trumbore, and many others, is well known to Soviet investigators. It gives me pleasure to learn that Western readers will now have an opportunity to become acquainted with the work done in the USSR. This will be helped considerably by the excellent translation commissioned by Plenum Press, to whom I am greatly indebted.

January 8, 1968   V. Fistul'

Preface to the Russian Edition

An attempt is made in this monograph to summarize the experimental and theoretical data on heavily doped semiconductors. Investigations of these semiconductors have been stimulated by the great variety of semiconductor devices now being made of heavily doped crystals. It has been found that an increase of the impurity concentration makes it possible to study more deeply certain aspects of carrier behavior (scattering, optical transitions). Moreover, basically new physical phenomena have been observed in heavily doped semiconductors; these include: allowed states in the forbidden band, influence of the nature of impurities on the transport phenomena, polytropy (the existence of impurities in a crystal in several forms), etc. A treatise on the theoretical and experimental aspects of heavily doped crystals must include a comparison with lightly doped or pure semiconductors. For this reason, almost every chapter of the book starts with an introductory section, which summarizes briefly the relevant properties of impurity-free semiconductors.

Some chapters (e.g., Chap. 5) are controversial and the views expressed in them will inevitably become outdated. However, if the present monograph helps to accelerate the pace at which current theories are displaced by new and better ones, the author will be well satisfied.

In the author's opinion, a chapter on the preparation of heavily doped semiconductors was required. The author is indebted to M. G. Mil'vidskii, a recognized expert in this field, for writing this chapter (Chap. 6).

The author also cordially thanks his colleagues É. M. Omel'-yanovskii and D. G. Andrianov, who have made considerable contributions to the discussion of many problems dealt with in the present book.

The author is particularly grateful to Professor A. G. Samoilovich and Doctor of Chemical Sciences R. N. Rubinshtein for their valuable advice on many aspects of the preparation of heavily doped semiconductors. The author is also deeply indebted to Mrs. A. D. Fistul' for her generous help in the preparation of the manuscript.

All critical comments will be gratefully received.

V. I. Fistul'

Contents

Introduction

There are many examples in the history of science of researchers returning to a given problem after a considerable lapse of time. In every case, the new approach has been made at a higher level of understanding and knowledge. This is a manifestation of the dialectic nature of progress in general and of progress in science in particular.

This has happened in the physics of semiconductors, as typified by the renewed interest in heavily doped semiconductors, i.e., semiconductors with high impurity concentrations.

When the studies of semiconductors began in the 1930's (in the Soviet Union, they were initiated by A. F. Ioffe), physicists and engineers had to contend with impure semiconductors containing 10^{19}-10^{20} cm^{-3} of foreign impurities.

However, the necessary techniques for preparing very pure semiconducting single crystals were developed fairly rapidly, and the gradual reduction achieved in the impurity concentration was accompanied by the discovery of new properties of these remarkable crystals. These properties were immediately applied in practical devices.

During this period, the theory was based mainly on an ideal semiconductor model. Therefore, the verification of the theoretical conclusions also required pure and perfect single crystals. It is interesting to note that, even during this early period, theoreticians were close to the concept of a p−n junction, but because of

1

the absence of pure single crystals, p−n junctions were first produced as late as 1948, in a transistor. At the same time, improved
semiconductor diodes were developed and since then tetrodes, crystal photodiodes, and many other devices have appeared. The theory of p−n junctions has indicated a close relationship between
the electrical properties of semiconductor devices and the properties of the crystals used in these devices, particularly with the
concentration and nature of the impurity on both sides of a p−n
junction. It has become clear that the technology of semiconductor
devices requires not simply crystals of maximum purity but crystals with impurities introduced deliberately in precisely known
amounts. In fact, postwar semiconductor electronics is based on
the use of doped semiconductors.

Up to 1958, all semiconductor devices were made of lightly
doped semiconductors containing not more than 10^{16} impurities per
cm^3. Only after the appearance of the tunnel diode did semiconductor technology begin to return to crystals containing impurities
in amounts reaching 10^{20} cm^{-3}. In fact, the applications of such
heavily doped semiconductors go well outside the tunnel diode.
These semiconductors are used also in laser and thermoelectric
devices, in semiconductor catalysts, and in some active components of solid−state systems. The term "impurity" should not be
understood in its literal sense. An "impurity" may be any point
structure defect which disturbs the ideal periodicity of the internal
potential field of the lattice. One extreme case of heavily doped
semiconductors is that of disordered alloys. Therefore, heavily
doped semiconductors form only a part of the more general problem of disordered structures.

We must now define the concepts of lightly and heavily doped
semiconductors.

These concepts have two aspects: electronic and atomic.

The electronic aspect is associated with the fact that
when the degree of doping with donor and acceptor impurities is increased, the carrier density in the semiconductor also increases.

At low impurity concentrations, local energy states are
formed in the forbidden band of a semiconductor. Since the impurity atoms are far apart, there is no interaction between them.

When the numbers of carriers in the allowed bands are small, these carriers obey the Boltzmann statistics, which ignores the quantum indistinguishability of particles.

The level of doping at which there is practically no interaction between impurity atoms (and the carrier gas obeys the Boltzmann statistics) can be called a low degree of doping and the corresponding conductors are called lightly doped.

When the degree of doping is increased beyond this level, the average distance between the impurity atoms decreases and consequently these atoms begin to interact, i.e., the wave functions of the electrons of neighboring impurity centers begin to overlap. In this case, local energy levels broaden out to form an impurity band. This process is accompanied by a decrease in the ionization energy of the impurity atoms. Such a degree of doping may be called moderate. In this case, nothing definite can be said a priori about the laws governing the energy distribution of carriers in the impurity band and in the allowed bands because these laws depend on the structure of the impurity band and on the relative number of carriers in the allowed states.

When the impurity concentration is reduced still further, the ionization energy falls to zero, i.e., the impurity band merges with the conduction band (in the case of donor impurities) or with the valence band (in the case of acceptor impurities). Thus, a single allowed band is formed in a crystal. This is the case of a high doping level and the corresponding crystals are called heavily doped.

The high density of carriers in heavily doped crystals requires the application of the Fermi−Dirac statistics. Since a gas of particles obeying such statistics is known as degenerate, the term "heavily doped semiconductors" is frequently identified with "degenerate semiconductors." This is not quite correct because, for example, a crystal may contain a number of impurities such that the electron gas is degenerate at room temperature but at high temperatures the degeneracy is lifted because of the appearance of intrinsic conduction.

The atomic aspect. An increase in the interaction between impurity atoms at high doping levels has other consequences, apart from the formation of an impurity band. The interaction can be more "chemical," i.e., it may result in the formation of aggregates of impurity atoms or of impurity atoms and host matrix atoms, which may be precipitated as second-phase occlusions provided the concentration is sufficiently high. If the concentration is insufficient for such precipitation, the aggregates present in a single-phase solution may have a considerable influence on the properties of a crystal due to the interaction structure defects or due to the formation of electrically inactive centers (electronic aspect).

Dopant atoms are fully ionized in the heavy doping case and the large number of ions affects the behavior of electrons and holes – if only by scattering.

Thus, the electronic and atomic aspects are closely related and are frequently manifested simultaneously. The electronic aspect has been treated more thoroughly in the published literature.

The present author has investigated both aspects of heavily doped semiconductors, which is reflected in the present monograph.

Chapter 1

Energy Spectrum of Electrons in Heavily Doped Semiconductors

§1.1. Preliminary Remarks

Schrödinger's Equation. In an isolated atom, such as hydrogen, only the valence electron and the nucleus interact. Therefore, Schrödinger's equation, which expresses essentially the law of conservation of energy, has the following simple form:

$$\frac{\hbar^2}{2m}\frac{d^2\psi(r)}{dr^2} + V_0(r)\psi(r) = E\psi(r). \qquad (1.1.1)$$

The first term on the left-hand side of this equation represents the kinetic energy of the interaction between the electron and the nucleus and the second term represents the potential energy. The sum of these energies is equal to the total energy of the system $E\psi(r)$. The solution of Eq. (1.1.1) yields plane waves of the type

$$\psi(r) = e^{\pm\, ikr}, \qquad (1.1.2)$$

where $\mathbf{k}$ is the wave vector of the electron, equal to $\mathbf{k}=\mathbf{p}/\hbar$, where $\mathbf{p}$ is the electron momentum.

Substituting Eq. (1.1.2) into Eq. (1.1.1), we can easily obtain the value of the kinetic energy of the electron $E_{kin} = E - V_0$, i.e., its energy spectrum.

A semiconductor crystal is a system of many electrons and nuclei. The fullest information on the properties of such a system, including its energy spectrum, can be obtained by solving Schrödinger's equation for the stationary states of this system. However, we then have to deal with a large number of interactions: electrons with other electrons, nuclei with other nuclei, and electrons with nuclei. Therefore, Schrödinger's equation should be written in the following form:

$$\left\{ -\frac{\hbar^2}{2m} \sum_i \frac{\partial^2}{\partial r_i^2} - \frac{\hbar^2}{2M} \sum_j \frac{\partial^2}{\partial R_j^2} + \sum_{j<n} \frac{Z_j Z_n e^2}{R_{jn}} + \sum_{i<k} \frac{e^2}{r_{ik}} - \sum \frac{Z_j e^2}{r_{ij}} \right\} \times$$

| kinetic energy of electrons | kinetic energy of nuclei | potential energy of interaction between nuclei | potential energy of interaction between electrons | potential energy of interaction between electrons and nuclei |

$$\times \psi(r_i, R_j) = W\psi(r_i, R_j), \qquad (1.1.3)$$

where m, M are, respectively, the electron and nuclear masses; r_i, R_j are the radius vectors of the i-th electron and j-th nucleus; Z_j, Z_n are the atomic numbers of nuclei; R_{jn}, r_{ik}, r_{ij} are the distances between the corresponding nuclei and electrons. It is impossible to obtain an exact solution of Eq. (1.1.3) because the wave function ψ depends on an enormous number ($\sim 10^{23}$) of independent variables. Two approximations can be used to simplify the situation: 1) the adiabatic approximation, in which it is assumed that all the nuclei are fixed relative to the electrons because $M \gg m$ – this allows us to drop the second and third terms from Eq. (1.1.3); 2) the one-electron approximation, in which all the electrons, except one, are replaced by an effective charge – in this case the potential energy of all the remaining electrons [the fourth and fifth terms in Eq. (1.1.3)] is replaced by some external self-consistent field $V(r)$, which gives the best description of the average interaction of all the remaining electrons with the one electron being considered.

Thus, Eq. (1.1.3) reduces to an equation which is of the same form as Eq. (1.1.1): the only difference between them is that $V_0(r)$ is replaced with a potential function $V(r)$. The aim of the theory is to obtain the best form for this function.

We shall deal with a one-dimensional crystal resulting from the condensation (mutual approach) of isolated atoms (Fig. 1.1a),

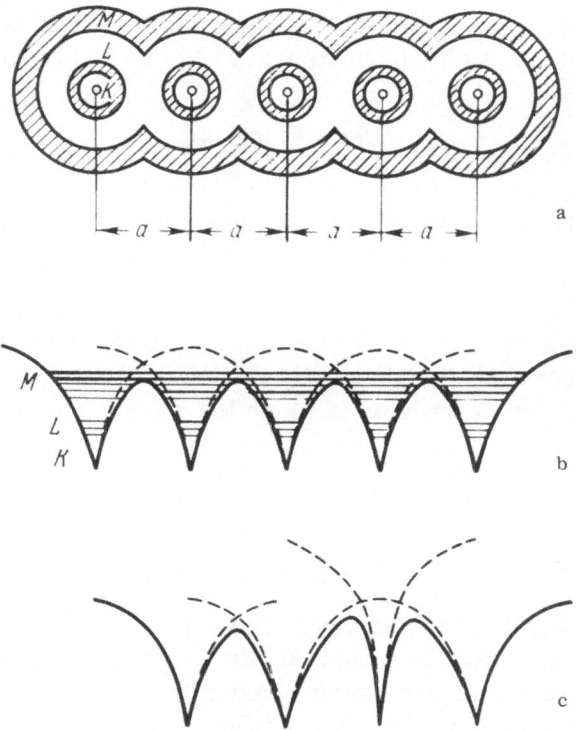

Fig. 1.1. Graphical representation of the potential energy of
an electron in a crystal. a) Distribution of atoms; b) potential energy; c) local departure from a periodic field in
a crystal (the dashed curves represent the potential energies
of free atoms).

separated from one another by the same distances, equal to the lattice parameter a. Simple summation of the values of the potential
functions of individual atoms gives the functions V(r), whose distinguishing property is its periodicity (Fig. 1.1b), i.e.,

$$V(r) = V(r + a).$$

In this case, the solution of Schrödinger's equation (1.1.1)
gives, instead of Eq. (1.1.2),

$$\psi(r) = U_k(r)\, e^{\pm ikr}, \tag{1.1.4}$$

i.e., the solution is in the form of plane waves modulated by a function $U_k(r)$, which is also periodic and whose period is equal to the lattice constant:

$$U_k(r) = U_k(r+a).$$

This result is frequently called the Bloch theorem [1] and the functions (1.1.4) are known as the Bloch functions.

Brillouin Zone. We can show that the electron states with wave vectors k and $k' = k + 2\pi/a$ are physically indistinguishable. Therefore, all the physically distinguishable states of an electron are included in a region in which the wave vector k is given by

$$-\frac{\pi}{a_\alpha} \leqslant k_\alpha \leqslant +\frac{\pi}{a_\alpha},$$

where the subscript "α" is used to denote x, y, z.

This region in the k space containing all k, which represent different states of an electron, is known as the first Brillouin zone. When the wave vector has values ranging from $-2\pi/a$ to $-\pi/a$ and from $+\pi/a$ to $+2\pi/a$, we have the Brillouin zone, etc.

For a cubic lattice, the first Brillouin zone is cube shaped. The vector $1/a$ is frequently called the reciprocal lattice vector.

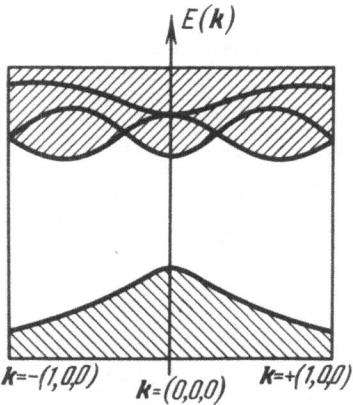

Fig. 1.2. Energy band structure of germanium along the [100] axis in the Brillouin zone.

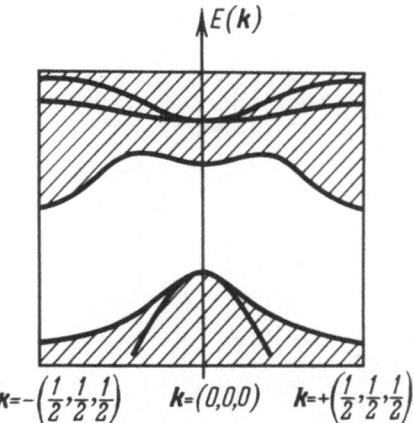

$$K=-\left(\frac{1}{2},\frac{1}{2},\frac{1}{2}\right) \qquad K=(0,0,0) \qquad K=+\left(\frac{1}{2},\frac{1}{2},\frac{1}{2}\right)$$

Fig. 1.3. Energy band structure of germanium
along the [111] axis in the Brillouin zone.

Energy Bands in Crystals. To find the energy spec-
trum of electrons, i.e., the eigenvalues of E, it is necessary to solve
Eq. (1.1.1) containing a definite function $V(\mathbf{r})$. Various methods for
solving this equation have been proposed. There is the method of
weakly bound electrons [2], in which the periodic potential is regarded as a small perturbation of the free motion of an electron.
The method of tightly bound electrons [3] is based on the use of
the wave functions of the electrons of an isolated atom and the influence of all the other electrons is allowed for in the form of a perturbation. The method of orthogonalized plane waves (OPW) [4]
postulates a function $V(\mathbf{r})$ which represents an almost ideal plane
wave in the space between the ions in the lattice and simultaneously
allows for the rapid fall of $V(\mathbf{r})$ near an ion. Thus, in the OPW
method, we use a periodic potential function with properties intermediate between the properties of the functions used in the methods
of weakly and tightly bound electrons.

Investigations carried out using all these methods show that
the energy spectrum of electrons is complex and represents alternating regions of allowed and forbidden energy values, which are
known as bands. The band pattern depends on the crystallographic
direction in a crystal.

The best results, confirmed experimentally, are given by the
OPW method. Thus, Herman has calculated the energy bands of
germanium and silicon [5-7]. By way of example, we shall consider
the solution for germanium (Figs. 1.2 and 1.3).

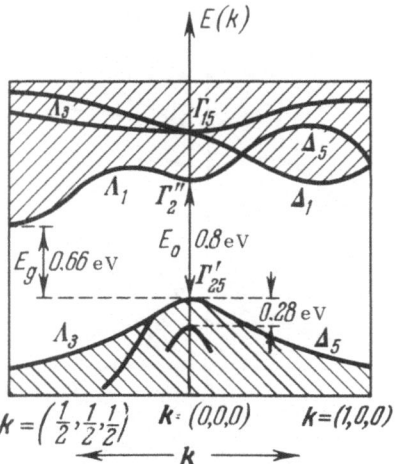

Fig. 1.4. Energy band structure of germanium calculated by the OPW method. The values of the energy gaps are experimental.

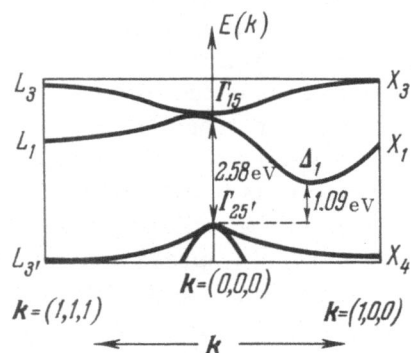

Fig. 1.5. Energy band structure of silicon, calculated by the OPW method. The values of the energy gaps are experimental.

Since the $E(\mathbf{k})$ curves are symmetrical for $0 \leq k \leq +\pi/a$ and $0 \geq k \geq -\pi/a$, it is usual to show $E(\mathbf{k})$ for the [111] axis in the Brillouin zone in the left-hand quadrant and for the [100] axis in the right-hand quadrant. Thus, instead of Figs. 1.2 and 1.3, we obtain one figure (Fig. 1.4). Similar representations of the energy spectra of silicon and $A^{III}B^{V}$ semiconductors are shown in Figs. 1.5

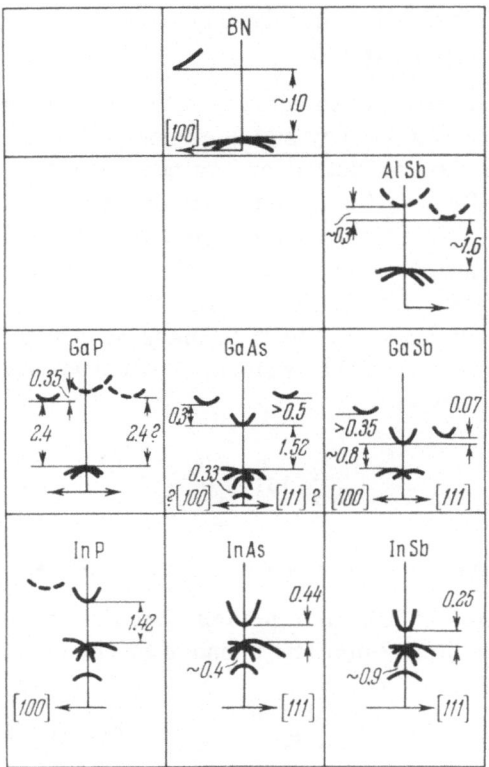

Fig. 1.6. Energy band structure of $A^{III}B^V$ compounds.
The values of the energy gaps are experimental [17].

and 1.6.* The numerical values of the energy gaps in these figures are taken from experimental data.

Energy Spectrum of Electrons in Ge and Si.
It is evident from Figs. 1.4 and 1.5 that the absolute minima of the conduction bands of Ge and Si do not lie at the point k = 0. Silicon has six symmetrically distributed energy minima at points k lying on the [100] axes. Germanium has eight equivalent minima located along the [111] directions on the Brillouin zone boundary. Since these minima lie on the Brillouin zone boundary, they naturally be-

*The letters Γ, Λ, Δ, X, and L in Figs. 1.4 and 1.5 represent points lying on the axes and planes of symmetry in the Brillouin zone of the diamond-like lattice.

long to two Brillouin zones simultaneously so that we may assume that one zone has four minima.

The dependences of the energy on the momentum and the shapes of the constant-energy surfaces near the energy extrema are different for the conduction and valence bands. When we consider the actual form of these dependences, it is convenient to use the concept of the effective mass of electrons in a solid. We may show that the equation of motion of an electron in the perfectly periodic field of the ideal lattice is similar to the equation of motion of a free electron if the electron mass m is replaced by a tensor quantity, known as the reciprocal effective mass tensor. This quantity is defined as

$$\frac{1}{m^*_{\alpha,\beta}} = \frac{1}{\hbar^2}\frac{\partial^2 E\,(\boldsymbol{k})}{\partial k_\alpha \partial k_\beta}, \tag{1.1.5}$$

where α and β are the directions of the principal axes of the tensor.

An expansion of $E(\boldsymbol{k})$ as a series near the energy extrema gives a quadratic dispersion law because the function $E(\boldsymbol{k})$ is even:

$$E\,(\boldsymbol{k}) = E\,(\boldsymbol{k}_0) \pm \frac{1}{2}\sum \frac{\partial^2 E\,(\boldsymbol{k})}{\partial k_\alpha \partial k_\beta}\bigg|_{k\,=\,k_0}(k_\alpha - k_{0\alpha})\,(k_\beta - k_{0\beta}). \tag{1.1.6}$$

The sign "+" designates an expansion near an energy minimum and the sign "−" designates an expansion near a maximum.

In this case, the reciprocal effective mass tensor

$$\frac{1}{m^*_{\alpha,\beta}} = \frac{\partial^2 E\,(\boldsymbol{k})}{\hbar^2 \partial k_\alpha \partial k_\beta}\bigg|_{E\,=\,0} \tag{1.1.7}$$

is independent of k, i.e., for an arbitrary value of k the energy $E(\boldsymbol{k})$ is governed by the same parameters $m^*_{\alpha,\beta}$.

In the general case of an arbitrary dependence $E(\boldsymbol{k})$, the parameters $m^*_{\alpha,\beta}$ are different.

Reducing the tensor $1/m^*_{\alpha,\beta}$ to the principal axes, we obtain from Eq. (1.1.6)

$$E\,(\boldsymbol{k}) - E\,(\boldsymbol{k}_0) = \pm\frac{\hbar^2}{2}\sum_\alpha \frac{(k_\alpha - k_{0\alpha})^2}{m^*_\alpha}. \tag{1.1.8}$$

Thus, in the absence of electron-state degeneracy, the constant-energy surfaces $E(\mathbf{k}) = $ const near the energy extrema are ellipsoids. In the special case of a completely isotropic scalar effective mass, Eq. (1.1.8) becomes

$$E(\mathbf{k}) = \frac{\hbar^2 k^2}{2m^*},\tag{1.1.9}$$

i.e., the ellipsoid becomes a sphere.

The constant-energy surfaces of the conduction bands of germanium and silicon are ellipsoids of revolution with symmetry axes directed, respectively, along the [111] and [100] axes

$$E(\mathbf{k}) = \frac{\hbar^2}{2}\left|\frac{k_x^2 + k_y^2}{m_\perp^*} + \frac{k_z^2}{m_\parallel^*}\right|,\tag{1.1.10}$$

where $m_\parallel^*$ is the component of the effective mass tensor along a symmetry axis, $m_\perp^*$ is the component of the effective mass tensor at right angles to such an axis.

The energy maxima of the valence bands of germanium and silicon lie in the center of the Brillouin zone at the point $k = 0$. Since this point is doubly degenerate, the dependence $E(\mathbf{k})$ is more complex than that given by Eq. (1.1.8):

$$E_{1,2}(\mathbf{k}) = -\frac{\hbar^2}{2m_0}[Ak^2 \pm \sqrt{B^2 k^4 + C^2(k_x^2 k_y^2 + k_x^2 k_z^2 + k_y^2 k_z^2)}],\tag{1.1.11}$$

where A, B, and C are constants. The expression with the positive square root describes the spectrum of the heavy holes in the first valence band, and the expression with the negative root describes the spectrum of the light holes in the second valence band.

The constant-energy surfaces corresponding to Eq. (1.1.11) are no longer ellipsoids. In the case of germanium, the sections of such surfaces cut, for example, by a (100) plane have the form shown in Fig. 1.7. Writing Eq. (1.1.11) in spherical coordinates and averaging over angles, we can obtain

$$E_{1,2}(\mathbf{k}) \approx -\frac{\hbar^2 k^2}{2m_0}\left[A \pm \sqrt{B^2 + \frac{C^2}{5}}\right],$$

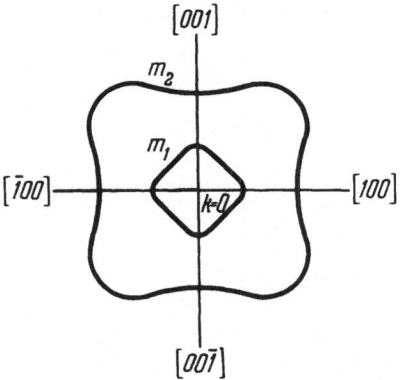

Fig. 1.7. Constant-energy surfaces for the
two upper valence bands of germanium,
shown as sections made by (100) plane in
the k space; m_1 denotes the light-hole band,
m_2 the heavy-hole band.

or

$$
\begin{aligned}
E_1(\boldsymbol{k}) &\approx -\frac{\hbar^2 k^2}{2m_1^*}, \\
E_2(\boldsymbol{k}) &\approx -\frac{\hbar^2 k^2}{2m_2^*},
\end{aligned} \right\}
$$

(1.1.12)

where $m_{1,2}^*$ are the scalar effective masses of the light and heavy
holes, defined as

$$
m_{1,2} = \frac{m_0}{A \pm \sqrt{B^2 + \dfrac{C^2}{5}}}.
$$

In the third valence band, split-off by the spin-orbital interaction, the dispersion law is quadratic and isotropic

$$
E_3(\boldsymbol{k}) = -\Delta E - \frac{\hbar^2 k^2}{2m_3^*},
$$

where ΔE is the energy gap between the third valence band and the
bands 1 and 2 at the point $k = 0$.

TABLE 1.1

	$m^*_\parallel/m_0$	$m^*_\perp/m_0$	$K_m = \dfrac{m^*_\parallel}{m^*_\perp}$	m^*_1/m_0	m^*_2/m_0
Germanium	1.58	0.08	19.3	0.33	0.042
Silicon	0.98	0.19	5.15	0.49	0.16
	m^*_1/m^*_2	m^*_3	A	B	C
Germanium	7.9	0.077	13.1 ± 0.4	8.3 ± 0.6	12.5 ± 0.5
Silicon	3.1	0.245	4.0 ± 0.1	1.1 ± 0.4	4.1 ± 0.4

We shall also quote an expression of the type

$$E = E_c + \frac{\hbar^2 k^2}{2m^*}[1 - \alpha k^2], \qquad (1.1.13)$$

which describes the energy near a minimum of a "nonstandard" band. At the bottom of such a band, the effective mass is very low but since the energy increases more slowly than the square of the wave vector, the states with higher energies correspond to a larger effective mass.

Thus, Eq. (1.1.13) represents an isotropic but nonparabolic dispersion law.

The value of α gives the degree of deviation of the conduction band from parabolicity.

Experimental Data on the Band Structure. The main experiments which have confirmed the correctness of theoretical calculations of the energy band structures of germanium and silicon are the cyclotron resonance measurements [8, 9], which have given the values of the effective masses of electrons and holes as well as the parameters A, B, and C. The results of these experiments are presented in Table 1.1.

The existence of two types of hole in the valence bands of germanium was discovered by Willardson, Harman, and Beer [10] in an investigation of the Hall effect.

Also the value of the anisotropy coefficient K_m has been determined in an investigation of the magnetoresistance [11, 12]. Comparison of the experimental and theoretical data on the absorption in the infrared range has yielded values of the spin-orbital splitting ΔE of the third valence band at the point $k = 0$. These quantities are shown in Figs. 1.4 and 1.5). To determine the thermal (i.e., minimal) and optical (k = const) widths of the forbidden band, a series of experiments has been carried out, in which the most accurate values have been obtained by Macfarlane and his colleagues [13, 14]. These values are also given in Figs. 1.4 and 1.5.

Energy Structure of $A^{III} B^V$ Semiconductors. The energy band structure of $A^{III}B^V$ semiconducting compounds has been investigated less fully than the structures of germanium and silicon. However, many properties of semiconductors depend only on the shape of the conduction band in the region of a minimum and the shape of the valence band in the region of a maximum, so even limited information is of great importance.

Theoretical investigations of the band structures of $A^{III}B^V$ compounds have been usually carried out using Herman's perturbation method [15]. This method makes it possible to obtain the principal characteristics of the band structure of $A^{III}B^V$ crystals, using the band structure of a diamond-like elemental semiconductor, occupying an intermediate position between the components A and B of an $A^{III}B^V$ compound belonging to the same period in Mendeleev's table. Thus, the band structure of GaAs can be calculated by the perturbation method from the band structure of Ge, which occupies an intermediate position between Ga and As. Silicon can be used in the same way for AlP. However, if an $A^{III}B^V$ compound has components which belong to different periods, for example, AlSb, its band structure must be calculated using an imaginary component SiSn.

TABLE 1.2. Temperature Coefficients of the Forbidden Band Widths of $A^{III}B^V$ Compounds [17] (100-300°K)

Compound	InSb	InAs	InP	GaSb	GaAs	GaP	AlSb
$-\dfrac{dE_g}{dT} \cdot 10^4$ eV	2.9	3.5	4.5	4.3	4.3	5.4	4.1

TABLE 1.3. Values of Effective Carrier Masses
in $A^{III}B^V$ Compounds near Allowed Band Extrema

Compound	InSb	InAs	InP	GaSb	GaAs	GaP	AlSb
m_n^*/m_0	0.014	0.02	0.073	0.047	0.078	0.12	0.09
m_{p1}^*/m_0	0.6?	0.41		$\sim$0.5	$\sim$0.5		0.4
m_{p2}^*/m_0	0.18	0.025			0.01		

The perturbation method shows that the internal potential $V^{III-V}(\mathbf{r})$ consists of the internal potential representing a diamond-like semiconductor $V^{IV}(\mathbf{r})$ and a perturbing correction $\Delta V^{III-V}(\mathbf{r})$:

$$V^{III-V}(\mathbf{r}) = V^{IV}(\mathbf{r}) + \Delta V^{III-V}(\mathbf{r}).$$

Using the symmetry of the perturbation potential together with the experimental data, Herman has determined [16] the main parameters of the band structures of some $A^{III}B^V$ compounds. Thus, Herman's method is semiempirical.

The accumulated theoretical and experimental data make it possible to predict the probable band structures of cubic $A^{III}B^V$ compounds in the form shown in Fig. 1.6. The values of the energy gaps given in this figure refer to 0°K. The value of E_g at 300°K can be easily found using the coefficient dE_g/dT, whose values are listed in Table 1.2.

Very little is known about the existence or positions of the bands shown dashed in Fig. 1.6.

Comparison of the band structures shown in Fig. 1.6 with the band structures of germanium and silicon (Figs. 1.4 and 1.5) shows that they are of the same type but the $A^{III}B^V$ compounds have a number of distinguishing features. The valence bands of the $A^{III}B^V$ compounds, like those of germanium and silicon, are degenerate at the center of the Brillouin zone and consist of a heavy-hole band, a light-hole band, and a third band, split-off by the spin-orbital interaction. However, the absolute maxima of the valence bands do not lie in the center of the Brillouin zone but at a small distance from it. In the case of the $A^{III}B^V$ compounds with relatively large average atomic numbers (GaAs, GaSb, InAs, InP, InSb), the absolute minima of the conduction bands lie at the center of the Brillouin zone at the point k(0, 0, 0). In the case of $A^{III}B^V$ materials with a small value of the average atomic number (AlSb, GaP), the absolute minima of the conduction bands lie along the [111] and [100] axes in the Brillouin zone.

The effective carrier masses at the bottom of the conduction band or at the top of the valence band are given in Table 1.3 [17, 18].

$A^{III}B^V$ compounds with large atomic numbers, have nonparabolic conduction bands because the forbidden band is narrow and therefore the valence and conduction bands interact.

Density of States in the Allowed Bands [19]. This density, i.e., the number of allowed states per unit energy interval, is an important characteristic of a crystal. It can be used to find the carrier density in the allowed bands, the average carrier energy, etc.

The number of allowed states in an infinitely small energy interval $(E, E + \Delta E)$ is given by the expression

$$\rho(E)\,dE = \frac{V}{(2\pi)^3} \sum_n \int\int\int_{E \leq E(k) \leq E + \Delta E} dk_x dk_y dk_z. \qquad (1.1.14)$$

The summation is carried out over all the energy bands. An element of integration volume in the k space can be represented in the form

$$dk_x dk_y dk_z = d\sigma\,dk_\perp,$$

where $d\sigma$ is an element of the constant-energy surface and $dk_\perp$ is an increment in the wave vector $\mathbf{k}$ along the normal to this surface. Since $|\mathrm{grad}_k E| = \partial E / \partial k_\perp$, it follows that

$$\rho(E) = \frac{V}{(2\pi)^3} \sum_n \int \frac{d\sigma}{|\mathrm{grad}_k E|}. \qquad (1.1.15)$$

We shall now consider various dependences $E(\mathbf{k})$.

1. Simple parabolic and isotropic dispersion law, defined by Eq. (1.1.9).

In this case, the constant-energy surface is a sphere with a surface area $\sigma = 4\pi k^2$. Then, the density of states per 1 cm^3 ($V = 1$ cm^3) is found from Eq. (1.1.15):

$$\rho(E) = \frac{1}{\sqrt{2}\,\pi^2}\,\frac{m^{*3/2}\sqrt{E}}{\hbar^3}. \tag{1.1.16}$$

2. The dispersion law is parabolic but anisotropic [Eq. (1.1.10)] and the constant-energy surfaces are ellipsoids of revolution. This case represents the structure of the conduction bands of germanium and silicon.

In this case, we shall use new variables

$$k_x' = \frac{k_x}{\sqrt{m_\perp^*}}\,; \qquad k_y' = \frac{k_y}{\sqrt{m_\perp^*}}\,; \qquad k_z' = \frac{k_z}{\sqrt{m_\parallel^*}}\,,$$

and we thus transform Eq. (1.1.10) to a form similar to that of Eq. (1.1.9):

$$E(\boldsymbol{k}) = \frac{\hbar^2\,(k')^2}{2}.$$

Moreover, since $dk_x dk_y dk_z = (m_\perp^2\,m_\parallel)^{1/2} dk_x'\,dk_y'\,dk_z'$, it follows that

$$\rho(E) = \frac{\nu}{\sqrt{2}\,\pi^2}\,\frac{(m_\perp^2\,m_\parallel)^{1/2}}{\hbar^3}\,\sqrt{E}, \tag{1.1.17}$$

where ν is the number of equivalent minima (4 for germanium and 6 for silicon).

The expression (1.1.17) can be written also in the same form as Eq. (1.1.16) by introducing the concept of the effective density-of-states mass m_N^*, which is defined as

$$m_N^* = \nu^{2/3}\,(m_\perp^2 m_\parallel)^{1/3}$$

[we must mention that sometimes m_N^* is defined simply as $(m_\perp^2 m_\parallel)^{1/3}$ without the factor $\nu^{2/3}$].

3. The dispersion law is isotropic but nonparabolic, as in Eq. (1.1.13). This case represents the conduction band of InSb. The

density of states is found approximately using an expression derived in [25]:

$$\rho(E) \approx 4\pi \left(\frac{2m^*}{\hbar^2}\right)^{3/2} V \bar{E} \left(1 + \frac{5am^*E}{\hbar^2}\right). \tag{1.1.18}$$

The condition of the validity of this expression is the inequality

$$E \ll \frac{\hbar^2}{am^*}.$$

§1.2. Energy Spectrum of Electrons in a Doped Semiconductor (Low Doping Level)

Impurity Levels. In a perfect infinite crystal, there are no electron energy levels in the forbidden band. However, in real crystals there are always some departures from a perfect periodic field, such as those shown in Fig. 1.1c. Such local departures may be due to a great variety of causes. One of the simplest departures is in the form of a foreign atom with which the crystal lattice is doped; for example, atoms of elements of groups III and V in the periodic table may be introduced into germanium or silicon and atoms of groups II and VI may be introduced into $A^{III}B^V$ semiconductors. These atoms form substitutional solid solutions with the host semiconductors.

By way of example, we shall consider a substitutional solid solution of As in Ge. Arsenic is pentavalent, i.e., its outer shell has five electrons. With the exception of the fifth electron, an As ion is very similar to a Ge ion, so that in the first approximation we may assume that the structure of the energy levels of the whole crystal remains practically unchanged if Ge atoms are replaced with As atoms at only very few positions. The excess – fifth – electron of As, which does not take part in the four covalent bonds in the lattice, is acted upon by the periodic field of the lattice and by the Coulomb field of the As^+ ions. The potential of this Coulomb field is weaker than the usual potential of ions in vacuum, because of the influence of the crystal as a whole. In a rough approximation, an impurity atom may be considered as being immersed in a medium of permitivity $\varkappa$. Then, the ion potential is

$$U(r) = \frac{e^2}{\varkappa r}.$$ (1.2.1)

Since the Coulomb potential varies fairly slowly at distances of the order of the lattice parameter, the problem of the energy spectrum of electrons in such a potential can be solved in the effective mass approximation, i.e., we can solve Schrödinger's equation (1.1.1) in which the potential function $V_0(\mathbf{r})$ is replaced with Eq. (1.2.1). If we also add the condition that the constant-energy surfaces are spherical, we find that the solution of this equation yields wave functions and energy values similar to the wave functions and energy values of a free hydrogen atom except that the electron mass and charge should be replaced with their effective values m* and e* $= e/\sqrt{\varkappa}$.

If the electron is bound near an impurity ion, its energy is [20]:

$$E_d = \frac{m^* e^{*4}}{2\hbar^2 n},$$

where n is the principal quantum number.

Bearing in mind that the first ionization potential of a free hydrogen atom is

$$U_i = \frac{m_0 e^4}{2\hbar^2 n^2} = 13.5 \text{ eV},$$

we obtain:

$$E_d = -13.5 \left(\frac{m^*}{m_0}\right) \frac{1}{\varkappa^2}.$$ (1.2.2)

Since the electron energy is usually measured from the bottom of the conduction band, a level with negative energy E_d must lie in the forbidden band below the conduction band.

If the effective mass is assumed to be equal to the free-electron mass, it follows from Eq. (1.2.2) that the energy necessary for the ionization of an As atom should be 0.0508 eV in germanium and 0.0908 eV in silicon.

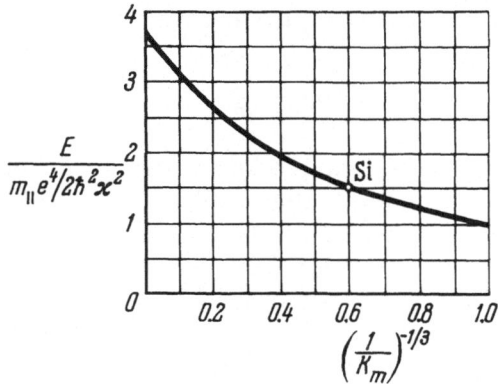

Fig. 1.8. Dependence of the impurity ionization energy
on the anisotropy factor of the effective electron mass.

A similar result is obtained in the case of acceptor impurities which are typically atoms of group III introduced into germanium or silicon. These atoms form local energy levels lying in the forbidden band above the top of the valence band. To ionize such acceptors we require an energy E_a, which – according to Eq. 1.2.2 – should be equal to E_d.

A more exact theory, which makes it possible to calculate quantitatively the impurity ionization energies E_d and E_a, has been developed allowing for the nonsphericity of the constant-energy surfaces. Kittel and Mitchell [21], and Kohn and Luttinger [22–25] have solved this problem for ellipsoidal constant-energy surfaces of germanium and silicon. They have shown that there is a simple function F

$$F = (\pi a^2 b)^{-1/2} \exp\left[-\left(\frac{x^2+y^2}{a^2} + \frac{z^2}{b^2}\right)^{1/2}\right], \qquad (1.2.3)$$

which, for suitably selected parameters a and b, gives a very good approximation for all values of $K_m = m_\parallel/m_\perp$ in the range $0 \le K_m^{-1} \le 1$. Using this function, Kohn and Luttinger [25] have obtained a dependence shown in Fig. 1.8.

For silicon, $E_d = -0.029$ eV when $K_m^{-1} = 0.19$, and for germanium, $E_d = -0.008$ eV when $K_m^{-1} = 0.05$. In the case of GaAs and InSb, the value of E_d is given by the simple expression (1.2.2) be-

TABLE 1.4. Values of Donor Level
Energies in Si (Measured from $E = E_C$)

States	$E \cdot 10^2$, eV		
	P	As	Sb
Ground	—4.4	—4.9	—3.9
	—3.2	—3.3	—3.1
	—1.13	—1.13	—1.13
Excited	—1.06	—1.11	—0.94
	—0.93	—0.95	—0.90
	—0.59	—0.59	—0.59
	—0.57	—0.57	—0.57

cause cyclotron resonance experiments have indicated no effective anisotropy in these compounds [17].

A similar problem has been solved by Kohn and Schechter [26] for the acceptor levels in germanium and the value of E_a has been found to be $+0.0089$ eV, if the energy is measured from the top of the valence band.

There is as yet no satisfactory theory of acceptor levels in silicon since the acceptor ionization energy in this material is of the same order of magnitude as the value of the spin-orbital splitting.

Excited States. In addition to the ground-energy levels of impurities, excited levels can also exist in the forbidden band of a semiconductor. From general considerations, it follows that such levels lie above the ground donor level in an n-type semiconductor and below the acceptor level in a p-type crystal.

A theoretical calculation of the excited states in silicon [25] has yielded a donor level scheme given in Table 1.4.

Excited states in Si were observed experimentally [27] in an investigation of the long-wavelength infrared absorption at liquid helium temperature. The positions of the excited levels are in very good agreement with the theory.

The ionization energies of the ground donor states in germanium are considerably lower than in Si. Therefore, the excited states in germanium are separated by $\sim 10^{-4}$ eV from the bottom of the conduction band. To observe them experimentally, it would be necessary to use radiation of millimeter wavelengths, a technique which meets with considerable difficulties. The extremely small

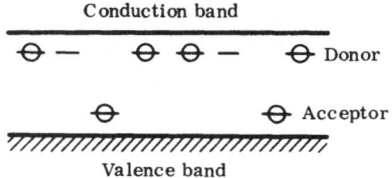

Fig. 1.9. Energy diagram of a semiconductor with impurities.

values of the energy gaps between the conduction band of germanium and the excited states of impurities allow us to ignore them altogether. However, in the case of silicon, we must remember the excited states in the interpretation of some of the physical phenomena.

D e e p   L e v e l s .   Donors and acceptors with low ionization energies form systems of "shallow" levels. Real crystals also contain also "deep" levels. The latter are observed in those cases when the impurity ion potential varies so rapidly that we may assume it to differ from zero only over a small region in space, for example, within one cell in a crystal. Thus, the impurity ion potential is in the form of a local perturbation and it can no longer be regarded as a Coulomb field. In this case, we cannot use the effective mass method and the problem is not hydrogen-like.

A theory of deep impurity levels has not yet been developed.

A full summary of the known impurity levels in Ge, Si, and one $A^{III}B^V$ compound, taken from [18], is given in Table 1.5. It is evident from this table that the experimental values of the ionization energies are generally in agreement with the theoretical values, but in some cases, the former are lower than the latter.

I m p u r i t y   C o n d u c t i o n .   This type of conduction can be observed even at low doping levels and low temperatures, when there are no free carriers in the allowed bands. This conduction is due to the formation of an impurity band. To understand this phenomenon we must look at Fig. 1.9. This figure shows the energy band structure of an n-type semiconductor, containing a certain number of acceptors. These acceptors are capable of capturing electrons not from the valence band but from some donors. Consequently, electrons are capable of traveling from an occupied donor to a vacant donor center (Fig. 1.9).

TABLE 1.5. Ionization Energies of Impurities in Ge, Si, and GaAs (d = donor, a = acceptor, "−" represents measurements from E_c, "+" represents measurements from E_v)

Group	Impurity	Ionization energy, eV		
		Ge	Si	GaAs
I	Li	d −0.0093 d −0.045 a +0.32 a −0.22	d −0.033 d +0.24 a +0.49	a +0.063
	Ag	a +0.13 a −0.29 a −0.09		
	Au	d +0.05 a +0.16 a −0.20 a −0.04	d +0.35 a − 0.54	
II	Be	a +0.07		
	Zn	a +0.03 a +0.09	a +0.31 a +0.55	a +0.26
	Cd	a +0.06 a +0.2		a +0.04
III	B	a +0.0104	a +0.046	
	Al	a +0.0102	a +0.057	
	Ga	a +0.0108	a +0.065	
	Tl	a +0.01	a +0.26	
	In	a +0.0112	a +0.16	
V	P	d −0.0120	d −0.044	
	As	d −0.0127	d −0.049	
	Sb	d −0.0096	d −0.039	
	Bi	−	d −0.069	
VI	Cr	a +0.07 a +0.12		
	O	d −0.01	d −0.03	d −0.0015
	S	d −0.18	d −0.18	d −0.03
	Se	d −0.14 d −0.28	d −0.37	d ∼ −0.004
	Te	d −0.11 d −0.30		d ∼ −0.004
VII	Mn	a +0.16 a −0.37	d −0.53	a +0.095
	Fe	a +0.34 a −0.27	d +0.40 d −0.55	a +0.36 a +0.59
	Co	d +0.09 a +0.25		a +0.54
	Ni	a −0.31		a +0.53

Such impurity conduction was first observed by Busch and Labhart [28] in silicon carbide. Since then, it has been found in a large number of n- and p-type semiconductors. The literature on this subject and a detailed theory are given in [29].

From our point of view, impurity conduction will be important only if a semiconductor contains one impurity partly compensated with another impurity.

Since impurity conduction is a tunnel effect without any activation energy, we may assume (incorrectly) that local impurity states form a band in which electrons travel and have a different effective mass from that in the conduction band. This assumption is used to explain impurity conduction by the presence of an "impurity band" although the values of E_d and E_a are not affected.

§1.3. Energy Spectrum of Electrons in a Doped Semiconductor (High Doping Level)

So far, we have assumed that the interaction between impurities may be neglected. This is true only if the impurity concentration is extremely low. If there is a small but finite overlap of the electron wave functions of neighboring impurity centers, individual electron levels broaden to form a real "impurity band." For this to happen, no impurity compensation is necessary. A level

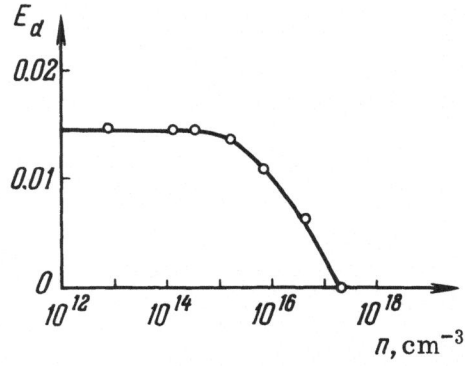

Fig. 1.10. Dependence of the ionization energy of donors in germanium on their concentration.

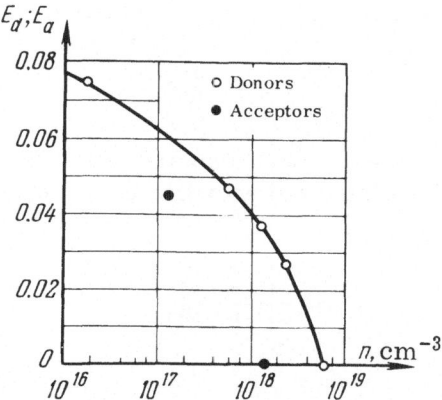

Fig. 1.11. Dependence of the ionization energies of donors and acceptors in silicon on their concentrations.

broadens symmetrically above and below its original position and thus the values of E_d and E_a become smaller.

Such a reduction in these energies, found experimentally [30, 31], is shown in Figs. 1.10 and 1.11. The ionization energies vanish in Ge and Si at different concentrations. This is because the Bohr radius of electrons at impurity centers in Si is approximately one-third of the radius in Ge, i.e., in Si the impurity states are more highly localized. Consequently, a greater impurity concentration is required in silicon for the same wave overlap as in germanium.

The problem of impurity bands can be approached in several ways. Of the earlier approaches, we shall consider the methods of Baltensperger [32] and of Lax and Phillips [33]. A more recent method will be considered by examining the work of Bonch–Bruevich [34, 35].

Impurity Band Model for a Regular Impurity Distribution. In this model it is assumed that impurity atoms form a regular lattice, consisting of hydrogen-like centers. This approach is artificial because there is, in fact, no such regular impurity distribution in real crystals.

Solving Schrödinger's equation within a sphere of radius r_s, where r_s is the average distance between impurities, which is related to the concentration of impurity atoms N by

$$\frac{4}{3} \pi r_s^3 = \frac{1}{N},$$
(1.3.1)

and assuming that the wave functions are of the Bloch type, Baltensperger [32] has calculated the energies corresponding to the edges of an impurity band ΔE_n:

$$\Delta E_n = \frac{1}{0.171} \frac{m^*}{m^+} \frac{e^2}{2\varkappa^2 a} \left(\frac{\varkappa a}{r_s}\right)^2,$$
(1.3.2)

where m^* and m^+ are, respectively, the effective masses of an electron in the conduction band and in the impurity band; a is the Bohr radius defined as

$$a = \frac{\hbar^2}{m^* e^2}.$$

An important aspect of this result is that an impurity band has sharp boundaries, similar to the boundaries of the conduction or valence bands.

Modified Baltensperger Model. Brody [36] attempted to allow for the broadening of an impurity band due to departures from a regular impurity distribution. He used a Gaussian curve, with its maximum at the initial position of an impurity level, to represent the density of states within an impurity band. The justification for this approach is provided by the calculations, reported in [37] for a one-dimensional random lattice, which give a density-of-states curve of nearly Gaussian form. At first sight, it would seem that a Gaussian curve may apply also in the three-dimensional case.

To determine the parameters of the selected Gaussian function, it is necessary to normalize the area bounded by this function by making it equal to the total number of states 2N. The standard deviation, defined by this Gaussian curve, should be related to the impurity band width ΔE_n given by Baltensperger's equation (1.3.2). This relationship may be twofold.

First, the value of ΔE_n may be assumed to be equal to two standard deviations σ. Secondly, we can assume, for example, that

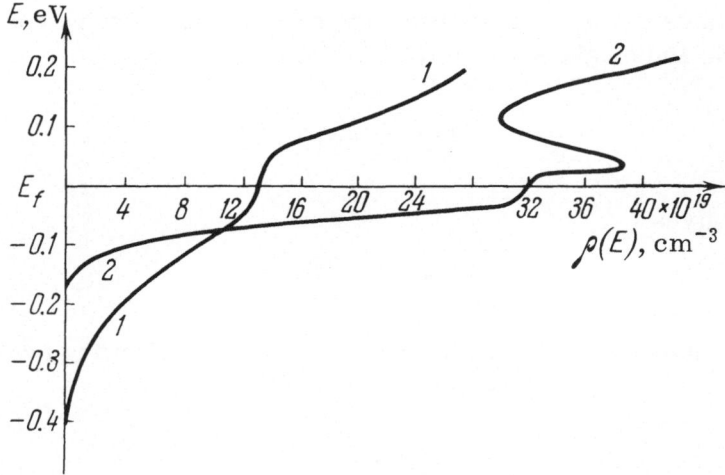

Fig. 1.12. Energy dependence of the density of states for overlapping impurity and conduction bands: 1) on the assumption of Eq. (1.3.3) for $N_i = 2 \cdot 10^{19}$ cm^{-3}; 2) on the assumption of Eq. (1.3.4) for $N_i = 2 \cdot 10^{19}$ cm^{-3}.

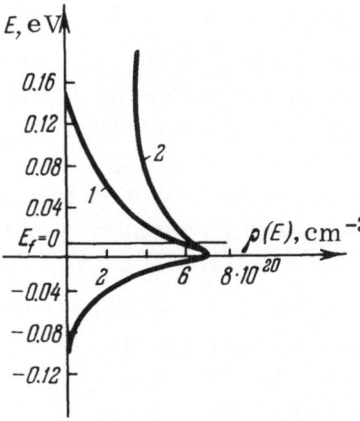

Fig. 1.13. Energy dependence on the density of states based on the Lax and Phillips impurity band model for $N_i = 3.4 \cdot 10^{19}$ cm^{-3} and $T = 300°K$: 1) in an impurity band; 2) for overlapping impurity and conduction bands.

all the states in the impurity band are enclosed in a square box of width ΔE_n and height equal to $N/(2\pi)^{1/2}\sigma$, i.e.,

$$\frac{\Delta E_n N}{(2\pi)^{1/2}\sigma} = 2N,$$

and this gives

$$\Delta E_n = (8\pi)^{1/2}\sigma.$$

From the normalization conditions, we obtain the following two density-of-states functions:

$$\rho_1(E) = \frac{2\sqrt{2}\,N}{\sqrt{\pi}\,\Delta E_n}\,e^{-2\left(\frac{E}{\Delta E_n}\right)^2}, \tag{1.3.3}$$

$$\rho_2(E) = \frac{4N}{\Delta E_n}\,e^{-4\pi\left(\frac{E}{\Delta E_n}\right)^2}. \tag{1.3.4}$$

All these considerations apply only to an impurity band consisting of ground impurity states. Bands representing excited states will fully overlap the conduction band. This follows from the estimates of Baltensperger [32] and of Conwell [38].

If, in the first approximation, the impurity bands of excited states are ignored, then by combining the expression for the density of states in a standard conduction band [Eq. (1.1.16)] either with Eq. (1.3.3) or with (1.3.4), we obtain a general curve, representing the energy dependence of the density of states in the overlapping conduction and impurity bands. Examples of such curves are given in Fig. 1.12.

These calculations should be regarded only as qualitative illustrations of the situation obtaining after the merging of the impurity and conduction bands. The main result is the existence of a "tail" of allowed states below the Fermi level.

Impurity Band for a Random Impurity Distribution. A tail in the density of states has been deduced more rigorously in [33, 37, 39-41] by considering a one-dimensional chain of randomly distributed impurity atoms. The potentials of the impurity atoms have been approximated by δ functions.

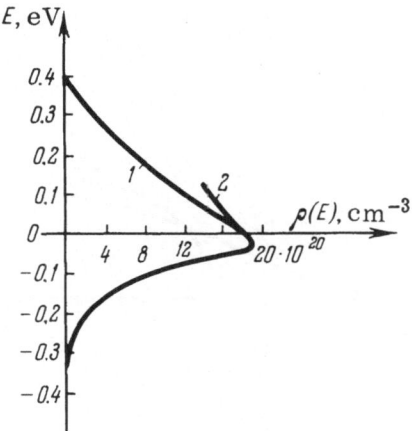

Fig. 1.14. Energy dependence of the density
of states based on the Lax and Phillips impurity band model for $N_i = 1.3 \cdot 10^{20}$ cm^{-3}
and $T = 300°K$: 1) in an impurity band; 2)
for overlapping impurity and conduction bands.

The problem has been solved by the perturbation-theory
method. The results of such calculations, carried out using the
method of Lax and Phillips [33], are shown in Figs. 1.13 and 1.14
for two donor concentrations ($3.4 \cdot 10^{19}$ and $1.3 \cdot 10^{20}$ cm^{-3}) in germanium. These figures show the curves representing the total
density of states for an impurity band overlapping a standard conduction band.

Review of Earlier Investigations of Impurity
Bands [34]. The "impurity band" concept must not be taken
too literally. Thus, for example, when the impurity concentration
is increased considerably, the density of free carriers increases
as well and, consequently, the screening of the impurity centers by
electrons becomes stronger. As a result of the screening, the impurity levels may disappear altogether without the formation of an
impurity band [42]. In germanium, this may happen at carrier
densities somewhat lower than 10^{18} cm^{-3}. Under these conditions,
the usual statement of the impurity band problem becomes mean-
ingless. However, this does not mean that there are no impurity
bands at all. We simply regard an impurity band as a range of
values of E in the forbidden band in which the density of states $\rho(E)$
differs from zero.

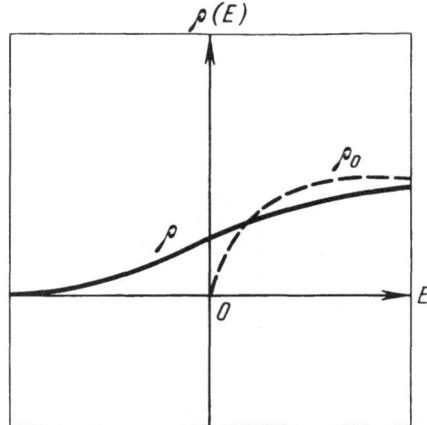

Fig. 1.15. Density of states in a pure (ρ_0) and
heavily doped (ρ) semiconductor. The zero
level represents the bottom of the conduction
band of a pure semiconductor.

One limitation of the earlier investigations [33, 39-41, 43]
has been the use of a one-dimensional model. Such a model gives
only qualitative results.

Another shortcoming of these investigations is making the
allowance for the interaction between electrons only through the
screening of the impurity potential. It is known [44] that such an
interaction alters the dispersion law of the density of states. A
serious deficiency of these early studies [33, 39, 40] is the use of
a δ-type screened impurity potential. Such an approximate treatment is justified if the screening radius R_0 is small compared with
the average distance between impurity atoms $r_s \sim N^{-1/3}$. In heavily
doped semiconductors, R_0 may be larger than r_s. In fact, if we
use the Debye approximation to estimate the screening radius, we
find that

$$R_0 \approx \frac{1}{2} \sqrt{\frac{a}{N^{1/3}}}. \tag{1.3.5}$$

When $N \geq 10^{19}$ cm^{-3}, the value of R_0 is not small compared with r/s,
estimated from Eq. (1.3.1).

TABLE 1.6. Changes in the Lattice Parameter
and Forbidden Band Width of Silicon
Due to Doping with Boron

Boron content, cm^{-3}	Lattice parameter a, Å	Change in forbidden band width, ΔE_g, eV
$1 \cdot 10^{14}$	5.4295	0.0
$1 \cdot 10^{17}$	–	–
$1 \cdot 10^{18}$	–	–
$1 \cdot 10^{19}$	5.4291	$0.3 \cdot 10^{-3}$
$2.6 \cdot 10^{19}$	5.4282	$1.2 \cdot 10^{-3}$
$5.2 \cdot 10^{19}$	5.4281	$1.3 \cdot 10^{-3}$
$1.5 \cdot 10^{20}$	5.4249	$4.25 \cdot 10^{-3}$
$2.0 \cdot 10^{20}$	5.4241	$5.0 \cdot 10^{-3}$

On the other hand, it has been demonstrated in [34] that the actual form of the expression used to describe the density of states is quite sensitive to the form of the screened potential. Consequently, the δ-approximation is unsatisfactory.

Modern Theory of Impurity States. The three-dimensional problem has been solved by Bonch−Bruevich [34] in the limiting case of high-impurity concentrations. Bonch−Bruevich not only has allowed for the interaction of electrons through the screened potential but also for the Coulomb interaction.

The problem has been solved using the Green functions. It has been found that the density of states differs from zero everywhere in the forbidden band. Near the Fermi level, the density of states differs little from the density of states in an ideal Fermi gas, but at the bottom of the conduction band* the density changes substantially, decreasing at first linearly when $E < 0$ but more rapidly in the middle of the band (Fig. 1.15). However, the explicit form of the $\rho(E)$ asymptote for $E \to -\infty$ (a factorially [34] or exponentially [45] decreasing function) has not yet been established finally because it is sensitive to the selected form of the screened impurity potential.

A discussion of the band structure of a heavily doped semiconductor with two interacting bands [35] has shown that only un-

─────────

*We mean here the conduction band of an impurity-free semiconductor.

important quantitative corrections must be made to the one-band case.

It has been found that the compensation decreases the density of states in the "tail" (in the case of total compensation, the density of states is reduced to zero).

Experimental Investigations of the Band Structure of Heavily Doped Semiconductors. Experimental studies of the band structure of heavily doped semiconductors can be divided into two groups.

The first group comprises investigations of energies in the band structure mainly due to distortions of the crystal lattice of a semiconductor.

The second group comprises studies of the density-of-states tail, extending into the forbidden band.

Changes in the lattice parameter, which are expected in the heavy-doping case, may affect the band structure of a crystal in the same way as omnidirectional compression.

If we denote the compressibility of a semiconductor by χ (cm^2/kg) and the change in the forbidden band width under uniform compression by dE_g/dP (eV $\cdot$ cm^2 $\cdot$ kg^{-1}), the absolute change in the forbidden band width ΔE_g of a crystal with a cubic lattice can be defined as

$$\Delta E_g = \frac{3\frac{\Delta a}{a}}{\chi}\frac{dE_g}{dP}, \qquad (1.3.6)$$

where $\Delta a/a$ is a relative change in the lattice parameter.

The results of an experimental determination of the lattice parameter of silicon doped with boron [46, 47] are given in Table 1.6. The same table includes values of ΔE_g calculated using Eq. (1.3.6).

Comparing the tetrahedral radii of phosphorus (1.10 Å), arsenic (1.18 Å), and boron (0.88 Å), we may conclude that changes in the lattice parameter (and small changes in the forbidden band width) should have smaller absolute values in the case of doping with phosphorus and arsenic than in the case of doping with boron (this has not yet been checked experimentally).

X-ray diffraction investigations of germanium doped with arsenic, carried out by the present author, have shown no changes in the lattice parameters right up to an arsenic concentration of $4.5 \cdot 10^{19}$ cm^{-3}.

The purpose of the second group of studies has been to investigate the density-of-states tail. However, no investigations have yet been published which give more accurately the form of the curve in Fig. 1.15. In the majority of these investigations, it has been assumed that because of the very rapid decrease of the $\rho(E)$ curve in the forbidden band, we can still speak of an effective forbidden band width, obtained by truncating the tail in some suitable manner.

The value of such an effective forbidden band width may be obtained from optical investigations (absorption of light and recombination radiation). These methods have, in fact, been used in the majority of such investigations [48-52]. These investigations will be discussed in detail in Chap. 4, but here we shall mention only that a considerable narrowing of the forbidden band (by 0.06 eV) has been reported for heavily doped germanium [48]. An exactly opposite result has been reported in [49]: it has been found that the energy gap between the valence and conduction bands has become greater in the heavily doped material.

Some narrowing (~0.03 eV) of the forbidden band in heavily doped n-type germanium has also been reported in [50].

Dubrovskii [47, 52] has investigated heavily doped n- and p-type silicon but has found no appreciable reduction in the value of E_g.

Finally, Sommers [53] has analyzed the results of several experiments on heavily doped germanium and has concluded that the narrowing of the energy gap E_g hardly exceeds 0.03 eV at an impurity concentration of $2 \cdot 10^{19}$ cm^{-3}. It is worth mentioning that Sommers has also used experimental data obtained by methods other than the optical absorption.

The influence of heavy doping on the forbidden band width has been investigated also for some $A^{III}B^V$ compounds. The narrowing of the forbidden band due to an increase in the dopant concentration has been observed for InAs [54, 55], InSb [55, 56], and InP [57]. For an impurity concentration of $2 \cdot 10^{19}$ cm^{-3}, the narrowing amounts to 0.078 eV for InSb and 0.059 eV for InAs.

The narrowing of the forbidden band has also been observed in uncompensated doped InSb [58].

In all these investigations of $A^{III}B^{V}$ compounds, the narrowing of the forbidden band has been attributed to the Coulomb interaction of electrons. However, a more probable mechanism of such a strong reduction in the forbidden band may be a change in the lattice parameter. Thus, for InSb the value of $\Delta E_g/dP$ is $15.5 \cdot 10^{-6}$ eV $\cdot$ cm$^2 \cdot$ kg^{-1} [59], which is an order of magnitude larger than $\Delta E_g/dP$ for Si.

The change in the lattice parameter $\Delta a/a$, necessary to reach the value of ΔE_g reported in [55], represents — according to Eq. (1.3.6) — 0.1% for an impurity concentration of 10^{19} cm^{-3}.

A direct proof of the existence of a density-of-states tail in the forbidden band of heavily doped GaAs has been obtained by the present author and Agaev [60] in an investigation of the excess current in tunnel diodes. However, the results of this investigation were still insufficient to determine the form of the curve in Fig. 1.15.

Chapter 2

Statistical Physics of Carriers in Heavily Doped Semiconductors

§2.1. Preliminary Remarks

Distribution-Function Concept. In a system of N particles there are some whose coordinates lie within the intervals

$$x = x \pm \Delta x, \quad y = y \pm \Delta y, \quad z = z \pm \Delta z, \tag{2.1.1}$$

and whose momentum components are, respectively,

$$p_x = p_x \pm \Delta p_x, \quad p_y = p_y \pm \Delta p_y, \quad p_z = p_z \pm \Delta p_z. \tag{2.1.2}$$

If the number of such particles is ΔN, they represent a fraction $\Delta N/N$ of the total number of particles. The fraction of such particles will increase or decrease proportionally to the variation of the intervals Δx, Δy, Δz, and Δp_x, Δp_y, Δp_z. The coefficient of proportionality may be a function of coordinates and momenta, as well as of the time t:

$$\frac{\Delta N}{N} = f(x, y, z, p_x, p_y, p_z, t) \, \Delta x \, \Delta y \, \Delta z \, \Delta p_x \, \Delta p_y \, \Delta p_z. \tag{2.1.3}$$

Since $\Delta x \, \Delta y \, \Delta z = \Delta V$ is an element of volume in the coordinate space, $\Delta p_x \, \Delta p_y \, \Delta p_z = \Delta \omega$ is an element of volume in the momentum space, and $\Delta V \Delta \omega = \Delta \gamma$ is an element of volume in the phase space, it fol-

37

lows that

$$\frac{\Delta N}{N} = f\Delta\gamma.$$

Thus, the function f has a simple physical meaning: it represents the relative concentration of particles contained in a given phase interval, calculated per unit phase volume. In other words, the function f is the probability of finding a particle in a given phase interval, calculated per unit phase volume.

Since the total of all probabilities is the certainty of an event, it follows that

$$\sum \frac{\Delta N}{N} \approx \frac{1}{N} \int_{\Gamma} \Delta N = 1,$$

or

$$\sum_i f\Delta\gamma_i \approx \int_{\Gamma} fd\gamma = 1, \tag{2.1.4}$$

in which integration is carried out over the whole phase volume Γ. Equation (2.1.4) is the normalization condition of the function f. We can easily see that f has the dimensions of density. Therefore, we shall call it the distribution density. The form of this function represents a distribution of particles over states, where a state is understood to be a phase interval defined by Eqs. (2.1.1) and (2.1.2).

Similarly, if we consider not N particles but M systems, each of which consists of many particles, the corresponding distribution density is

$$w = \frac{\Delta N}{M}. \tag{2.1.5}$$

Liouville Theorem. The positions of particles in the phase space cannot be determined exactly since their degree of localization is determined by an element of the phase space, i.e., the quantity $\Delta\gamma$ is also a measure of the indeterminacy of the positions of the particles in the phase space. It is found that this indeterminacy is independent of time. This is the meaning of the Liouville theorem, the derivation of which will be found in any statistical physics course [1-3].

However, it also follows from this theorem that the distribution density is also independent of time.

In fact, if $\Delta\gamma =$ const and the number of systems M and the number particles ΔN are independent of time, it follows from Eq. (2.1.5) that w = const. This means that the distribution density is a function of the coordinates and momenta which does not change during the motion of the particles. Such functions, which remain invariant during the motion of a body, are known as integrals of motion.

Integrals of Motion. These integrals include, for example, the total energy of a free particle E, its momentum P, its angular momentum J, etc. Consequently, we can represent the dependence of the distribution density w on coordinates and momenta in the form of a dependence of the integrals of motion

$$w = w(E, \ \boldsymbol{P}, \ \boldsymbol{J}, \ \dots). \qquad (2.1.6)$$

We shall consider two elements of the phase space $\Delta\gamma_1$ and $\Delta\gamma_2$. The number of systems within these volumes will be denoted by ΔM_1 and ΔM_2. Then the probabilities of finding the systems in $\Delta\gamma_1$ and $\Delta\gamma_2$ are:

$$w_1 = w_1(E_1, \ \boldsymbol{P}_1, \ \boldsymbol{J}_1, \ \dots) \Delta\gamma_1 \ \text{ and } \ w_2 = w_2(E_2, \ \boldsymbol{P}_2, \ \boldsymbol{J}_2, \dots) \Delta\gamma_2.$$
$$(2.1.7)$$

Since we are assuming that the systems do not interact with one another, their presence in different volumes $\Delta\gamma$ are independent events. Therefore, the probability w of finding simultaneously the first system in $\Delta\gamma_1$ and the second system in $\Delta\gamma_2$ is equal to the product of the probabilities of the separate events:

$$w = w_1(E_1, \ \boldsymbol{P}_1, \ \boldsymbol{J}_1, \ \dots) \, w_2(E_2, \ \boldsymbol{P}_2, \ \boldsymbol{J}_2, \ \dots). \qquad (2.1.8)$$

The two systems can be regarded as one complex system and the state of this new system can be described in a phase space in which an element $\Delta\Gamma$ is equal to the product $\Delta\gamma_1 \cdot \Delta\gamma_2$. The probability of finding such a complex system in $\Delta\Gamma$ is:

$$w = w(E_1 + E_2, \ \boldsymbol{P}_1 + \boldsymbol{P}_2, \ \boldsymbol{J}_1 + \boldsymbol{J}_2, \ \dots). \qquad (2.1.9)$$

Equating Eqs. (2.1.8) and (2.1.9) and taking the logarithms, we obtain:

$$\ln w\,(E_1 + E_2,\ \boldsymbol{P}_1 + \boldsymbol{P}_2,\ \boldsymbol{J}_1 + \boldsymbol{J}_2,\ \dots\,) =$$
$$= \ln w_1\,(E_1,\ \boldsymbol{P}_1,\ \boldsymbol{J}_1,\ \dots\,) + \ln w_2\,(E_2,\ \boldsymbol{P}_2,\ \boldsymbol{J}_2,\ \dots\,).$$

Here, we see that a logarithm of the function w is an additive quantity and is a sum of the logarithms of the probability densities of separate systems, i.e.,

$$\ln w = \ln w_1 + \ln w_2. \qquad (2.1.10)$$

If we bear in mind that the integrals of motion E, $\boldsymbol{P}$, $\boldsymbol{J}$ are also additive quantities, we find that when the distribution function depends only on the total energy of the system and not on the energies of individual particles, Eq. (2.1.10) is satisfied only by a linear function of the integrals of motion, i.e.,

$$\ln w = a + bE + c\boldsymbol{P} + d\boldsymbol{J}, \qquad (2.1.11)$$

where a, b, c, d are constant coefficients.

Gibbs Distribution. This discussion applies to a system with a constant number of particles. The number of particles N is also an integral of motion and is also additive, like E, $\boldsymbol{P}$, and $\boldsymbol{J}$. Using this property, we can rewrite Eq. (2.1.11) for a system with a variable number of particles:

$$\ln w = a + bE + c\boldsymbol{P} + d\boldsymbol{J} + eN, \qquad (2.1.12)$$

where e is a constant coefficient.

If all the particles considered are not in translational or rotational motion, it follows that c = 0 and d = 0. In fact, if the coefficient c were greater than zero, the probability of motion of the whole system of particles to the right $e^{c_x P_x}$ would have exceeded the probability of its motion to the left $e^{-c_x P_x}$. Thus, we would have observed a macroscopic motion of a system of particles. We can easily show that if the particles are at rest the components c_x, c_y, c_z and d_x, d_y, d_z are all equal to zero. Then, Eq. (2.1.12) becomes

$$\ln w = a + bE + eN, \qquad (2.1.13)$$

and hence

$$w = \exp \frac{\Omega - E + \mu N}{\theta}, \qquad (2.1.14)$$

where $\theta = 1/b$, $\Omega = a\theta$, $\mu = e\theta$.

The distribution (2.1.14) is known as the grand canonical distribution or the Gibbs distribution [4]. We shall now consider the physical meaning of the quantities occurring in this expression.

The Distribution Modulus θ. Assuming that the quantities a, e, and N in Eq. (2.1.13) are constant, we find that:

$$w = A \exp (bE). \qquad (2.1.15)$$

Since the probability of finding states with an infinitely large energy should be infinitely small, the quantity b should be basically negative. Therefore, we shall assume that $b = -\theta^{-1}$, where the distribution modulus θ has the dimensions of energy and gives the average energy of the particles, i.e., the quantity θ is the temperature expressed in some "natural" units. It is related to the absolute temperature T by $\theta = kT$, where k is Boltzmann's constant.

Thermodynamic Potential Ω. We shall write Eq. (2.1.14) in the form

$$w = \exp \frac{F - E}{\theta}, \qquad (2.1.16)$$

where

$$F = \Omega + \mu N. \qquad (2.1.17)$$

We shall substitute Eq. (2.1.16) into a normalization condition analogous to Eq. (2.1.4), and we shall differentiate it with respect to $1/\theta$. Since E is a function only of coordinates and momenta, and therefore is independent of $1/\theta$, the differentiation yields

$$\int \left(F + \frac{1}{\theta} \frac{\partial F}{\partial \left(\frac{1}{\theta} \right)} - E \right) e^{\frac{F-E}{\theta}} \, d\gamma = 0,$$

whence — using Eq. (2.1.4) as well as the fact that F and $1/\theta$ are independent of integration variables — we find that:

$$F + \frac{1}{\theta} \frac{\partial F}{\partial \left(\frac{1}{\theta} \right)} - \int E e^{\frac{F-E}{\theta}} \, d\gamma = 0,$$

and hence

$$\int E e^{\frac{F-E}{\theta}} \, d\gamma = F - \theta \frac{\partial F}{\partial \theta} = F - T \frac{\partial F}{\partial T}.$$

The general rules for the calculation of averages [5] show that the integral in the above expression is simply the average energy of the system $\overline{E}$. Hence,

$$\overline{E} = F - T \frac{\partial F}{\partial T}. \tag{2.1.18}$$

On the other hand, there is a similar thermodynamic expression [4] relating the energy of a system E with that part of the energy F, which gives the ability of the system to carry out work, i.e., the free energy. Consequently, Ω [cf. Eq. (2.1.17)] represents the possibility of a change in the free energy of the system. For this reason it is called the thermodynamic potential.

Chemical Potential μ. We shall assume that a system of particles is at absolute zero. Then the change in the energy of the system dE can take place only by a change in the free energy and, therefore, from Eq. (2.1.18), we obtain

$$dE|_{T \to 0} = dF. \tag{2.1.19}$$

However, the energy of the system changes if we add (or take away) some particles which belong to the system. Consequently, Eq. (2.1.19) should include the change in the number of particles

$$dE|_{T \to 0} = dF + \mu \, dN. \tag{2.1.20}$$

We shall assume that $dE|_{T=0} = 0$, i.e., we shall assume that there is no change in the energy. Then, the quantity μ is defined thus:

$$|\mu| = \frac{dF}{dN}. \tag{2.1.21}$$

It follows that the quantity μ represents the energy per particle. In other words, μ indicates the ability of a system to change its energy by exchanging particles with an external medium (or another system). Therefore, the quantity μ is called the chemical potential.

The phrase "chemical" potential stresses the fact that we are considering particles capable of moving in accordance with the laws of mechanics. In particular, we shall not consider yet the case of charged particles, which can also move under the action of electric forces.

§2.2.   Fermi — Dirac Statistics for Free Electrons

Quantum Indistinguishability of Electrons. We must now specify exactly what we mean by a "particle." First of all, we shall divide all particles into "macroparticles" and "microparticles." The former are those whose motion is sufficiently well described by the laws of classical mechanics. The latter obey the laws of quantum mechanics. They include the particles in the "microworld," e.g., electrons.

Microparticles cannot be distinguished in the quantum sense. This means that the macroproperties of a body are not affected by an interchange of particles between two microstates. To make this clearer, we shall consider a simple example. Let us assume that we have two identical particles presented by the circles in Fig. 2.1a. Each of these particles is in a definite state represented by a square. When they are interchanged, the situation is still the same (Fig. 2.1b). If these two particles are distinguishable, then an interchange will produce a different distribution (Figs. 2.1c and 2.1d).

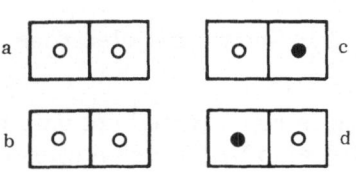

Fig. 2.1. Possible distributions of two particles.

Hence, we see that in the second case we may speak of two states, which differ by the interchange of the two particles, whereas in the first case there is only one state. Generally speaking, in the case of macrobodies, we can speak of "N!" states, differing by the interchange of N particles; but in the case of atomic particles, we should say: "a state in which there are ΔN particles," i.e.,

$$
\left.
\begin{array}{lll}
\Delta N_1 & \text{particles in state} & E_1, \\
\Delta N_2 & \text{particles in state} & E_2, \\
. \; . \; . \; . \; . \; . \; . \; . \; . \; . \; . \; . \\
\Delta N_k & \text{particles in state} & E_k.
\end{array}
\right\}
\qquad (2.2.1)
$$

This has been our approach in the derivation of the Gibbs distribution. Consequently, the expression (2.1.14) represents the distribution of the probabilities for quantum-indistinguishable particles.

We note that for distinguishable particles Eq. (2.1.14) would have had a pre-exponential factor, equal to 1/N! [6].

Pauli Exclusion Principle. In the case of classical particles, we can have an infinite number of particles in any given state. Some quantum-indistinguishable particles are, in this respect, similar to classical particles. They include light quanta and many atoms and molecules.

Electrons belong to a different class of quantum-indistinguishable particles which obey an additional selection rule: there cannot be more than one electron in any given state. This rule is known as the Pauli exclusion principle. In general, elementary particles have spin, i.e., a mechanical angular momentum similar to that of ordinary hard spheres rotating about their axes. The spin S can be represented by

$$ S = \pm s\hbar, $$

where s is the spin quantum number. It is known that electrons have spin quantum numbers of $\pm 1/2$.

In the case of electrons, the state is represented not only by the motion of a particle (momentum) but also by the orientation of its spin. The number of particles which can have a given value of the momentum is

$$\Delta N = 2s + 1, \tag{2.2.2}$$

i.e., two electrons can have the same momentum.

Fermi — Dirac Distribution. We shall now find the electron energy distribution.

The energy of a system of particles is found by adding the energies of individual particles

$$E = \sum_0^k \Delta N_k E_k.$$

The total number of particles in such a system is

$$N = \sum_0^k \Delta N_k. \tag{2.2.3}$$

Using these expressions we shall represent the Gibbs distribution of Eq. (2.1.14) in the form

$$w = \exp\left[\frac{\Omega + \sum_0^k (\mu - E_k)\,\Delta N_k}{\theta}\right]. \tag{2.2.4}$$

We shall find the $w_i(\Delta N_i)$ particle distribution function at the i-th level

$$w_i(\Delta N_i) = \sum_{\Delta N_j\,(j \neq i)} w(\Delta N_1, \ldots, \Delta N_i, \ldots). \tag{2.2.5}$$

This relationship can be rewritten so that all the particles associated with the i-th level can be distinguished by a separate factor

$$w_i(\Delta N_i) = C \exp\left[\frac{(\mu - E)\,\Delta N_i}{\theta}\right], \tag{2.2.6}$$

where C is the remaining sum.

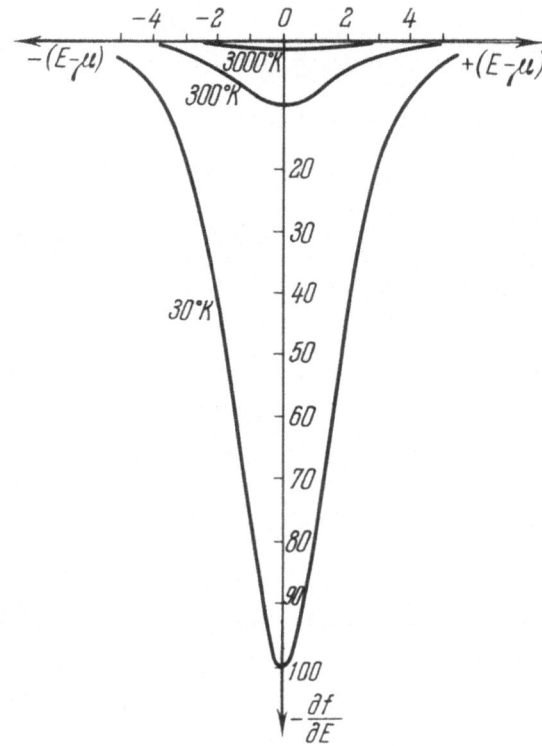

Fig. 2.2. Derivative of the Fermi–Dirac distribution function.

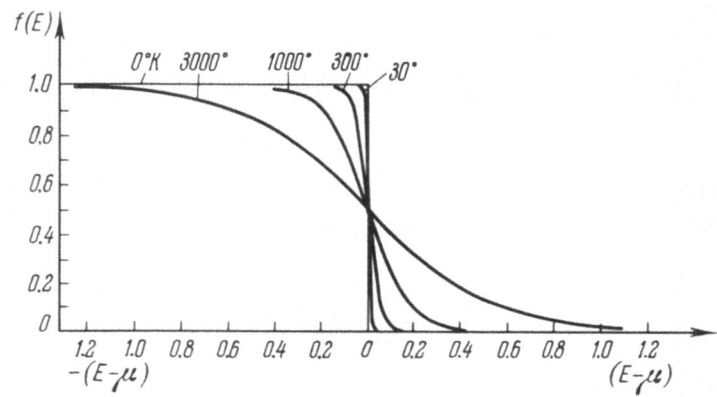

Fig. 2.3. Fermi–Dirac distribution function.

The distribution of particles over all the levels is

$$f = \frac{\sum\limits_{\Delta N} \Delta N_i w_i}{\sum\limits_{\Delta N} w_i} \tag{2.2.7}$$

Substituting Eq. (2.2.6), we obtain (after summation and allowance for the fact that $\Delta N_i = 0, 1$ for electrons) the following function, known as the Fermi–Dirac distribution:

$$f = \frac{1}{1 + \exp\left(\dfrac{E - \mu}{kT}\right)}. \tag{2.2.8}$$

Properties of the Fermi – Dirac Function.
To analyze the Fermi–Dirac distribution (2.2.8), we shall consider its derivative with respect to the energy:

$$\frac{\partial f}{\partial E} = -\frac{1}{kT} \frac{\exp\left(\dfrac{E - \mu}{kT}\right)}{\left[1 + \exp\left(\dfrac{E - \mu}{kT}\right)\right]^2}. \tag{2.2.9}$$

If we multiply the numerator and denominator of Eq. (2.2.9) by $\exp[-2(E - \mu)/kT]$, which does not alter $\partial f / \partial E$, we obtain

$$\frac{\partial f}{\partial E} = -\frac{1}{kT} \frac{\exp\left(-\dfrac{E - \mu}{kT}\right)}{\left[1 + \exp\left(-\dfrac{E - \mu}{kT}\right)\right]^2}. \tag{2.2.10}$$

Comparing these two expressions, we find that the derivative remains constant when the sign of the quantity $(E - \mu)/kT$ is reversed, i.e., the function $\partial f / \partial E$ is even, and this means that its values are distributed symmetrically with respect to the point $(E - \mu)/kT = 0$ or, which is equivalent, with respect to the energy $E = \mu$. This can be seen clearly in Fig. 2.2. We can show easily that at this point

$$\frac{\partial f}{\partial E} = -\frac{1}{4kT}, \tag{2.2.11}$$

and since this derivative is equal to tan α, where α is the angle of the slope of the tangent to the curve $f(E)$ at the point $E = \mu$, it follows that at this point the function f decreases. It also follows from Eq. (2.2.11) that the function $f(E)$ at the point $E = \mu$ becomes less steep when the temperature is lowered, and at absolute zero $\partial f/\partial E$ becomes $-\infty$, i.e., the tangent becomes perpendicular to the abscissa and the function $f(E)$ acquires the step-like form shown in Fig. 2.3.

If we assume that the energy E is the reference energy in the Fermi–Dirac distribution, we can conveniently represent Eq. (2.2.8) in the form

$$f = \frac{1}{1 + e^{-\mu^*}}. \tag{2.2.12}$$

where the quantity $\mu^* = \mu/kT$ is dimensionless and is known as the reduced electrochemical potential or the reduced Fermi level. The values of $f(\mu^*)$ are tabulated in Appendix A.1.

Electrochemical Potential. We shall consider once again the physical meaning of the quantity μ. The components of a system are electrons which, in addition to their mass, also have an electric charge. Consequently, going back to Eq. (2.1.20), we should include in that equation a term representing the possibility of a change in the energy due to a change in the amount of charge:

$$dE\,|_{T-0} = dF + \mu\,dN + \psi\,dq. \tag{2.2.13}$$

In this expression, the coefficient

$$\psi = \frac{\partial\,|\,E\,|_{T-0}}{\partial q}\,\Big|_{\substack{\Omega\,=\,\text{const} \\ N\,=\,\text{const}}}$$

is known as the electrostatic potential and represents that fraction of the work which has been done on a unit charge. The total charge dq is equal to the product of the electron charge and the number of particles; therefore,

$$dE\,|_{T-0} = dF + (\mu - e\psi)\,dN. \tag{2.2.14}$$

The new coefficient of dN now represents the algebraic sum of the chemical and electrostatic potentials and is known as the electrochemical potential. Usually, the electrochemical potential is denoted by the same symbol μ as the chemical potential. We shall also use the same notation.

It is evident from Fig. 2.3 that when $E - \mu = 0$, the value of $f(E)$ is equal to $1/2$, irrespective of the temperature. Consequently, we may identify the electrochemical potential with some energy E_F, for which the probability of occupation is equal to $1/2$. We can identify E_F and the electrochemical potential only when the electron system is under equilibrium conditions. Under nonequilibrium conditions, $E_F \neq \mu$.

Boltzmann Distribution. When

$$e^{-\mu^*} \gg 1 \qquad (2.2.15)$$

the Fermi—Dirac distribution function of Eq. (2.2.12) can be approximated by a simpler formula

$$f = e^{\mu^*}, \qquad (2.2.16)$$

which is identical with the well-known Boltzmann distribution [5] and applies to particles obeying the laws of classical mechanics. At first sight, it is difficult to see how these quantum-indistinguishable particles, having spin and obeying Pauli's exclusion principle, can be described by a "classical" distribution. This apparent paradox can be understood if we recall the most important difference between classical and quantum particles. This difference is the continuity of the energy spectrum of classical particles and the discrete nature of the spectrum of quantum particles. The condition $e^{-\mu^*} \gg 1$ is equivalent to a transition from a discrete to a continuous spectrum. In fact, Eq. (2.2.15) is satisfied better at higher temperatures. When the temperature is high, the energy levels broaden and the distances between them become so small that the energy spectrum is practically continuous. Conversely, as the temperature is lowered, the distribution of the particles in an electron gas differs more and more from the Boltzmann distribution.

Thus, the Boltzmann distribution can be regarded as the limiting case of the Fermi—Dirac distribution.

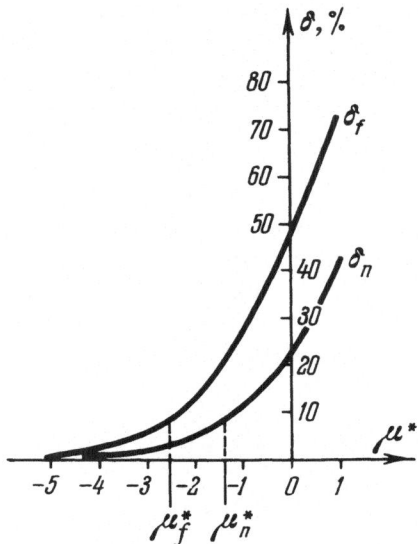

Fig. 2.4. Errors due to the replacement of
the Fermi—Dirac distribution function with
the Boltzmann distribution. The subscript f
refers to the function f and the subscript n
to the electron density.

Degeneracy Criterion. An electron gas obeying the
Fermi—Dirac distribution is known as degenerate and the value of
the reduced Fermi level μ^* is called the degree of degeneracy.
The division into degenerate and nondegenerate states is quite arbitrary because it is difficult to draw a definite line between these
states. This can be seen clearly from Fig. 2.4, which shows the
value of δ_f, representing the relative error in the calculation of
the function f using the Boltzmann rather than the Fermi—Dirac
distribution, i.e.,

$$\delta_f = 1 - \frac{e^{-\mu^*}}{1 + e^{-\mu^*}}. \qquad (2.2.17)$$

If we regard the value of the error δ_f of, for example, 8% as still
permissible, the degeneracy limit $\mu_f^* = -2$. When $\mu_f^* > -2$, the
electron gas is degenerate and when $\mu_f^* < -2$, this gas is nondegenerate.

We have used the notation δ_f and μ_f to indicate (by the subscript f) that the error δ and the degeneracy limit apply only to a particular distribution function.

We shall show later that different physical effects are affected in different ways by the nature of the distribution. Moreover, the values of δ will be different in different cases. Consequently, we shall ascribe different values of μ^* to the degeneracy limit.

§2.3. Statistics of Electrons and Holes

in Semiconductors

Electron Density in the Conduction Band. From the most general considerations it follows that the number of conduction electrons with energies in an interval dE is equal to the product of the number of states in this interval $\rho(E)dE$ and the probability of occupation of these states by electrons $f(E)$. The total number of electrons in the conduction band, n, is therefore:

$$n = \int_{E_c}^{\infty} f(E)\,\rho(E)\,dE. \tag{2.3.1}$$

The lower limit of integration is equal to the minimum electron energy, i.e., the energy corresponding to the bottom of the conduction band E_c. The upper limit of integration is, generally speaking, equal to the maximum possible value of the electron energy E_{max}, which may be assumed to be infinite.

Substituting into Eq. (2.3.1) the expressions obtained earlier for $f(E)$ [Eq. (2.2.8)] and $\rho(E)$ [Eq. (1.1.16)], which are valid for a parabolic conduction band, and allowing for the possibility of two values of the electron spin, we obtain

$$n = \int_{E_c}^{\infty} \frac{4\pi \left(\frac{2m_n^*}{h^2}\right)^{3/2} (E - E_c)\,dE}{1 + \exp\dfrac{E - \mu}{kT}}. \tag{2.3.2}$$

Measuring the energy from E_c and introducing the reduced (di-

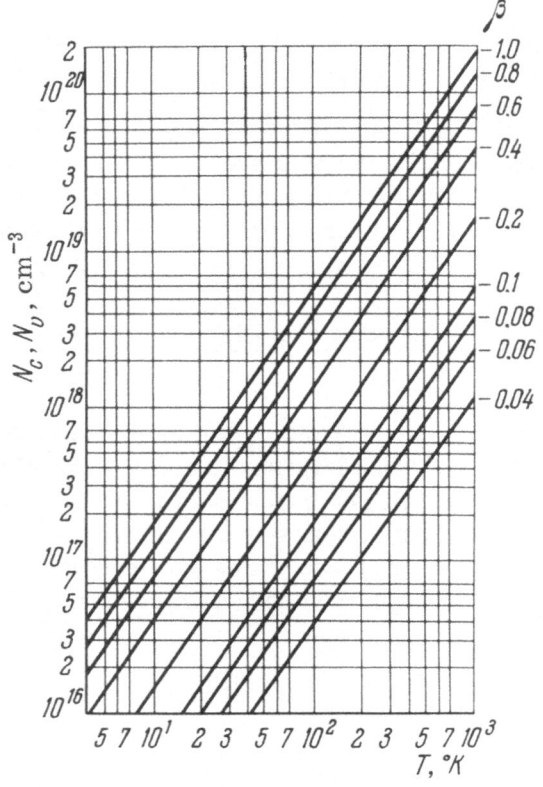

Fig. 2.5. Nomogram for the calculation of N_c and N_v.

mensionless) Fermi level μ^* and the reduced energy

$$\varepsilon^* = \frac{E - E_c}{kT},$$ (2.3.3)

we obtain:

$$n = N_c F_{1/2}(\mu^*),$$ (2.3.4)

where the integral

$$F_{1/2} = \int_0^\infty \frac{(\varepsilon^*)^{1/2}\, d\varepsilon^*}{1 + \exp(\varepsilon^* - \mu^*)}$$ (2.3.5)

is known as the Fermi integral with one-half index and the quantity

$$N_c = 4\pi \left(\frac{2m_n^* kT}{h^2}\right)^{3/2} \tag{2.3.6}$$

represents the effective density of states in the conduction band. In fact, in the case of a completely nondegenerate electron gas ($\mu^* \ll 0$) the Fermi integral in Eq. (2.3.5) becomes $\sqrt{\pi}\, e^{\mu^*}/2$, so the total number of electrons is then

$$n = \frac{\sqrt{\pi}}{2} N_c e^{\mu^*}, \tag{2.3.7}$$

and, consequently, the value of n/N_c is equal, to within the coefficient $\sqrt{\pi}/2$, the value of the distribution function Eq. (2.2.16), which gives the degree of occupation of the level E_c in the classical limit. Thus, the quantity N_c is equivalent to the density of all possible states in the conduction band of a completely nondegenerate semiconductor.

Substituting the numerical values of the constants into Eq. (2.3.6) we obtain:

$$N_c = 5.437 \cdot 10^{15} \cdot \beta^{3/2} T^{3/2}, \tag{2.3.8}$$

where $\beta = m_n^*/m_0$.

The dependence (2.3.8) is shown graphically in Fig. 2.5.

If the conduction band of a semiconductor is not parabolic, it is necessary to substitute into Eq. (2.3.1) the expression for $\rho(E)$ for the nonparabolic case: Eq. (1.1.18). This gives

$$n = N_c F_{1/2}(\mu^*) + \frac{5akT}{h^2} N_c F_{3/2}(\mu^*), \tag{2.3.9}$$

where

$$F_{3/2} = \int_0^\infty \frac{(\varepsilon^*)^{3/2}\, d\varepsilon^*}{1 + \exp(\varepsilon^* - \mu^*)}$$

which is also called the Fermi integral but this time with the three-halves index, and not the one-half index as in Eq. (2.3.5).

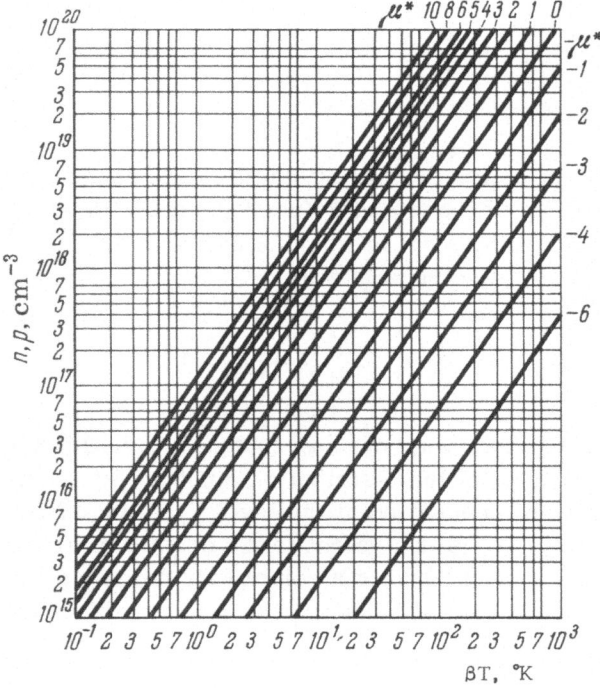

Fig. 2.6. Nomogram for the calculation of electron and hole
densities in allowed bands of a semiconductor.

Dividing this expression by Eq. (2.3.4), we can easily show
that

$$\frac{5akT}{\hbar^2}\frac{F_{3/_2}(\mu^*)}{F_{1/_2}(\mu^*)} = \frac{n'-n}{n} \qquad (2.3.10)$$

is simply the relative difference in the number of electrons in a
nonparabolic band compared with a parabolic band.

The dependence of the electron density on the parameter βT
for a parabolic band is shown graphically in Fig. 2.6.

Criterion of Degeneracy. As in the preceding sec-
tion, we shall find the relative error δ_n in the determination of the
number of electrons, using a formula valid in the classical case

[Eq. (2.3.7)] and a general formula [Eq. (2.3.4)]:

$$\delta_n = 1 - \frac{F_{1/2}(\mu^*)}{\frac{\sqrt{\pi}}{2}\exp\mu^*}. \tag{2.3.11}$$

The dependence of δ_n on the reduced Fermi level is also shown in Fig. 2.4. If an error of 8% is permissible, then the "limit" of degeneracy is $\mu_n^* = -1.4$. We can see that $\mu_f^* = \mu_n^*$.

The usual degeneracy limit $\mu^* = 0$ gives rise to an error of ~20% (Fig. 2.4) in the determination of the electron density in the conduction band.

It is clear from Fig. 2.6 that the transition from the nondegenerate to the degenerate state is easier when the parameter βT is small, i.e., the smaller the value of $\beta T|_{\mu^* = 0}$, the lower are the densities at which degeneracy begins. This parameter may be called the reduced degeneracy temperature.

Usually, the degeneracy temperature is understood to be the true value of the temperature at which the electron gas in a given semiconductor (with a given value of β) becomes degenerate. This definition is valid only if the effective carrier masses are constant. If the value of β varies with temperature, as is the case in compounds of Bi and Te [29], or if it varies with the electron density, as in the case of InSb and GaAs [7], it is more convenient to use the reduced degeneracy temperature.

Density of Holes in the Valence Band. The probability that a state E is occupied by a hole is equivalent to the probability that this state is not occupied by an electron, i.e.,

$$f_p(E) = 1 - f(E) = \frac{1}{1 - \exp\frac{\mu - E}{kT}}. \tag{2.3.12}$$

Since the values of the energy in the valence band are measured from E_V downwards, the total number of holes per unit volume is given by:

$$p = \int_{-\infty}^{E_v} \rho_p(E) f_p(E) \, dE. \tag{2.3.13}$$

Substituting $\rho_p(E)$, similar to Eq. (1.1.16), and $f_p(E)$ from Eq. (2.3.12), we obtain

$$p = 4\pi \left(\frac{2m_p^*}{h^2}\right)^{3/2} \int_{-\infty}^{E_v} \frac{(E_v - E)^{1/2}\, dE}{1 + \exp\dfrac{\mu - E}{kT}}. \qquad (2.3.14)$$

We shall also introduce a dimensionless energy for the valence band

$$E_p^* = \frac{E_v - E}{kT}.$$

Comparing this expression with Eq. (2.3.3), we easily find that:

$$E_p^* = -\, \varepsilon^* - E_g^*, \qquad (2.3.15)$$

where E_g^* is the reduced forbidden band width, equal to

$$E_g^* = \frac{E_c - E_v}{kT}.$$

Using this notation, the expression for the density of holes [Eq. (2.3.14)] can be rewritten in a form similar to the expression for the electron density:

$$p = N_v \cdot F_{1/2}(-E_g^* - \mu^*), \qquad (2.3.16)$$

where

$$N_v = 4\pi \left(\frac{2m_p^* kT}{h^2}\right)^{3/2} \qquad (2.3.17)$$

is the effective density of states in the valence band and $F_{1/2}(-E_g^* - \mu^*)$ is an integral fully analogous to that given by Eq. (2.3.5).

We can determine N_v using the same graph as for N_c (Fig. 2.5) if we define β as the ratio m_p^*/m_0. Similarly, we can use Fig. 2.6 to calculate the density of states in the valence band.

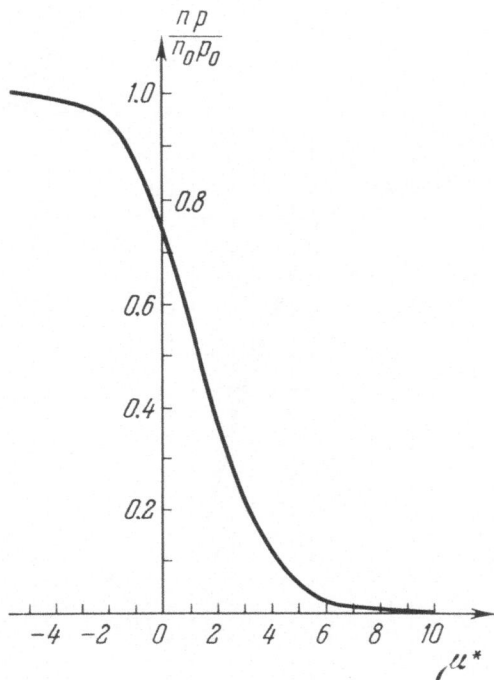

Fig. 2.7. Dependence of the product np on the degree of
degeneracy of the electron gas in a crystal (at T = 300°K).

Law of Mass Action for Carriers.   From an elementary theory of semiconductors [8], it is known that a nondegenerate carrier gas obeys the relationship

$$n_0 p_0 = n_i^2 = \frac{\pi}{4} N_c N_v e^{-\frac{E_g}{kT}}, \tag{2.3.18}$$

which depends only on the properties of a semiconductor (m_n^*, m_p^*,
and E_g) if the temperature is constant. The subscript "0" indicates densities in the nondegenerate case.

Multiplying Eq. (2.3.4) by Eq. (2.3.16), we obtain the corresponding expression for any degree of degeneracy:

$$np = N_c N_v F_{1/2}(\mu^*) F_{1/2}(-\mu^* - E_g^*). \tag{2.3.19}$$

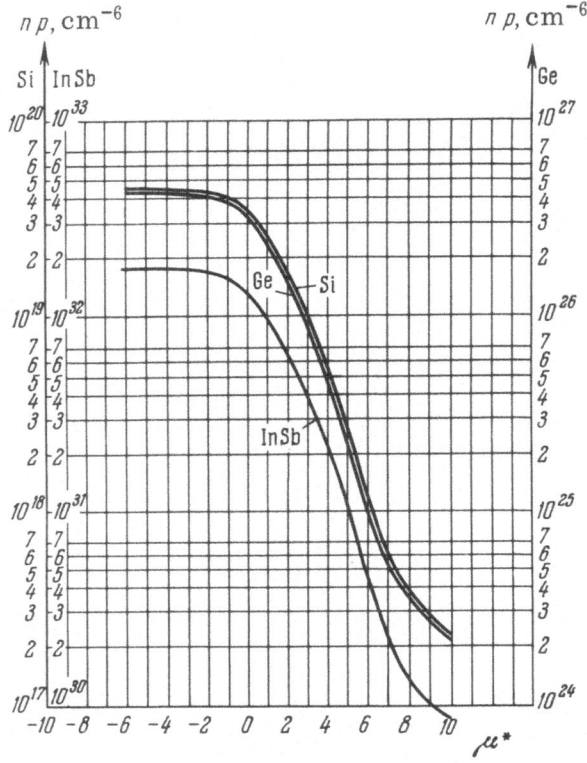

Fig. 2.8. Absolute values of the product np as a function of the
degree of doping of Ge, Si, and InSb at various temperatures.

We can easily show that in the nondegenerate case ($\mu^* \ll 0$)
this expression does indeed reduce to Eq. (2.3.18) since both integrals are then approximated by exponential functions of the type
$\sqrt{\pi} e^{\mu^*}/2$ and $\sqrt{\pi} e^{-\mu^* - E_g^*}/2$ (cf. §2.5). By way of illustration,
the ratio

$$\frac{np}{n_0 p_0} = \frac{4}{\pi} \frac{F_{1/2}(\mu^*) F_{1/2}(-\mu^* - E_g^*)}{\exp(-E_g^*)} \qquad (2.3.20)$$

is plotted in Fig. 2.7.

We can easily show that in practice the curve in Fig. 2.7 is
the same for all common semiconductors because the high value
of E_g^* compared with μ^* allows us to write Eq. (2.3.20) in an ap-

proximate form:

$$\frac{np}{n_0 p_0} \approx \frac{2}{\sqrt{\pi}} e^{-\mu^*} F_{1/2}(\mu^*). \qquad (2.3.21)$$

Comparing it with Eq. (2.3.11), we see that

$$\frac{np}{n_0 p_0} \approx 1 - \delta_n,$$

i.e., the difference between the products np and $n_0 p_0$ is equal to the error in the calculation of the electron density using the Boltzmann distribution rather than the Fermi distribution.

In general, this result is trivial because in the case of carrier gas degeneracy in one of the bands, the carrier gas of the opposite sign remains nondegenerate.

The absolute values of np for various semiconductors are naturally different (Fig. 2.8).

Density of Electrons at Local Levels. In principle, a local level, for example at a donor impurity center, can also be occupied by two electrons with oppositely directed spins. Actually, there are either no electrons at such a center or only one electron, because the second electron cannot be captured by such a level in view of the strong electrostatic repulsion between electrons. In other words, in allowed bands we have treated electrons as noninteracting particles but at local levels electrons must be treated as strongly interacting (repulsing) particles. This condition affects considerably the form of the distribution function for localized electrons. We shall not prove this [9, 10]. It follows from Eq. (2.2.4):

$$e^{-\frac{\Omega_l}{kT}} = \sum_{\Delta N_1, \Delta N_2} \exp\left[\frac{\mu(\Delta N_1 + \Delta N_2) - E_1 \Delta N_1 - E_2 \Delta N_2 - A \Delta N_1 \Delta N_2}{kT}\right].$$

$$(2.3.22)$$

Here, Ω_l refers to a local center; the energy of an impurity level can be written in the form:

$$E = \sum_k E_k \Delta N_k = E_1 \Delta N_1 + E_2 \Delta N_2 + A \Delta N_1 \Delta N_2,$$

where $E_1 = E_2 = -E_d$ is the ionization energy of a donor center and

A is the energy of interaction between two electrons at one impurity center.

The quantities ΔN_1 and ΔN_2 can assume the following values:

Center free of electrons: $\Delta N_1 = 0, \Delta N_2 = 0;$
Center occupied by first electron: $\Delta N_1 = 1, \Delta N_2 = 0;$
Center occupied by second electron: $\Delta N_1 = 0, \Delta N_2 = 1;$
Center occupied by both electrons: $\Delta N_1 = 1, \Delta N_2 = 1.$

Using these values, we can sum Eq. (2.3.22) to obtain:

$$e^{-\frac{\Omega_l}{kT}} = 1 + 2\exp\left(\frac{\mu + E_d}{kT}\right) + \exp\frac{2(\mu + E_d) - A}{kT}. \qquad (2.3.23)$$

Since the energy A is large because of the strong repulsion between electrons, the last term on the right-hand side of Eq. (2.3.23) is much less than unity and it can be neglected. Then, after taking the logarithms of this expression, we obtain the thermodynamic potential

$$\Omega_l = -kT\ln\left[1 + 2\exp\left(\frac{\mu + E_d}{kT}\right)\right]. \qquad (2.3.24)$$

On the other hand, the following thermodynamic relationship applies to a local center and to any other energy level:

$$\mu\,dn = dF + S\,dT,$$

which yields the following expression at constant temperature when F is replaced by the expression in Eq. (2.1.17):

$$n = -\left(\frac{\partial\Omega}{\partial\mu}\right)_T, \qquad (2.3.25)$$

i.e., to find the average number of localized electrons at a donor center, it is necessary to differentiate Eq. (2.3.24):

$$n_l = \frac{1}{1 + \frac{1}{2}\exp\left(-E_d^* - \mu^*\right)}. \qquad (2.3.26)$$

If the total number of such local centers is N_d, the density of all localized electrons is given by:

$$n_d = \frac{N_d}{1 + \frac{1}{2} \exp\left(-E_d^* - \mu^*\right)}. \tag{2.3.27}$$

If the impurity levels are of the acceptor type, we can similarly obtain the average density of holes p_a at these levels:

$$p_a = \frac{N_a}{1 + 2 \exp\left(-E_a^* + \mu^* + F_g^*\right)}. \tag{2.3.28}$$

§2.4.   Determination of the Fermi Level

Neutrality Equation. Statistical physics methods allow us to find an equation for the determination of the chemical potential in each special case. We shall consider the case of an extrinsic n-type semiconductor with donor local centers. As in the preceding sections, we shall regard the energy of interaction between conduction electrons as very large. Then, the thermodynamic potential of the whole system Ω is the sum of the potentials of the conduction band Ω_c, of the valence band Ω_v, and of the local donor levels Ω_d:

$$\Omega = \Omega_c + \Omega_v + \Omega_d. \tag{2.4.1}$$

Differentiating Ω with respect to μ^* we obtain – in accordance with Eq. (2.3.25) – the following equation:

$$n = n_c + n_v + n_d.$$

On the other hand, $n = N_0 + N_d$ (N_0 is the number of states in the valence band), so that we obtain:

$$n_c = (N_d - n_d) + p. \tag{2.4.2}$$

This equation shows that the number of conduction electrons is equal to the total number of holes at the impurity levels, $(N_d - n_d)$, and in the valence band (p). Therefore, this equation is known as the equation of neutrality for an extrinsic n-type semiconductor.

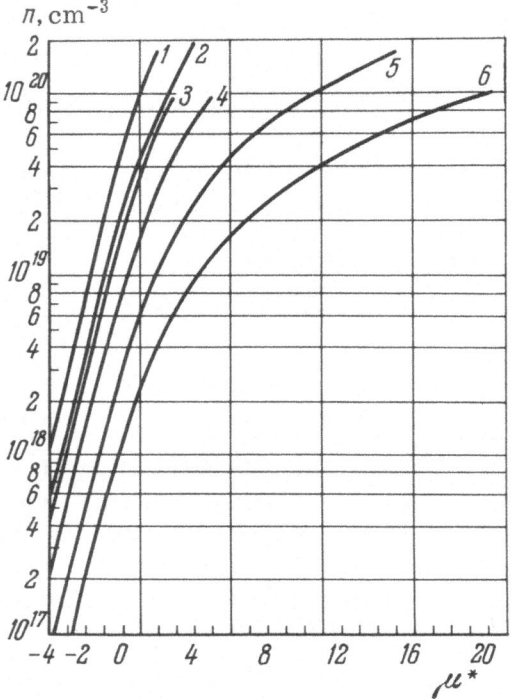

Fig. 2.9. Solution of the neutrality equation for Ge and Si containing one type of donor. Si: 1) 500°K; 2) 300°K; 5) 78°K. Ge: 3) 500°K; 4) 300°K; 6) 78 °K.

Substituting into Eq. (2.4.2) the values of n_c, n_v, and n_d from Eqs. (2.3.4), (2.3.16), and (2.3.27), we obtain the neutrality equation in its explicit form

$$4\pi\left(\frac{2m_n^*kT}{h^2}\right)^{3/2} F_{1/2}(\mu^*) = \frac{N_d}{2e^{E_d^*+\mu^*}+1} + 4\pi\left(\frac{2m_p^*kT}{h^2}\right)^{3/2} F_{1/2}(-E_g^*-\mu^*).$$

$$(2.4.3)$$

In the case of an extrinsic p-type semiconductor, the neutrality equation is

$$p = (N_a - p_a) + n_c. \qquad (2.4.4)$$

Substituting into the above expression Eqs. (2.3.4), (2.3.16), and

(2.3.28), we obtain

$$4\pi \left(\frac{2m_p^* kT}{h^2}\right)^{3/2} F_{1/2}(- E_g^* - \mu^*) = \frac{N_a}{\frac{1}{2} e^{E_a^* - E_g^* - \mu^*} + 1} + 4\pi \left(\frac{2m_n^* kT}{h^2}\right)^{3/2} F_{1/2}(\mu^*).$$

$$(2.4.5)$$

In the presence of donors and acceptors in a semiconductor, the neutrality equation becomes

$$(N_d - n_d) + p = (N_a - p_a) + n_c, \tag{2.4.6}$$

whose explicit form the reader can work out for himself.

These neutrality equations are the starting points in the determination of the chemical potential and its dependence on the carrier density and temperature. However, the neutrality equations contain the Fermi integrals and, therefore, they can be solved analytically in the complete absence of degeneracy but only in the simplest cases.

Heavily Doped Semiconductors. When the concentration of impurities (for example, the concentration of donors) is high, we may assume that the impurities are ionized over a wide range of temperatures. This condition is satisfied for donor concentrations exceeding $1.5 \cdot 10^{17}$ cm^{-3} in germanium and $5 \cdot 10^{18}$ cm^{-3} in silicon [11, 12].

We shall consider only temperatures at which the intrinsic conduction is still unimportant. When this restriction is made, the values of p and n_d can be assumed to be equal to zero and Eq. (2.4.2) simplifies:

$$n_c = N_d$$

or

$$N_c F_{1/2}(\mu^*) = N_d. \tag{2.4.7}$$

Figure 2.9 shows the solution of this equation for germanium and silicon.

The solution given by Eq. (2.4.7) is not valid at impurity concentrations in germanium lower than $1.5 \cdot 10^{17}$ cm^{-3}, at very low temperatures when not all the impurities are ionized, or at very high temperatures when the intrinsic conduction cannot be neglected.

In such cases, we can find μ^* either by numerical solution or by developing special approximate methods, which can be divided into two groups: graphical and analytic.

Graphical Methods for Determining the Chemical Potential. Various graphical methods for determining μ^* have been proposed [13-15]. They all have one common deficiency: they are either applicable in certain special cases or they ignore the electron gas degeneracy. A general method is described in [16] but this method is very cumbersome and is not used much by experimenters.

A simple method is described in [17]. It is based on a suitable selection of the scale which permits the writing of the neutrality equation in such a form that only two universal curves need be used for any value of E_g, E_d, E_a, N_a, N_d, m_n^*, m_p^*, and T.

We shall illustrate the use of this method in the case of impurity conduction in a semiconductor with one donor impurity.

Equation (2.4.2) becomes:

$$n_c = N_d - n_d,$$

or, in the explicit form

$$C \, (\beta_n T)^{3/2} F_{1/2} \, (\mu^*) = \frac{N_d}{1 + 2e^{\mu^* + E_d^*}}, \qquad (2.4.8)$$

where

$$C = 4\pi \left(\frac{2k m_0}{h^2}\right)^{3/2}, \qquad \beta_n = \frac{m_n^*}{m_0}.$$

Taking logarithms of Eq. (2.4.8), we obtain:

$$\log C F_{1/2} \, (\mu^*) + \frac{3}{2} \, (\log T + \log \beta_n) = \log N_d - \log \, [1 + 2e^{\mu^* + E_d^*}]. \qquad (2.4.9)$$

We note that the second term on the right-hand side of this equation depends on μ^* in the same way as log $(1+x)$ on 2.3 log x, provided a shift is made along the abscissa. In fact, if we assume that

$2e^{\mu*} + E_d^* = x$, then

$$2.3 \log x - 2.3 \log 2 - E_d^* = \mu^*.$$

Hence, we see that $\log (1+x)$ depends in the same way on $2.3 \log x$ as $\log (1 + 2e^{\mu*} + E_d^*)$ on μ^* if the origin of coordinates is shifted along the abscissa to the left by an amount $- (2.3 \log 2 + E_d^*) = - (0.7 + E_d^*)$. Thus, in Eq. (2.4.9), the functions depending on μ^* do not include the parameters of a semiconductor. The left-hand side of Eq. (2.4.9) is given by a graph of $CF_{1/2} (\mu^*) = \vartheta (\mu^*)$ if the origin is shifted along the ordinate by an amount $A = (^3/_2) (\log T - \log \beta_n)$ in the upward direction for $A > 0$ or in the downward direction for $A < 0$.

The right-hand side of Eq. (2.4.9) is given by a graph of $\varphi(\mu^*)$, which represents the dependence of $\log (1+x)$ on $2.3 \log x$, if the origin is shifted along the abscissa to the left by $(0.7 + E_d^*)$ and along the ordinate by $\log N_d$. The value of μ^* is found from the intersection of the curves $\vartheta (\mu^*)$ and $\varphi(\mu^*)$ when the origin of coordinates is shifted in this way (Fig. 2.10). In practice, it is more convenient to use a curve $[18 - \varphi(\mu^*)]$ rather than $\varphi(\mu^*)$; the former curve is shown in Fig. 2.10 and the origin of coordinates is shifted along the ordinate by $\log N_d/10^{18}$. The origin of the curve $[18 - \varphi(\mu^*)]$ can be placed right at the start at a point -0.7 and then the curve needs to be shifted by only E_d^* in the determination of μ^*. The coordinate grid on which these curves are plotted with suitable shifts is shown in Fig. 2.11.

By way of example, we shall find μ^* for germanium containing 10^{19} cm^{-3} of arsenic, at 500°K. We shall assume that $E_d = 0.01$ eV and $\beta_n = 1/2$.

First, we calculate the value of A and find it to be 3.6. We superimpose on the grid the curve $\vartheta(\mu^*)$ of Fig. 2.10, shifting it along the ordinate by $+ 3.6$ (Fig. 2.12). Then, we superimpose on the same grid the curve $[18 - \varphi(\mu^*)]$, shifting it along the abscissa to the left by an amount $E_d^* = E_d/kT = 0.01/0.043 = 0.23$ and along the ordinate by $\log (N_d/10^{18}) = + 1$, which is also shown in Fig. 2.12. Next, we find μ^* from the point of intersection of these curves.

This method becomes more complicated if intrinsic conduction is allowed for. However, basically, the procedure for finding μ^* remains the same. In the intrinsic conduction case a curve $\vartheta'(\mu^*)$ is not governed by the integral $F_{1/2}(\mu^*)$ but by the integral $F_{1/2}(-\mu^* - E_g^*)$. Therefore, it is obtained by mirror reflection of the previous curve, $\vartheta (\mu^*)$, and is shifted along the abscissa by

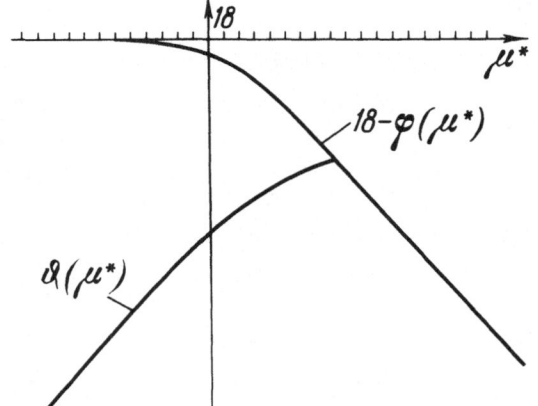

Fig. 2.10. Graphs of the functions $\vartheta(\mu^*)$ and $[18 - \varphi(\mu^*)]$.

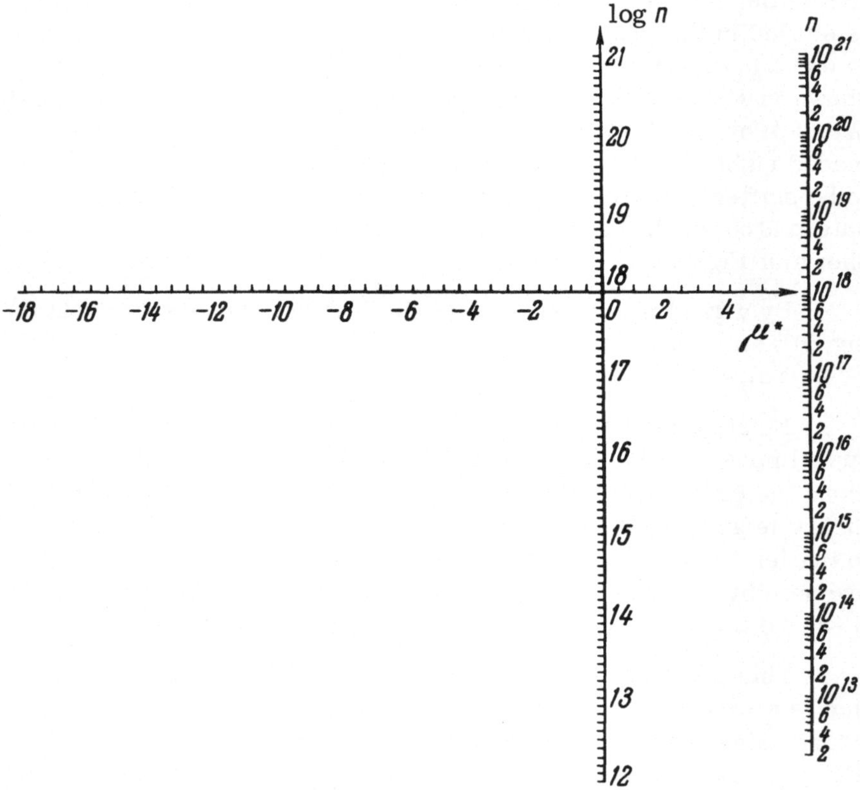

Fig. 2.11. Coordinate grid for the neutrality equation.

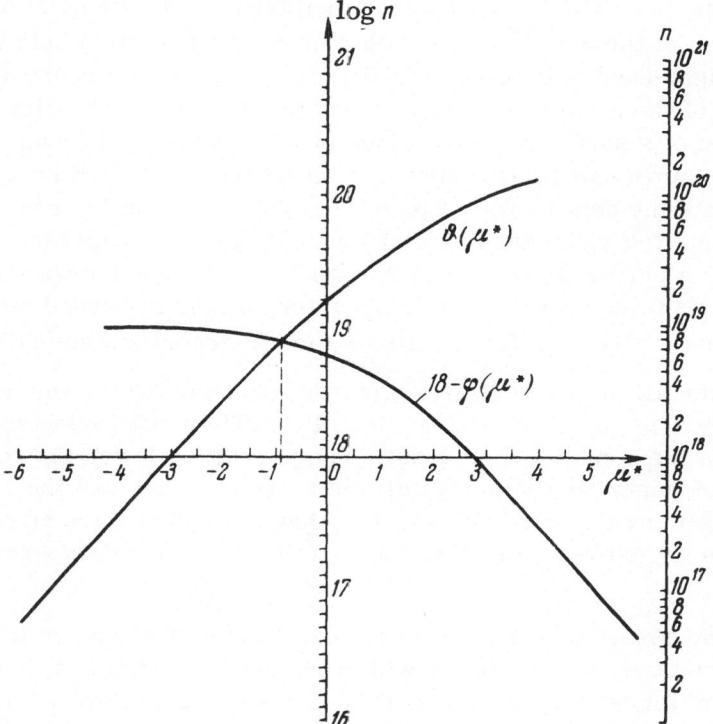

Fig. 2.12. Example of the application of the graphical method.

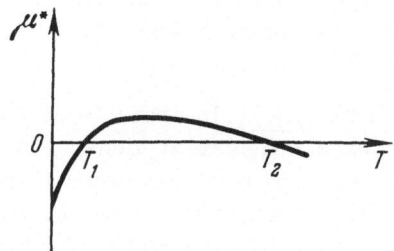

Fig. 2.13. Temperature dependence of the reduced Fermi level for a crystal with one type of donor center, on the assumption that E_d = const.

an amount E^*_g. The Fermi level is then found from the point of intersection of the $\vartheta'(\mu^*)$ curve with a curve representing both terms on the right-hand side of Eq. (2.4.3), i.e., with a curve representing the sum of the curves $[18 - \varphi(\mu^*)]$ and $\vartheta(\mu^*)$. However, with the exception of a small range of values of μ^*, where $p \approx N_d - n_d$, the summed curve coincides either with the dependence of $(N_d - n_d)$ on μ^* or with the dependence of p on μ^*. Thus, outside the region $p \approx N_d - n_d$, the right-hand side of Eq. (2.4.3) can be represented either by a universal graph of $[18 - \varphi(\mu^*)]$ or by a universal graph of $\vartheta'(\mu^*)$ shifted by E^*_g. A suitable interpolation is needed only in the region $p \approx N_d - n_d$; for details of such interpolation see [17].

Like all the approximate methods, this method for the determination of μ^* is of limited applicability. The most important limitation is the condition $E_d = \text{const}$. Therefore, the lower the concentration of impurities in a crystal, the more accurate are the values of μ^* found by this method. On the other hand, there are no temperature limitations in this method, even on the high-temperature side.

The assumption $E_d = \text{const}$ means that the dependence $\mu^*(T)$, considered over a sufficiently wide range of temperatures, has a maximum for an n-type crystal (Fig. 2.13) and a minimum for a p-type semiconductor. On the low-temperature side, there is a degeneracy temperature T_1 at which the carriers "freeze." On the high-temperature side, there is another degeneracy temperature T_2, at which the electrons transferred from the valence to conduction band begin to make an important contribution.

In fact, heavily doped semiconductors have no degeneracy temperature T_1 because they do not have shallow impurity levels [18]. Therefore, at low temperatures this method of finding μ^* should be used with caution.

Analytic Methods. In order to be able to describe analytically the temperature and carrier-density dependences of the Fermi level, it is necessary first to approximate the integral $F_{1/2}(\mu^*)$, occurring in the neutrality equation, with some simple expression which can be used to solve the neutrality equation for μ^*. The approximations usually employed will be considered in the next section. A detailed discussion of the analytic methods is of no great interest because they reduce to a simple algebraic solution of the neutrality equation. Moreover, the analytic solutions are less general than the graphical methods because all approximations limit their validity.

Corrections to the Fermi Level. In heavily doped semiconductors, when $\mu^* \gg 1$, the interaction between electrons themselves and between electrons and impurities alters somewhat the values of μ^* calculated in the manner described earlier. The interaction between electrons themselves is the more important [19, 20]. A general expression for the Fermi level μ, which includes corrections for both types of interaction, has the following form [19]:

$$\mu = \mu_0 \left[1 - \frac{2}{\pi} \mu_0^{-1/2} - \frac{\pi^2}{12} \left(\frac{\mu_0}{kT} \right)^{-2} - \pi^{-1} \mu_0^{-1/2} \left(\frac{\mu_0}{kT} \right)^{-2} \right], \qquad (2.4.10)$$

where μ_0 is the value of the Fermi level calculated without these corrections.

When the carrier density is increased, μ tends to μ_0. Therefore, the correction represented by Eq. (2.4.10) is not very important.

§2.5. Fermi Integrals

Numerical Values of the Integrals. The integrals $F_{1/2}(\mu^*)$ and $F_{3/2}(\mu^*)$ mentioned in the preceding sections are widely used. Tables of values of the Fermi integrals and approximation formulas are used in such cases. Fermi integrals with various subscripts will be used frequently in many problems discussed in the present monograph.

TABLE 2.1. Tabulated Values of the Fermi Integrals

Tabulated function	Values of n	Range of μ^*	Interval	Source
$\dfrac{1}{n!} F_n(\mu^*)$	—1; 0; 1; 2; 3; 4	$0 \leqslant \mu^* \leqslant 10$	0.1	[21]
$\dfrac{1}{n!} F_n(\mu^*)$	1; 2; 3; 4	$-4 \leqslant \mu^* \leqslant 0$	0.1	[22]
$F_n(\mu^*)$	1/2; 3/2	$-4 \leqslant \mu^* \leqslant 20$	0.1	[23]
$F_n(\mu^*)$	—1/2; 1/2; 3/2; 5/2; 7/2; 9/2; 11/2	$-4 \leqslant \mu^* \leqslant 20$	0.1	[24]
$\dfrac{1}{\Gamma(n+1)} F_n(\mu^*)$	—1; 0; 2; 3; 4; —3/2; —1/2; 1/2; 3/2; 5/2; 7/2	$-4 \leqslant \mu^* \leqslant 4$ $4 \leqslant \mu^* \leqslant 10$	0.1 0.2	[25]

TABLE 2.2

n	$-1/2$	0	$1/2$	1	$3/2$	2	$5/2$
$C(n)$	0.6	0.375	0.25	0.18	0.12	0.08	0.06

n	3	$7/2$	4	$9/2$	5	$11/2$	6
$C(n)$	0.04	0.03	0.02	0.015	0.01	0.0075	0.005

Table 2.1 shows where the numerical values of the Fermi integrals can be found.

Methods for the Calculation of the Fermi Integrals. Since the numerator in the integrand of a Fermi integral

$$F_n(\mu^*) = \int_0^\infty \frac{x^n dx}{1 + \exp(x - \mu^*)} \qquad (2.5.1)$$

remains finite when $x \to 0$, it follows that when $n < -1$, the integral in Eq. (2.5.1) diverges and, therefore, we can calculate it only when $n \geq -1$. In one special case, when $n = 0$, the Fermi integral can be expressed in an analytic form:

$$F_0(\mu^*) = \int_0^\infty \frac{dx}{1 + \exp(x - \mu^*)} = \int_0^\infty \frac{\exp(\mu^* - x)\, dx}{1 + \exp(\mu^* - x)} = \ln[1 + \exp(\mu^*)]. \qquad (2.5.2)$$

Two cases should be distinguished in the calculation of the integral in Eq. (2.5.1): a strongly degenerate and a completely nondegenerate electron gas, i.e., $\mu^* \gg 0$ and $\mu^* \ll 0$.

Completely Nondegenerate Electron Gas: $\mu^* \ll 0$. We shall write Eq. (2.5.1) in the form:

$$F_n(\mu^*) = \int_0^\infty \frac{x^n \exp(\mu^* - x)\, dx}{1 + \exp(\mu^* - x)} . \qquad (2.5.3)$$

Since $\mu^* < 0$, it follows that $\exp(\mu^* - x) < 1$ over the whole integration range. Therefore, the denominator in the integrand in Eq. (2.5.3) can be expanded as a series and we can write:

$$F_n(\mu^*) = \int_0^\infty \left\{ x^n \exp(\mu^* - x) \sum_{k=0}^\infty (-1)^k \exp[k(\mu^* - x)] \right\} dx =$$

$$= \sum_{k=0}^\infty (-1)^k \exp[(k+1)\mu^*] \cdot \int_0^\infty x^n \exp[-x(k+1)] \, dx.$$

In the above expression, we shall assume that $(k+1)\, x = u$, so that

$$F_n(\mu^*) = \sum_{k=0}^\infty (-1)^k \frac{\exp[(k+1)\mu^*]}{(k+1)^{n+1}} \int_0^\infty u^n \exp(-u) \, du,$$

or

$$F_n(\mu^*) = \sum_{k=0}^\infty (-1)^k \frac{\exp[(k+1)\mu^*]}{(k+1)^{n+1}} \Gamma(n+1). \tag{2.5.4}$$

where $\Gamma(n+1)$ is a gamma function whose values are tabulated, for example, in [26]. If we use only the first term in the series (2.5.4), we obtain:

$$F_n(\mu^*) \approx \Gamma(n+1) \exp(\mu^*). \tag{2.5.5}$$

This expression represents the most widely used approximation for the Fermi integrals in the nondegenerate case. Hence

$$
\left.
\begin{aligned}
&F_{-1/2}(\mu^*) = \sqrt{\pi}\exp(\mu^*), \quad F_3(\mu^*) = 6\exp(\mu^*), \\
&F_0(\mu^*) = \exp(\mu^*), \quad\quad F_{7/2}(\mu^*) = \frac{105}{16}\sqrt{\pi}\exp(\mu^*), \\
&F_{1/2}(\mu^*) = \frac{\sqrt{\pi}}{2}\exp(\mu^*), \quad F_4(\mu^*) = 24\exp(\mu^*), \\
&F_1(\mu^*) = \exp(\mu^*), \quad\quad F_{9/2}(\mu^*) = \frac{945}{32}\sqrt{\pi}\exp(\mu^*), \\
&F_{3/2}(\mu^*) = \frac{3\sqrt{\pi}}{4}\exp(\mu^*), \quad F_5(\mu^*) = 120\exp(\mu^*), \\
&F_2(\mu^*) = 2\exp(\mu^*), \quad F_{11/2}(\mu^*) = \frac{10395}{64}\sqrt{\pi}\exp(\mu^*), \\
&F_{5/2}(\mu^*) = \frac{15\sqrt{\pi}}{8}\exp(\mu^*), \quad F_6(\mu^*) = 720\exp(\mu^*).
\end{aligned}
\right\} \tag{2.5.6}
$$

TABLE 2.3

k	1	2	3	4
B_k	$1/6$	$1/30$	$1/42$	$1/30$
t_{2k}	$\dfrac{\pi^2}{12}$	$\dfrac{7\pi^4}{720}$	$\dfrac{31\pi^6}{30,240}$	$\dfrac{127\pi^8}{1,209,600}$

The approximations just given are most accurate for high absolute values of $|\mu^*|$, i.e., they represent the case of a completely non-degenerate electron gas. The limit of applicability of the expressions given in Eq. (2.5.6) can be easily found by calculating the error in the determination of the values of F_n found by means of these expressions.

According to our calculations, this upper limit is $\mu^* = -1$. When $\mu^* = -1$, the errors are:

Integral	$F_{-1/2}$	F_0	$F_{1/2}$	F_1	$F_{3/2}$	F_2	$F_{5/2}$	F_3	$F_{7/2}$
Error, %	25	18	12	9	7	5	3	2	1

For integrals with indices exceeding 7/2, the errors are even smaller and, therefore, they are not tabulated.

A higher-order approximation is easily obtained by taking the first two terms in the series of Eq. (2.5.4):

$$F_n(\mu^*) \approx \Gamma(n+1)\left[\exp(\mu^*) - \frac{\exp(2\mu^*)}{2^{n+1}}\right] = \frac{\Gamma(n+1)}{\exp(-\mu^*)}\left[1 - 2^{-n-1}\exp(\mu^*)\right]. \tag{2.5.7}$$

This approximation is fairly inconvenient and, therefore, Ehrenberg [27] has suggested replacing Eq. (2.5.7) with a simpler expression of the type

$$F_n(\mu^*) \approx \frac{\Gamma(n+1)}{\exp(-\mu^*) + C(n)}, \tag{2.5.8}$$

where C(n) replaces the second term in Eq. (2.5.7). The numerical values of C(n) are given in Table 2.2.

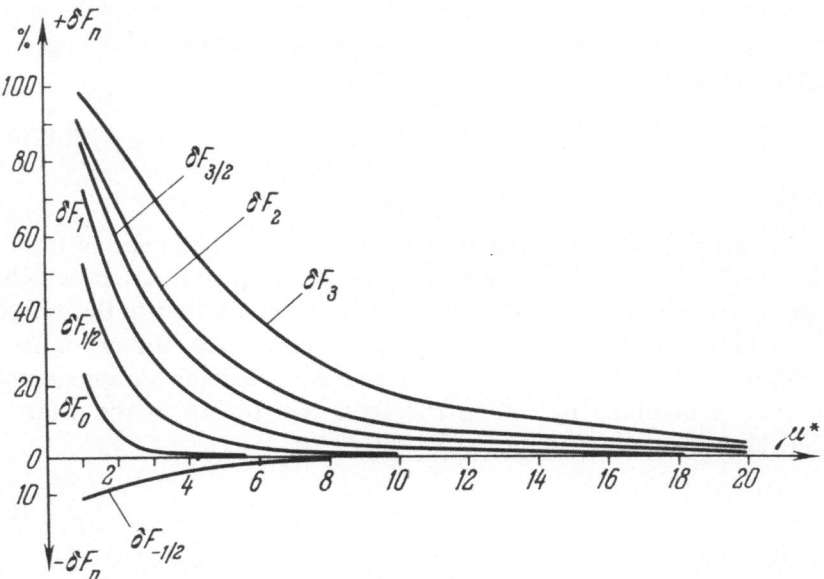

Fig. 2.14. Errors in the determination of the Fermi integrals using Eq. (2.5.11).

The approximation of Eq. (2.5.8) with C(n) listed in Table 2.2 gives an error of less than 5% in the range $\mu* < -1.8$.

Strongly Degenerate Electron Gas: $\mu* \gg 0$. Dingle [21] has deduced the following general formula for the calculation of $F_n(\mu*)$ in the case of high positive values of $\mu*$:

$$F_n(\mu*) = 2\Gamma(n+1)(\mu*)^{n+1} \sum_{k=0}^{\frac{n+1}{2}} \frac{t_{2k}(\mu*)^{-2k}}{\Gamma(n+2-2k)}. \qquad (2.5.9)$$

The upper limit of summation, $(n+1)/2$, shows that we must take the largest integer contained in $(n+1)/2$ (for example, the upper limit is 1 when $n=2$; 2 when $n=3$, etc.). The functions t_{2k} are defined thus:

$$t_0 = \frac{1}{2}; \qquad t_{2k} = \frac{1}{2}(2\pi)^{2k}[1 - 2^{(1-2k)}]\frac{B_k}{(2k)!}, \qquad (2.5.10)$$

where B_k are the Bernoulli numbers listed in Table 2.3. The same table includes the first four functions t_{2k}.

If we use only the first term in Eq. (2.5.9), we obtain the simplest approximation for the Fermi integrals valid in the strongly degenerate case:

$$F_n(\mu^*) = \frac{(\mu^*)^{n+1}}{n+1}.$$

(2.5.11)

Figure 2.14 shows the errors committed by the use of Eq. (2.5.11). We can see that the validity of this approximation, which we shall call the first-order approximation, is limited to the range of very high values of μ^*. As the index n increases, the error in the determination of F_n also increases. A second-order approximation can be obtained by taking the first two terms in the sum of Eq. (2.5.9):

$$F_n(\mu^*) = \frac{(\mu^*)^{n+1}}{n+1} + \frac{n\pi^2}{6}(\mu^*)^{n-1}.$$

(2.5.12)

Then, the integrals $F_n(\mu^*)$ assume the following form for $\mu^* \gg 0$:

$$
\left.
\begin{aligned}
F_{-1/2} &= 2\left((\mu^*)^{1/2} + \frac{\pi^2}{24}(\mu^*)^{-3/2}\right), \quad F_0 = \mu^*, \\
F_{1/2} &= \frac{2}{3}\left((\mu^*)^{3/2} + \frac{\pi^2}{8}(\mu^*)^{-1/2}\right), \\
F_1 &= \frac{1}{2}\left((\mu^*)^2 + \frac{\pi^2}{3}\right), \\
F_{3/2} &= \frac{2}{5}\left((\mu^*)^{5/2} + \frac{5\pi^2}{8}(\mu^*)^{1/2}\right), \\
F_2 &= \frac{1}{3}\left((\mu^*)^3 + \pi^2\mu^*\right), \\
F_{5/2} &= \frac{2}{7}\left((\mu^*)^{7/2} + \frac{35\pi^2}{24}(\mu^*)^{3/2}\right), \\
F_3 &= \frac{1}{4}\left((\mu^*)^4 + 2\pi^2(\mu^*)^2\right), \\
F_{7/2} &= \frac{2}{9}\left((\mu^*)^{9/2} + \frac{63\pi^2}{24}(\mu^*)^{5/2}\right), \\
F_4 &= \frac{1}{5}\left((\mu^*)^5 + \frac{10\pi^2}{3}(\mu^*)^3\right), \\
F_{9/2} &= \frac{2}{11}\left((\mu^*)^{11/2} + \frac{33\pi^2}{8}(\mu^*)^{7/2}\right), \\
F_5 &= \frac{1}{6}\left((\mu^*)^6 + 5\pi^2(\mu^*)^4\right), \\
F_{11/2} &= \frac{2}{13}\left((\mu^*)^{13/2} + \frac{143\pi^2}{24}(\mu^*)^{9/2}\right), \\
F_6 &= \frac{1}{7}\left((\mu^*)^7 + 7\pi^2(\mu^*)^5\right).
\end{aligned}
\right\}
$$

(2.5.13)

TABLE 2.4

n	0.0	0.5	1.0	1.5	2.0	3.0	4.0
$R(n)$	−1.645	−0.582	0.000	0.448	0.822	1.437	1.939

We can easily show that by dropping the second terms in these expressions, we obtain the first-order approximation represented by Eq. (2.5.11).

Moderate Degeneracy: $\mu* \approx 0$. The most difficult case to approximate is the intermediate case of small absolute values of $\mu*$. The approximation just given can be used also for $\mu* \approx 0$ but the accuracy is then very low (error of several tens percent). If a higher accuracy is required, it is necessary to take numerical values of $F_n(\mu*)$ from suitable tables.

Recurrence Formulas. In some cases, the value of a Fermi integral of a particular index can be found from the known value of a Fermi integral of another index. Differentiation gives

$$\frac{dF_{n+1}}{dA} = \int_0^\infty \frac{x^{n+1}e^x A^{-2}}{(A^{-1}e^x + 1)^2}\, dx,$$

(2.5.14)

where $A = e^{\mu*}$. We can easily show that

$$\frac{d}{dx}\left(A^{-1}e^x + 1\right)^{-1} = -\frac{A^{-1}e^x}{(A^{-1}e^x + 1)^2}.$$

Substituting this expression into the integrand in Eq. (2.5.14), we obtain:

$$\frac{d}{dA}F_{n+1} = -A^{-1}\int_0^\infty x^{n+1}\frac{d}{dx}(A^{-1}e^x + 1)^{-1}\, dx,$$

which − after integration by parts − gives the following expression for n > −1:

$$\frac{d}{dA} F_{n+1} = (n + 1) A^{-1} F_n.$$

Therefore,

$$F_{n+1} = (n + 1) \int_0^A A^{-1} F_n dA. \qquad (2.5.15)$$

If we now use the expression (2.5.8) for $F_n(\mu^*)$, then

$$F_{n+1} = \Gamma(n + 2) \int_0^A \frac{dA}{1 + AC(n)} = \frac{\Gamma(n + 2)}{C(n)} \ln[1 + AC(n)]. \quad (2.5.16)$$

In exactly the same way, we can find that

$$F_{n+2} = \frac{\Gamma(n + 3)}{C(n)} R(n) [1 + AC(n)], \qquad (2.5.17)$$

where R(n) is a function sufficiently fully tabulated in [28]. Some values of this function are given in Table 2.4.

Chapter 3

Transport Phenomena in Heavily Doped Semiconductors

§3.1. Preliminary Remarks

Transport Equation. So far we have considered carriers under statistical equilibrium conditions. We shall now deal with the processes taking place under the action of external forces (electric and magnetic fields, temperature, etc.) – processes of great practical and theoretical interest. In these processes, carriers are no longer under equilibrium conditions. We thus have phenomena associated with the ordered motion of carriers and known as transport or transfer phenomena.

There are several possible theoretical approaches to transport phenomena. The most widely used is the method of Boltzmann's transport equation [1-3], which can be employed to calculate the carrier mobility, the electrical conductivity, the Hall coefficient, the thermoelectric power, the thermal conductivity, and other transport coefficients representing the behavior of a semiconductor when it is subjected to external forces.

Other well-known methods for the determination of the transport coefficients are: the solution of the equation of motion of carriers, averaged out by a "weighting function" [4, 5]; the density matrix method of Kubo [6]; and the Green's functions method [7].

We shall consider only the transport equation method because it yields, relatively simply, some valuable information on the transport processes, and the results obtained by this method are frequently identical with those found using the other methods [8].

   We shall restrict our discussion to weak external forces which alter (perturb) only the particle distribution function but do not change the total number of particles in a crystal.

   Thus, carriers are acted upon by external forces which tend to order their motion, as well as by collisions which tend to disorder this motion. The transport equation expresses the fact that the dependence of the distribution function of particles $f(\mathbf{k}, \mathbf{r}, t)$ on time is governed by the net effect of these two competing processes:

$$\frac{\partial f}{\partial t} = \left(\frac{\partial f}{\partial t}\right)_\mathbf{t} + \left(\frac{\partial f}{\partial t}\right)_c .$$

(3.1.1)

The first term on the right-hand side of this equation represents the change in the function f due to transport phenomena under the action of external forces; the second term represents the change in the same function due to random collisions between carriers. In this general case, the transport equation can be easily obtained as a consequence of the Liouville theorem (cf. §1 in Chap. 2).

   We shall be interested in steady-state transport phenomena in which the experimentally observed macroscopic effect is the result of numerous microscopic collisions of the particles. For such a state of the system, we may assume that $\partial f/\partial t = 0$. Consequently, Eq. (3.1.1) becomes

$$\left(\frac{\partial f}{\partial t}\right)_\mathbf{t} + \left(\frac{\partial f}{\partial t}\right)_c = 0.$$

(3.1.2)

   A detailed derivation of the transport equation can be found elsewhere (for example in [3]); here we shall give the expression for its components:

$$-\left(\frac{\partial f}{\partial t}\right)_\mathbf{t} = \frac{1}{\hbar}\,(F\,\nabla_k f) + (v\,\nabla f),$$

(3.1.3)

where $\mathbf{F}$ is a generalized external force acting on a carrier in a crystal; $\mathbf{v}$ is the velocity of carriers; $\nabla_\mathbf{k}$ and ∇ are gradients in the wave-number and coordinate spaces, respectively;

$$-\left(\frac{\partial f}{\partial t}\right)_c = \frac{f - f_0}{\tau} = \frac{f_1}{\tau}, \tag{3.1.4}$$

where f_1 is a correction which transforms the original equilibrium distribution function f_0 to a new distribution function f valid under the action of perturbing forces; τ is a constant with the dimensions of time (its physical meaning will be discussed later). Since we are dealing with weak perturbing forces, we may additionally assume that

$$f_1 \ll f_0. \tag{3.1.5}$$

Therefore, the new distribution function f can be represented in the form of an expansion

$$f(\mathbf{k}) = f_0(\mathbf{k}) + \alpha \frac{\partial f_0}{\partial E} + \beta \frac{\partial^2 f_0}{\partial E^2}, \tag{3.1.6}$$

where α and β are coefficients of this expansion. We must bear in mind that $f_0(\mathbf{k}) \equiv f_0(E)$, since

$$f_0 = [1 + \exp(\varepsilon^* - \mu^*)]^{-1} \text{ and } E = \hbar^2 k^2 / 2m^*$$

(we shall consider a parabolic band).

We shall take only the first two terms of the expansion in Eq. (3.1.6) and we shall write the correction to the original distribution function in the form:

$$\alpha \frac{\partial f_0}{\partial E} = f_1(\mathbf{k}) = - C(E) \lambda(\mathbf{k}) \frac{\partial f_0}{\partial E}, \tag{3.1.7}$$

i.e., in the form of a product of certain functions of $\mathbf{k}$ and of E. This is justified because the perturbed distribution function should be governed not only by the energy of the particles but also by their velocity. The problem is now to find explicitly the form of $C(E)$ and $\lambda(\mathbf{k})$. To do this, we shall assume first that a crystal is subject only to an external force $e\mathcal{E}$ due to a uniform electrostatic field $\mathcal{E}$.

The transport equation then becomes:

$$\frac{e}{\hbar} \mathcal{E} \nabla_k f = - \frac{C(E) \lambda(\mathbf{k})}{\tau(\mathbf{k})} \frac{\partial f_0}{\partial E}. \tag{3.1.8}$$

In order to calculate the left-hand side of this expression with an accuracy to the first power of f_1, it is sufficient to assume – in accordance with Eq. (3.1.5) – that $f = f_0$. Then

$$\nabla_k f \approx \nabla_k f_0 = \frac{\hbar^2}{m^*} \, \boldsymbol{k} \, \frac{\partial f_0}{\partial E}, \qquad (3.1.9)$$

and Eq. (3.1.8) becomes:

$$\frac{\hbar e}{m^*} \, \mathscr{E} \boldsymbol{k} \, \frac{\partial f_0}{\partial E} = - \frac{C(E) \, \lambda(\boldsymbol{k})}{\tau(\boldsymbol{k})} \, \frac{\partial f_0}{\partial E}. \qquad (3.1.10)$$

Hence, we see that

$$\lambda(\boldsymbol{k}) \equiv \boldsymbol{k} \qquad (3.1.11)$$

and

$$C(E) = - \frac{e\hbar}{m^*} \tau(\boldsymbol{k}) \, \mathscr{E}. \qquad (3.1.12)$$

Thus, we obtain

$$f_1(\boldsymbol{k}) = - C(E) \, \boldsymbol{k} \, \frac{\partial f_0}{\partial E}, \qquad (3.1.13)$$

$$f(\boldsymbol{k}) = f_0(\boldsymbol{k}) - C(E) \, \boldsymbol{k} \, \frac{\partial f_0}{\partial E}. \qquad (3.1.14)$$

We can show that the identity $\lambda(\mathbf{k}) \equiv \mathbf{k}$ applies to any force. The function $C(E)$ depends strongly on the nature of the force acting on the carriers in a crystal. It is this function that determines a particular transport coefficient.

The aim of the transport equation method is to express the current density j_e and the heat flux density $\mathbf{q}$ in terms of the function f_1, i.e., in terms of $C(E)$.

We shall illustrate this in the case of the electrical conductivity.

It is known that the current density is

$$j_c = e \int \boldsymbol{v} \, dn,$$

where dn is the number of carriers in an element of the wave-number space $(dk_x dk_y dk_z = d^3k)$, which is given by

$$dn = \frac{2}{(2\pi)^3} f(\boldsymbol{k}) \, d^3k.$$

Consequently,

$$j_e = \frac{2e}{8\pi^3} \int \boldsymbol{v} f(\boldsymbol{k}) \, d^3k = -\frac{2e\hbar}{(2\pi)^3 \, m^*} \int \boldsymbol{k} f(\boldsymbol{k}) \, d^3k. \qquad (3.1.15)$$

Substituting Eq. (3.1.14) and bearing in mind that in the absence of external forces the current is equal to zero, i.e.,

$$\int \boldsymbol{k} f_0(\boldsymbol{k}) \, d^3k = 0,$$

we obtain:

$$\bar{j}_e = -\frac{2e\hbar}{(2\pi)^3 \, m^*} \int \boldsymbol{k} C(E) \, \boldsymbol{k} \frac{\partial f_0}{\partial E} d^3k.$$

Making the substitution

$$\frac{\partial f_0}{\partial E} = \frac{m^*}{k\hbar^2} \frac{\partial f_0}{\partial k}$$

and integrating in terms of the spherical coordinates k, θ, φ, where

$$k_x = k \sin\theta \cos\varphi,$$
$$k_y = k \sin\theta \sin\varphi,$$
$$k_z = k \cos\theta,$$
$$d^3k = k^2 \sin\theta \, d\theta \, d\varphi \, dk,$$

we find that

$$j_e = -\frac{e}{3\pi^2\hbar} \int_0^\infty k^3 C(E) \frac{\partial f_0}{\partial k} \, dk. \qquad (3.1.16)$$

TABLE 3.1. Expressions for Generalized Forces
F_i in Boltzmann's Transport Equation

Transport effect	Acting force	Generalized force
Electrical conductivity	$\mathscr{E}$	$-e\,\mathscr{E}$
Hall effect	$\mathscr{E}, H$	$-e\left(\mathscr{E}+\dfrac{1}{c}\left[v,\,H\right]\right)$
Thermoelectric power	∇T	$-e\nabla\left(\varphi-\dfrac{\mu}{e}\right)$
Thermal conductivity	∇T	$-e\nabla\left(\varphi-\dfrac{\mu}{e}\right)$
Nernst−Ettingshausen effect	$\nabla T, H$	$-e\left\{\nabla\left(\varphi-\dfrac{\mu}{e}\right)+\dfrac{1}{e}\left[v,\,H\right]\right\}$
Magnetoresistance	$\mathscr{E}, H$	$-e\left(\mathscr{E}+\dfrac{1}{c}\left[v,\,H\right]\right)$

Notation used: $\mathscr{E}$ is an external electric field; **H** is an external magnetic field; ∇T is a temperature gradient; φ is an electrostatic potential; c is the velocity of light; e is the electron charge; **v** is the electron velocity.

If we substitute $C(E)$ from Eq. (3.1.12) into the above expression, we find that

$$j_c = \frac{e^2\mathscr{E}}{3\pi^2 m^*}\int_0^\infty k^3\tau\,(k)\,\frac{\partial f_0}{\partial k}\,dk. \qquad (3.1.17)$$

Comparing this expression with the well-known equation

$$j_e = \sigma\mathscr{E},$$

we obtain an expression for the electrical conductivity

$$\sigma = \frac{e^2}{3\pi^2 m^*}\int_0^\infty k^3\tau\,(k)\,\frac{\partial f_0}{\partial k}\,dk. \qquad (3.1.18)$$

Similarly, we can obtain expressions for other transport coefficients when a crystal is subject to different forces. Thus, if a crystal is subjected to electric and magnetic fields, then, instead of Eq.

(3.1.12), we must find

$$C(E) = C(\mathscr{E}, H)$$

and we can thus find an expression for the Hall coefficient.

Clearly, in general, $C(E)$ is a function of the applied forces.

Table 3.1 shows the connection between external forces and effects they produce. The majority of these effects will be considered in later sections of the present chapter.

R e l a x a t i o n   T i m e . The expression (3.1.18) for the electrical conductivity includes a quantity τ, which also occurs in the formulas for other transport coefficients. This is because τ determines the nature of the collision term (3.1.14) in the transport equation. If this term is integrated, we can see from

$$f - f_0 = |f - f_0|_{t=0} \, e^{-\frac{t}{\tau}}$$

that τ is the time during which the difference $(f - f_0)$ decreases by a factor of e compared with its initial value. Consequently, the time τ represents the tendency of particles to return to equilibrium and it is called the relaxation time.

It follows from this definition that the relaxation time τ is a function of the particle velocity, $\tau(\mathbf{k})$, and of the particle energy, $\tau(E)$.

The use of the relaxation time concept represents a certain approximation which is rigorously justified in only two cases: if the collisions of the particles are elastic, i.e., if the change in the carrier energy in collisions is small ($\Delta E \ll E$), or if the carrier velocities obey a random distribution law, i.e., the scattering is isotropic, which means that it is equiprobable along any direction.

Studies of transport phenomena show that the first approximation is practically always satisfied. The second assumption is satisfied when the constant-energy surfaces are spherical. If these surfaces are ellipsoidal, then, strictly speaking, we must allow for the scattering anisotropy, and the relaxation time is no longer a scalar quantity but a tensor of the second rank [9, 10]. If the dispersion law is even more complex, such as that given by Eq. (1.1.11) — valid for p-type Ge and Si — the relaxation time concept cannot be used rigorously at all. For this reason, the theory of transport phenomena in such cases is much more complex [3].

Validity of the Transport Equation. The transport equation approach is valid if the concept of a mean free path of particles is meaningful, i.e., if the duration of a collision Δt is small compared with the time interval between two consecutive collisions τ. A criterion expressing this condition can be obtained easily from Heisenberg's indeterminacy principle, according to which $\Delta E \Delta t \sim \hbar$, where ΔE is the energy indeterminacy and Δt is the duration of the perturbation.

We have shown in Chap. 2 that the distribution function varies very rapidly near the Fermi level μ. The inequality $\Delta E < kT$ must be satisfied, because otherwise the values of the distribution function would be indeterminate. Thus, $\hbar/\Delta t < kT$. If Δt is replaced with τ, we find that:

$$\tau > \frac{\hbar}{kT}. \tag{3.1.19}$$

This criterion is derived more rigorously in [11, 12]. It can be written in a different form, by expressing the value of τ in terms of the experimentally determined value of the mobility

$$u > \frac{e\hbar}{m^*kT}. \tag{3.1.20}$$

If we now use the values of m^*, for example, the values for Ge and Si, we obtain, respectively,

$$u > \frac{1.1 \cdot 10^5}{T} \text{ and } u > \frac{5.15 \cdot 10^4}{T}.$$

Comparing these inequalities with the experimentally observed mobilities, we reach the unhappy conclusion that the transport equation is applicable to Ge and Si only for impurity concentrations of the order of 10^{17} cm^{-3} at 300°K. At low temperatures, the transport equation method is found to be unsuitable, even at lower impurity concentrations. We can see here the full analogy with metals, for which the electron relaxation time also does not satisfy the inequality (3.1.19) [12].

It has been shown [12] that in the case of metals Eq. (3.1.19) should be replaced with a different criterion

$$\tau > \frac{\hbar}{\mu}.$$ (3.1.21)

The correctness of this approach has been confirmed by other workers [13, 14].

Heavily doped semiconductors can be regarded quite satisfactorily as poorly conducting metals. In such metals and in heavily doped semiconductors, the Fermi level lies within the conduction band. Since only those electrons are scattered which have energies equal to the Fermi energy, it follows that the average energy of electrons in heavily doped semiconductors is equal to μ and not to kT. Thus, the criterion (3.1.21) is more suitable for heavily doped semiconductors than the expression (3.1.19).

The higher the doping level of a semiconductor, the stronger is the justification for the use of Eq. (3.1.21) as the criterion of the validity of the transport equation approach. There is some doubt about the applicability of the transport equation to semiconductors with moderately degenerate electron gas. In the case of germanium and silicon, this corresponds to carrier densities from 10^{17} cm^{-3} to $(3-5) \cdot 10^{18}$ cm^{-3}. However, it has been shown [15] that in the steady-state case the transport equation method is applicable to moderately degenerate semiconductors in spite of the fact that Eq. (3.1.21) is not satisfied. Therefore, we shall consider transport phenomena within the transport equation framework.

A different criterion for the validity of the transport equation method is obtained in the case of transport phenomena in magnetic fields. In this case, the transport equation is basically classical or semi-classical because the quantization of electron orbits in the presence of a magnetic field is ignored. This approximate approach is permissible only in weak fields, when the energy of thermal motion kT is greater than the distance between two neighboring discrete states in a magnetic field H [11]:

$$kT > \frac{1}{2} \frac{e\hbar}{m^*c} H.$$ (3.1.22)

We can show that this condition is easily satisfied for practically all semiconductors and at all temperatures provided $H < 10^4$ Oe.

$$\text{TABLE 3.2.} \quad \tau = \tau_0 \, (\varepsilon^*)^{r-1/2}$$

Scattering centers, r	τ_0	Notation used
Acoustical vibrations (phonon theory), r = 0	$\dfrac{9\pi}{4\sqrt{2}} \dfrac{\hbar^4 \omega^2 M}{C^2 a^3 (m^* kT)^{3/2}}$	ω — velocity of sound; M — atomic mass; C — Bloch constant; a — lattice parameter
Acoustical vibrations (deformation potential theory), r = 0	$\dfrac{\pi \hbar^4 C_{ll}}{\sqrt{2E_1^2}\,(m^* kT)^{3/2}}$	C_{ll} — elastic constant for longitudinal vibrations; $E_1 = \Omega_0 dE_0/d\Omega$; E_0 — energy of allowed band edge; Ω_0 — initial volume of unit cell before deformation
Optical vibrations $(T \ll \theta_D)$ in heavily doped crystals, r = $\frac{1}{2}$	$\dfrac{a^3 M}{2\pi \sqrt{2m^*}} \dfrac{(\hbar\omega_0)^{3/2}}{(\gamma Ze^2)^2} \times$ $\times \left[\exp\left(\dfrac{\hbar\omega_0}{kT}\right) - 1 \right](1-f_0)$	ω_0 — limiting frequency of longitudinal optical vibrations; Ze — ion charge; γ — factor representing the polarizability of ions; f — Fermi function; θ_D — Debye temperature
Optical vibrations $(T \ll \theta_D)$ in lightly doped crystals, r = $\frac{1}{2}$	$\dfrac{a^3 M}{2\pi \sqrt{2m^*}} \dfrac{(\hbar\omega_0)^{3/2}}{(\gamma Ze^2)^2} \times$ $\times \left[\exp\left(\dfrac{\hbar\omega_0}{kT}\right) - 1 \right]$	
Optical vibrations $(T \gg \theta_D)$ irrespective of degree of doping, r = 1	$\dfrac{a^3 M}{2\pi \sqrt{2m^* kT}} \left(\dfrac{\hbar\omega_0}{\gamma Ze^2}\right)^2$	
Impurity ions, r = 2	$\dfrac{\sqrt{2m^*}\chi^2}{\pi e^4 N_i g}$; $g = \ln(1+b) - \dfrac{b}{1+b}$; $b = \dfrac{8m^* E}{\hbar^2} R_0^2$	χ — permittivity; N_i — concentration of impurity ions
Neutral impurities, r = $\frac{1}{2}$	$\dfrac{(m^*)^2 e^2}{20\hbar^3 \chi N_n}$	N_n — concentration of neutral impurities

§3.2. Scattering Mechanisms

A system may return to equilibrium after the application of an external force only as a result of collisions between particles, i.e., as a result of scattering. Free carriers may interact ("collide") with a great variety of scattering centers. These centers may be impurity atoms (ionized or neutral), thermal vibrations (acoustical and optical), of the lattice atoms, structural defects (dislocations, vacancies), and other obstacles. The interaction between carriers and scattering centers determines the actual values of all the transport coefficients.

Scattering by the Acoustical Vibrations of the Lattice Atoms. There are two methods for solving the problem of the scattering of free carriers by thermal vibrations of the lattice atoms: the method based on making an allowance for the interaction between carriers and the phonon gas [16], and the deformation potential method [16].

In the first method, the lattice atom vibrations are represented by quasi-particles known as phonons.

In the second method, these vibrations are represented by longitudinal and transverse waves. When these waves are propagated along a crystal, an additional periodic potential is superimposed on the periodic potential of the internal field. This alters the energy structure of the crystal. A "ripple" appears at the zone boundaries. This is equivalent to an alternating potential energy of the carriers, known as the deformation potential.

The theory of the phonon approach is developed in [18] and the deformation potential theory is given in [19]. The expressions for the relaxation times obtained by these two methods are given in Table 3.2. These expressions are valid in the case of an impurity-free semiconductor. Comparison of the formulas for τ_L shows that both methods give the same type of dependence on the temperature and on the effective carrier mass. Moreover, a relationship between these two theories of scattering has been found in [20] and it has been shown that the constants occurring in the expression for τ_L are related by $E_1 = 2C/3$. The physical meaning of the constant C is that it represents the intensity of the interaction of electrons with the vibrations of the lattice atoms. The constant E_1 represents the shift of the bottom of a free band for unit volume deformation of a crystal.

We shall now consider the validity of the expressions given in Table 3.2 in the case of heavily doped crystals. It is sufficient to investigate any one of these expressions, for example, the first one. We must go back to the derivation of this expression. Details of this derivation are given in [21], but here we shall use an intermediate result, according to which the relaxation time is defined as

$$\frac{1}{\tau_L} = \frac{V}{8\pi^2} \frac{m^*}{\hbar^2 k^3} \int_{q_{min}}^{q_{max}} w(q)(2N_q + 1)q^3 \, dq, \qquad (3.2.1)$$

where V is the volume of the crystal; k is the electron wave number; q is the phonon wave number, w(q) is the probability of the emission (absorption) of a phonon during scattering; N_q is the average number of phonons with a wave number q, i.e., the phonon distribution function. It is important to note that the integration is carried out over the whole range of values of q. We shall determine the limits of this range. If the electron energy is $E = \hbar^2 k^2/2m^*$ and the frequency of the longitudinal acoustical waves is $\omega_q = vq$, where v is the velocity of these waves, it follows from the law of conservation of energy that:

$$\frac{\hbar^2(k \pm q)^2}{2m^*} = \frac{\hbar^2 k^2}{2m^*} \pm \hbar vq. \qquad (3.2.2)$$

The upper sign represents the absorption of a phonon and the lower sign the emission of a phonon. Hence,

$$q = \mp 2k\cos\vartheta \pm \frac{2m^* v}{\hbar}, \qquad (3.2.3)$$

where ϑ is the angle between the directions of the vectors **k** and **q**. We shall find the second term in Eq. (3.2.2), dividing it by k.

1. For pure semiconductors:

$$\frac{2m^* v}{\hbar k} = \frac{2m^* v}{p} \approx \frac{2m^* v}{\sqrt{m^* k_0 T}} = 2\sqrt{\frac{T_{cr}}{T}}, \qquad (3.2.4)$$

where $T_{cr} = m^* v^2/k_0$, and k_0 is Boltzmann's constant. Since electrons are distributed in accordance with Boltzmann's law, we can assume that the average value of p is $\sqrt{m^* k_0 T}$. If $v = 10^5$ cm/sec and $m^* = 0.5 \cdot 10^{-27}$ g, we find that $T_{cr} < 1°K$. At all temperatures $T \gg 1°K$, we can neglect the second term in Eq. (3.2.3) and, therefore,

$$q = \mp 2k \cos \vartheta, \qquad (3.2.5)$$

which shows that the range of integration with respect to q extends from $q_{min} = 0$ to $q_{max} = 2k$. The dropping of the second term in Eq. (3.2.3) is equivalent to the dropping of the second term in Eq. (3.2.2), i.e., equivalent to the rejection of the term describing the inelasticity of electron–phonon collisions.

Since the average value of k for electrons at 300°K is about 10^7 cm^{-1}, and according to the Debye theory [22] the value of q_0 is given by $q_0 = (6\pi^2/a^3)^{1/3} \approx 10^8$ cm^{-1}, it follows that $q_{max} \ll q_0$.

2. In heavily doped semiconductors, the electrons are distributed in accordance with the Fermi–Dirac law and not in accordance with Boltzmann's law. Therefore, instead of Eq. (3.2.4), we have

$$\frac{2m^* v}{\hbar k} \approx \frac{2m^* v}{\hbar k (\mu)}, \qquad (3.2.6)$$

where $k(\mu)$ is a wave vector which represents the electrons with the Fermi energy.

Thus a replacement is the consequence of the fact that even those small changes in the carrier energy which take place during scattering, cannot occur in the case of electrons with $E < \mu$, since all the states in this range of energies are fully occupied. Only in a narrow thermal ($\pm k_0 T$) layer near $E = \mu$ are there electrons whose energy can change a little and which, therefore, may be scattered. The wave vector of such electrons for $\mu^* \geq 2$ is

$$k(\mu) = \frac{1}{\hbar} \sqrt{2m^* \mu} > 10^8 \text{ cm}^{-1},$$

i.e., it is about an order of magnitude larger than in the case of pure semiconductors.

TABLE 3.3

Semi-conductor	$T\,°K$	μ^*_{lim}	Semi-conductor	$T\,°K$	μ^*_{lim}
Ge	100	81	Si	100	51
Ge	300	27	Si	300	17
Ge	500	16	Si	500	10

Thus, the relationship (3.2.6) is satisfied at all temperatures:

$$\sqrt{\frac{k_0 T_{cr}}{\mu}} \ll 1.$$

Hence, in heavily doped semiconductors electrons are also scattered elastically by the acoustical vibrations. This means that Eq. (3.2.5) remains valid. However, for pure semiconductors we have $q_{max} \ll q_0$ whereas for heavily doped materials we have $q_{max} = 2k \approx q_0$. Therefore, as shown in [3], we must distinguish two cases: $k(\mu) < q_0/2$ and $k(\mu) > q_0/2$. If the first of these conditions is satisfied, Eq. (3.2.1) should be integrated from zero to $2k(\mu)$ and the expression for τ_L remains the same as in Table 3.2. However, if the second condition is satisfied, the upper limit of integration in Eq. (3.2.1) is q_0 and the expression for τ_L becomes [3]:

$$\frac{2\sqrt{2}}{\pi^3} \frac{VMm^{*1/2}k_0\theta_D^3}{\hbar^4 C^2 T} E^{3/2}, \tag{3.2.7}$$

where θ_D is the Debye temperature given by $\theta_D = \hbar v q_0/k_0$.

Table 3.3 lists the limiting values of μ^*, needed to satisfy the condition $k(\mu) > q_0/2$, from which Eq. (3.2.7) begins to apply. We can see that Eq. (3.2.7) applies only to good metals. In the case of semiconductors, the expressions for τ_L are identical with the formulas given in Table 3.2 at all practically realizable carrier densities.

However, in principle, this does not mean that the numerical values of τ_L are independent of the degree of doping of a semiconductor. Thus, for example, we know nothing of how E_1 varies with the doping level. Equally, the constancy of the lattice parameter a or of the elastic modulus C_{ll} is not self-evident. There are as yet no theoretical predictions on this subject. Experimental results show that the lattice parameter may change in some cases (this problem has been considered in Chap. 1 in connection with the band structure of heavily doped semiconductors).

Scattering by the Optical Vibrations of the Lattice Atoms. The higher the "degree of ionicity" of a crystal the more likely will be the interaction between carriers and the optical vibrations of the lattice atoms. This type of scattering differs basically from the scattering by the acoustical phonons in respect of the absolute value of ΔE, which represents the change in the carrier energy during scattering. When an acoustical phonon is absorbed, or emitted, the value of $\Delta E = \pm \hbar vq$ is very small compared with E, and consequently the scattering is elastic. However, in the interaction with the optical phonons, $\Delta E = \pm \hbar \omega_0$ may be either smaller or larger than E, depending on the temperature. At high temperatures, when $\hbar \omega_0 \ll k_0 T$, the value of ΔE may be small. In this case, the scattering may be regarded as elastic. At low temperatures when $\hbar \omega_0 \gg k_0 T$, the electrons can absorb only phonons and this greatly alters their energy. Consequently, the concept of a relaxation time loses its meaning. However, if the interaction of a carrier with an optical phonon is regarded as taking place in two stages – an absorption of a quantum and an instantaneous emission of exactly the same quantum – the relaxation time can still be used [23]. In this approach, we neglect the time during which the electron is in a high-energy state. This is permissible because at low temperatures the probability of the emission of a phonon, proportional to $(N_q + 1)$, is much higher than the probability of absorption of a phonon, which is proportional to N_q. In fact, since N_q is defined by Planck's function

$$N_q = \left[\exp \left(\frac{\hbar \omega_0}{k_0 T} \right) - 1 \right]^{-1},$$

it follows that

$$\frac{N_q + 1}{N_q} \approx \exp \left(\frac{\hbar \omega_0}{k_0 T} \right) \gg 1.$$

Thus, we may assume that, even at low temperatures, carriers are scattered elastically. Such a simplification is fully justified, since the formulas obtained in this way (cf. Table 3.2) are identical with the expressions found by Sondheimer and Howarth [24], who have solved the transport equation by a variational method, and those found by Samoilovich et al. [8], who used the density matrix method.

The electron gas degeneracy affects the value of τ in different ways at low and high temperatures. The degeneracy increases the occupancy of the states to which electrons can be transferred after their interaction with phonons. Therefore, when the number of electrons is increased, the probability of electron transitions to states with lower energies becomes less.

Allowance for this circumstance leads to the appearance of a factor $(1 - f)$ in the collision integral of Eq. (3.1.4) [25]. Because of this, the solution is a Fermi function in the absence of external forces $[f(\mathbf{k}) = 0]$. Consequently, the final expression for τ given in Table 3.2 includes a factor $(1 - f_0)$. This expression is also valid at low temperatures. At high temperatures, the expression for τ, obtained for a nondegenerate electron gas (also given in Table 3.2), still remains valid [25]. This follows from the fact that τ is, in general, governed by the quantity $(N_q + 1)$ and the Fermi terms are due to unity in this factor [25]. At high temperatures $(\hbar\omega_0 \ll k_0 T)$, $N_q \approx k_0 T / \hbar\omega_0$ and there are no Fermi terms because unity in $(N_q + 1)$ can be ignored.

Scattering by Ionized Impurities. An ionized impurity center in the lattice of a semiconductor produces a long-range Coulomb field with a potential energy $U = e^2 / \varkappa r$, where $\varkappa$ is the macroscopic permittivity. The problem of the motion of a carrier in such a field is in many respects similar to the analogous problem of the Rutherford scattering of charged particles by nuclei [26], i.e., the relaxation time can be calculated using Rutherford's formula for the effective scattering cross section:

$$\sigma(\theta) = \frac{e^4}{(2\varkappa m^* v^2)^2} \operatorname{cosec}^4 \frac{\theta}{2}, \qquad (3.2.8)$$

where θ is the angle between carrier trajectories before and after collision with an ionized center.

The relationship (3.2.8) allows us to calculate the probability of scattering in a cell of the crystal momentum space. However, the integration of this expression over the whole space, which is necessary for the determination of the total scattering probability, shows that the integral diverges. This is because in a Coulomb field particles are scattered through small angles. To obtain the final result, it is necessary to truncate the Coulomb potential, as done by Conwell and Weisskopf [27]. They considered scattering only in a certain region R_0, where R_0 was defined as $R_0 = 1/(2N^{-1/3})$. Their calculation [27] yielded the following expression for the relaxation time in the case of scattering by impurity ions:

$$\tau_i = \frac{\sqrt{2m^*}\varkappa^2 \ (k_0 T)^{3/2} (\varepsilon^*)^{3/2}}{\pi N_i e^4 \ln\left[1 + \left(\dfrac{3\varkappa k_0 T}{e^2 N_i^{1/3}}\right)^2\right]} \cdot \tag{3.2.9}$$

   In contrast to this semiclassical solution, Brooks and Herring [28] developed a more rigorous theory which allows for the fact that free carriers may collect around a charged center and may partly screen its potential. The problem is to find a reasonable value for the distance at which one should truncate the potential of a partly screened impurity ion. To do this, it is necessary to solve Poisson's equation

$$\frac{1}{r}\frac{d^2(rV)}{dr^2} = \frac{4\pi q}{\varkappa}, \tag{3.2.10}$$

where $V = (1/e)U(r)$; $q = e(n' - n)$ is the space−charge density; n is the maximum density of free carriers at large distances r; n' is the varying density of carriers at short distances from the ion.

   The solution of Eq. (3.2.10) is

$$V(r) = \frac{e}{\varkappa r} e^{-\frac{r}{R_0}}, \tag{3.2.11}$$

where the screening radius for an arbitrary degree of degeneracy of the electron gas is given by the expression

$$\frac{1}{R_0^2} = \frac{16\pi^2 e^2 (m^*)^{3/2} (2k_0 T)^{1/2}}{h^3 \varkappa} F_{-1/2}(\mu^*). \tag{3.2.12}$$

We can easily show, using the approximation (2.5.6), that, in the absence of degeneracy, Eq. (3.2.12) becomes

$$R_0 = \frac{\varkappa^{1/2} (k_0 T)^{1/2}}{2\pi^{1/2} e n^{1/2}},$$ (3.2.13)

and in the strongly degenerate case ($\mu^* \gg 0$), we can use the approximation (2.5.11) to obtain

$$R_0 = \frac{\hbar \varkappa^{1/2}}{2e (m^*)^{1/2}} \left(\frac{3n}{\pi}\right)^{-1/6}.$$ (3.2.14)

According to [29] the total scattering cross section, calculated allowing for the screening of ions by electrons, has the following form:

$$\sigma = \frac{2\pi e^4}{\varkappa^2 (m^*)^2 v^2} \left[\ln (1 + b) - \frac{b}{1 + b}\right].$$ (3.2.15)

where

$$b = \frac{8m^* E}{\hbar^2} R_0^2.$$ (3.2.16)

(In this case, the expression for τ_i has the same form as in Table 3.2.) It is necessary to draw attention to the value of E in the above expression. If E is understood to be the average energy of scattered particles, then it should be defined as

$$E \approx \overline{E} = k_0 T \frac{F_{3/2} (\mu^*)}{F_{1/2} (\mu^*)}.$$ (3.2.17)

However, the closer the carrier distribution to the Fermi step, the less likely is the scattering of particles of energies lower than μ. Therefore, for an ideal Fermi gas we may assume

$$E = \mu.$$ (3.2.18)

The use of Eq. (3.2.18) is more justified at higher carrier densities and lower temperatures (i.e., when μ is larger). Conversely, the nearer the carrier distribution is to the Maxwellian type, the more justified we are in using Eq. (3.2.17). A simple calculation shows that $\mu^* \approx 3.5$ is the limit below which we must use Eq. (3.2.17) and above which we should employ Eq. (3.2.18).

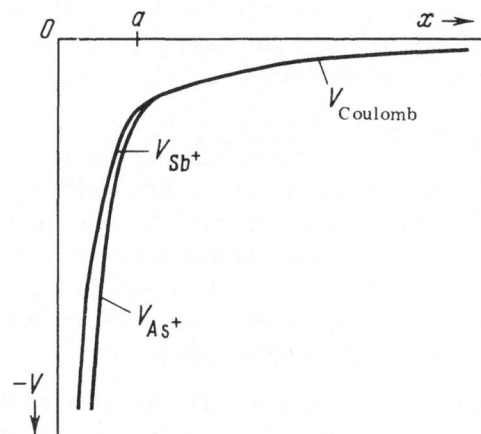

Fig. 3.1. Distribution of the potential at various
distances from an impurity ion.

Numerical comparison of Eq. (3.2.9) and the expression given
in Table 3.2 shows that the difference between them becomes very
large at high impurity concentrations. Therefore, we shall use
only the Brooks−Herring formula.

At very high impurity concentrations, the screening radius
decreases considerably, which can be seen clearly from Eq. (3.2.14).
The scattered particles may be affected not only by the Coulomb
forces but also by short-range forces [30]. In this case, the scattering laws are quite different. The effect of these forces is shown
schematically in Fig. 3.1, which indicates that the short-range
forces depend on the nature of the ion. Correspondingly, we should
observe experimental differences between the scattering of electrons by various impurities.

There is as yet no quantitative theory of scattering with allowance for the short-range forces.

Scattering by Neutral Impurities. The theory
described in [31-33] gives the following expression for τ_N:

$$\tau_N = \frac{(m^*)^2 e^2}{20 \hbar^3 \varkappa N_n}, \tag{3.2.19}$$

where N_n is the concentration of neutral impurities.

In the heavy doping case, this form of scattering is unimportant because, right down to the lowest temperatures, impurities in heavily doped semiconductors are fully ionized.

Scattering of Electrons by Electrons. In impurity-free semiconductors, the electron−electron interaction gives a correction up to 60% of τ_i [34], compared with τ_i calculated using the Brooks−Herring formula. Physically, the electron−electron interaction represents a redistribution between the electrons of that energy that they receive from the electric field. Consequently, the fast electrons acquire smaller momenta because some of the energy is transferred to the slower electrons.

In heavily doped semiconductors, the role of the electron−electron scattering should not be important since all the scattered electrons have the same velocity governed by the Fermi energy.

Scattering by Dislocations and Point Defects. It follows from the experimental data that the dislocation scattering is important only in very pure crystals at low temperatures, i.e., when all the other stronger scattering mechanisms are suppressed. In heavily doped crystals such conditions are practically impossible to satisfy.

Point defects result in scattering which is represented by a relaxation time of the following type [35]:

$$\tau = \frac{\pi \hbar^4}{(m^*)^{3/2} (2k_0 T)^{1/2} V_0 N_d} \left(\frac{k_0 T}{E}\right)^{1/2}, \qquad (3.2.20)$$

where N_d is the concentration of point defects; V_0 is a constant representing the amplitude of the δ-like potential: $V = V_0 \delta(r)$, which applies near a point defect.

The value of N_d in heavily doped crystals may reach very high values, comparable with the impurity ion concentration. Therefore, the mechanism of carrier scattering by point defects should be included in discussions of the experimental data.

Mixed Scattering. In real crystals, several scattering mechanisms usually act simultaneously. In fact, in every interpretation of the results obtained for a given transport phenomenon we are faced with the problem of the mixed scattering of carriers. To solve this problem, we must first identify the main scattering mechanisms and then, if possible, determine their contribution to the observed transport phenomenon. We must bear in mind that the contribution of each type of scattering varies very strongly with temperature and with the impurity concentration. A situation in which one of the scattering mechanisms is much more likely than all the others is very rare. This situation is encountered in impurity-free semiconductors but in heavily doped materials it is practically never realized. A much more likely situation involves two scattering mechanisms. These mechanisms are usually the scattering by ions and by acoustical phonons, or scattering by ions and optical phonons. The former pair of mechanisms is encountered in covalent semiconductors such as Ge and Si — the latter pair in ionic semiconductors, including $A^{III}B^{V}$ compounds.

Problems of this type have been considered for lightly doped semiconductors with a Boltzmann-type electron gas in calculations of the electrical conductivity, the Hall effect, and the thermoelectric power [36], in a calculation of the Hall coefficient [37], and in investigations of the Nernst–Ettingshausen effect [38, 39]. In all these investigations, only two mechanisms were considered: scattering by acoustical phonons and by impurity ions.

In connection with studies of the electrical conductivity [40] and mobility [41, 42], the same problem was considered for the same scattering mechanisms in heavily doped semiconductors with a Fermi electron gas.

We shall now give expressions for τ which apply when several scattering mechanisms act simultaneously. All the formulas for $\tau(E)$ can be expressed in the form

$$\tau(E) = \tau_0(T)(\varepsilon^*)^{r-1/2}, \qquad (3.2.21)$$

where $r = 0$ for the scattering by the acoustical phonons and by point defects; $r = 1/2$ or $r = 1$ for the scattering by the optical phonons below or above the Debye temperature, respectively; $r = 2$ for the scattering by impurity ions.

Applying the well-known addition rule to the reciprocals $1/\tau$ [43]:

$$\frac{1}{\tau} = \sum_{m=1}^{m} \frac{1}{\tau_m},$$

we obtain the following expression for the simultaneous scattering by the acoustical phonons and impurity ions:

$$\tau = \frac{\tau_{0L} \, (\varepsilon^*)^{3/2}}{(\varepsilon^*)^2 + a^2} , \qquad (3.2.22)$$

where

$$a^2 = \frac{\tau_{0L}}{\tau_{0i}} \qquad (3.2.23)$$

and

$$g(b) = \text{const},$$

i.e., it is assumed that the screening factor is independent of the energy of scattered particles or that at least this dependence can be neglected. We shall show later (§3.3) that, strictly speaking, we cannot ignore this dependence, particularly in the case of heavily doped semiconductors, because it may give rise to considerable errors.

Similarly, for the simultaneous scattering by ions and by the optical phonons the relaxation time is obtained in the form:

$$\tau = \frac{\tau_{0L} \, (\varepsilon^*)^2}{(\varepsilon^*)^{3/2} + a^2} \qquad (3.2.24)$$

Anisotropy of the Relaxation Time. In Chap. 1, we have shown that the constant-energy surfaces of Ge and Si are ellipsoids of revolution. This means that the effective masses and the relaxation times are anisotropic. A detailed theoretical analysis, carried out by Herring [44], has shown that the concept of the relaxation time, depending only on the energy and independent of the position on the constant-energy surface, can be used satisfactorily only to describe the scattering by the neutral impurities and by the optical phonons. The approximation of the scalar relaxation time is, strictly speaking, inapplicable to scattering by the acoustical phonons and particularly by impurity ions. In other words, in every analysis of the experimental results on the scattering it is necessary to allow for the anisotropy of the relaxation time, which implies that the probability of a transition $w_{kk'}$ (where k and k' are the wave vectors of the scattered particle before and after collision) depends not only on the scattering angle but also on the orientation of the vector k in a crystal.

A theory of transport phenomena allowing for the anisotropy of τ was developed by Herring and Vogt [9]. Considering the case when the principal axes of the mass ellipsoid coincide with the principal axes of the relaxation time tensor, they showed that the scattering process for each ellipsoidal constant-energy surface may be described by means of three relaxation times, which are the principal components of the relaxation time tensor. Herring and Vogt [9] investigated the anisotropy of τ when carriers are scattered by acoustical phonons. Thus, their theory is valid only for sufficiently pure undoped germanium and silicon crystals.

The theory of transport phenomena has been developed further by Samoilovich et al. [10, 45-47]. An effective method for solving the transport equation in the case of anisotropic scattering of carriers is given in [10], where it is applied to two scattering mechanisms acting independently: the acoustical phonons and impurity ions. This method gives expressions for the relaxation time

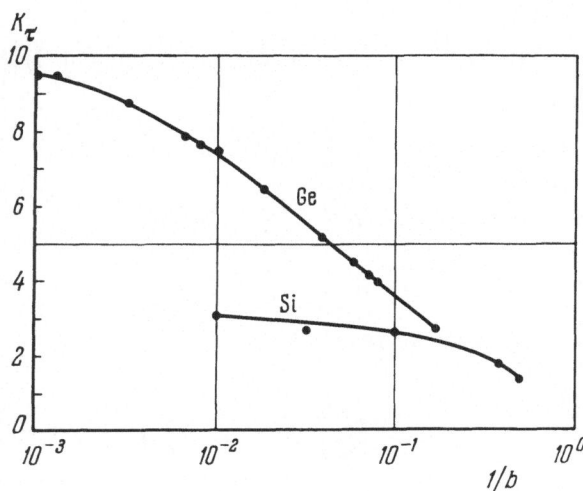

Fig. 3.2. Dependence of the degree of anisotropy of the relaxation times of Ge and Si on the carrier density.

components in both cases. The relationships found for τ_i^L in the case of scattering by the acoustical phonons are very similar to those obtained by Herring and Vogt [9]. The slight difference between the corresponding expressions is due to a slightly different allowance for the phonon spectrum anisotropy.

Relationships for τ in the impurity scattering case were obtained in [45, 46] for relatively lightly doped semiconductors. The impurity concentration criterion in these calculations was the parameter $1/b$, defined by Eq. (3.2.16) and Eq. (3.2.12). In these investigations, the calculations were limited to values of $1/b$ from 10^{-6} to 10^{-2}, which correspond to impurity concentrations in Ge and Si not less than 10^{17} cm^{-3}.

The validity of the results obtained in [45, 46] has been confirmed by obtaining the well-known Brooks—Herring formula for τ_i as a special case of anisotropic τ.

Finally, more general formulas for the relaxation tensor components in the impurity scattering case have been obtained in [48-49] for any value of $1/b$ and, consequently, for any impurity concentration:

$$\frac{1}{\tau_i^\parallel} = 2C\left[\frac{\pi - 2\chi}{2}\ln\frac{\beta\rho + 1}{\beta\rho - 1} - \left(2a + \frac{\pi}{2}\right)\ln 2 + L(\chi + a) - \right.$$

$$\left. - L(\chi - a) + L\left(\frac{\pi}{2} - 2a\right) + 2L(a) - \frac{\beta}{1 + \beta^2}\ln(b + 1)\right], \quad (3.2.25)$$

$$\frac{1}{\tau_i^\perp} = C(\beta^2 - 1)\left[\frac{\pi - 2\chi}{2}\ln\frac{\beta\rho + 1}{\beta\rho - 1} - \left(2a + \frac{\pi}{2}\right)\ln 2 + \right.$$

$$+ L(\chi + a) - L(\chi - a) + L\left(\frac{\pi}{2} - 2a\right) + 2L(a) +$$

$$\left. + \frac{\beta}{1 + \beta^2}\ln(b + 1) - \frac{\pi - 2\chi}{\rho(\beta^2 - 1)}\right], \quad (3.2.26)$$

where the following notation is used:

$$\beta^2 = \frac{m_\parallel - m_\perp}{m_\perp}; \qquad C = \frac{3\pi n e \sqrt[4]{2m_\parallel}}{8\beta^3 \varkappa^2 E^{3/2} m_\perp}; \qquad \rho = \frac{1}{\beta}\sqrt{1 + \frac{1 + \beta^2}{b}};$$

$a = \tan^{-1}\beta$; $\chi = \tan^{-1}\rho$; L is the Lobachevskii function tabulated in [55]. A numerical calculation of $\tau_i^\parallel$ and $\tau_i^\perp$ for Ge and Si with high impurity concentrations has been carried out in [50]. The results of this calculation can be represented conveniently in the form of a dependence of $K_T = \tau_i^{\parallel}/\tau_i^{\perp}$ on $1/b$ (Fig. 3.2). As the impurity concentration is increased $(1/b \to 1)$, the value of K_T decreases and at high values of $1/b$ it does not differ greatly from unity. This allows us to assume that impurity scattering in very heavily doped samples (particularly in Si) is almost isotropic.

§3.3.   Electrical Conductivity and Hall Effect

Electrical Conductivity σ.   In the first section of this chapter we obtained a general expression for the electrical conductivity [Eq. (3.1.18)]. We shall now analyze this expression. First, we shall integrate it by parts:

$$\sigma = - \frac{\epsilon^3}{3\pi^2 m^*} \left\{ \mid k^3 \tau f_0 \mid_0^\infty - \int_0^\infty f_0 \frac{d}{dk} (k^3 \tau) \, dk \right\}.$$

The first term in curly brackets (braces) vanishes when the integration limits are substituted. Integrating the second term by parts, replacing k with E (using the relationship $E = \hbar^2 k^2 / 2m^*$), and employing a dimensionless energy ε^*, we obtain:

$$\sigma = \frac{8\pi e^2 (2m^* k_0 T)^{3/2}}{3m^* h^3} \int_0^\infty \frac{\partial f_0}{\partial \varepsilon^*} (\varepsilon^*)^{3/2} \tau (\varepsilon^*) \, d\varepsilon^*, \qquad (3.3.1)$$

and since

$$\sigma = neu_\sigma,$$

[u_σ is the drift mobility of electrons and n is the electron density, given by Eq. (2.3.4)], the expression for the drift mobility becomes:

$$u_\sigma = \frac{2e}{3m^*} \frac{\displaystyle\int_0^\infty \frac{\partial f_0}{\partial \varepsilon^*} (\varepsilon^*)^{3/2} \tau \, d\varepsilon^*}{F_{1/2}(\mu^*)}. \qquad (3.3.2)$$

We shall substitute into these expressions the relaxation time τ for the mixed scattering mechanism [Eq. (3.2.22)], and for f_0 we shall substitute the Fermi function of Eq. (2.2.12). Then

$$\sigma = \frac{8\pi e^2}{3m^* h^3} (2m^* k_0 T)^{3/2} \tau_{0L} \Phi_3 (\mu^*, a),$$ (3.3.3)

$$u_\sigma = \frac{2e}{3m^*} \tau_{0L} \frac{\Phi_3 (\mu^*, a)}{F_{1/2} (\mu^*)},$$ (3.3.4)

where the integral $\Phi_3(\mu^*, a)$ has the form

$$\Phi_3 (\mu^*, a) = \int_0^\infty \frac{(\varepsilon^*)^3 \exp (\varepsilon^* - \mu^*) \, d\varepsilon^*}{[(\varepsilon^*)^2 + a^2][1 + \exp (\varepsilon^* - \mu^*)]^2}.$$ (3.3.5)

The numerical values of this integral are given in Appendix A.3 and the corresponding family of curves is shown in Fig. 3.3. Analysis of this and similar integrals of other indices is given in Appendix A.5.

We shall now analyze Eqs. (3.3.3) and (3.3.4). Let us assume that $a^2 \ll (\varepsilon^*)^2$, which represents the case when the carriers are

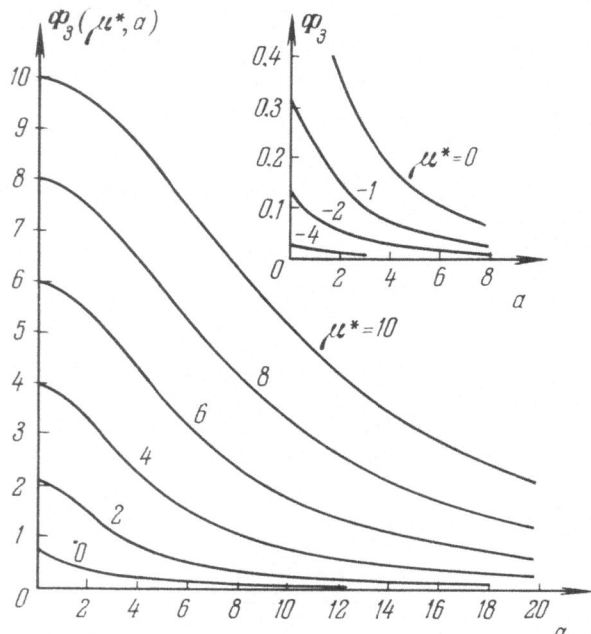

Fig. 3.3. Graphical representation of the integral $\Phi_3(\mu^*, a)$.

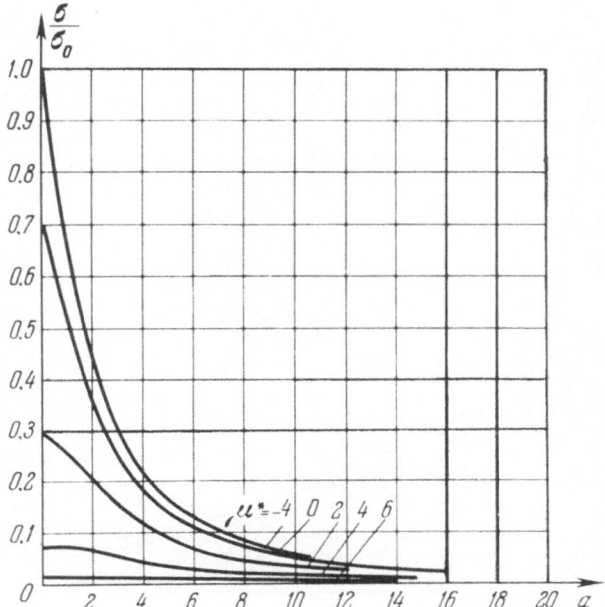

Fig. 3.4. Calculated electrical conductivity for the mixed
scattering of carriers.

scattered mainly by the thermal vibrations of the lattice and the
contribution of scattering by the impurity ions is negligibly small.
In this case, the integral $\Phi_3(\mu^*, a)$ simplifies and transforms into
the Fermi integral $F_0(\mu^*)$. This can be demonstrated easily when
Φ_3 is integrated by parts. As a result of these operations, we obtain:

$$\sigma_L = \frac{8\pi e^2}{3m^*h^3} (2m^*k_0T)^{3/2} \tau_{0L} F_0(\mu^*) \tag{3.3.6}$$

and

$$u_{aL} = \frac{2e}{3m^*} \tau_{0L} \frac{F_0(\mu^*)}{F_{1/2}(\mu^*)}. \tag{3.3.7}$$

Here and later, we shall use an additional subscript L to denote
quantities referring solely to the case of scattering by the acoustical vibrations of the lattice atoms.

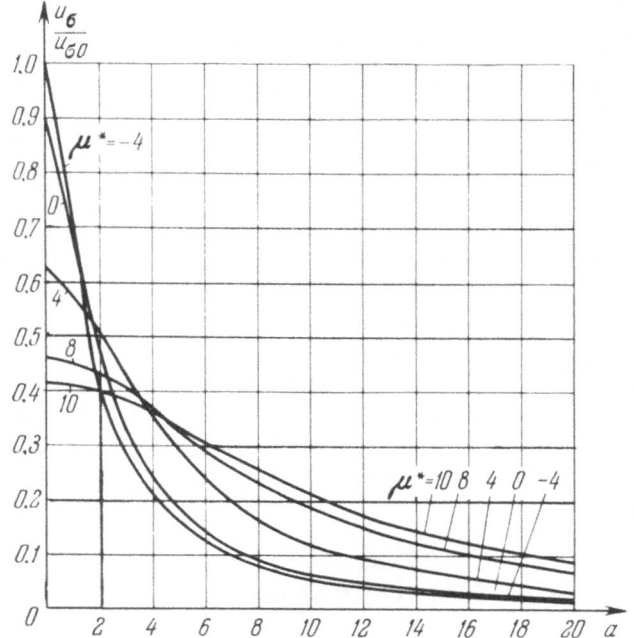

Fig. 3.5. Calculated drift mobility for the mixed scattering of electrons.

From Eqs. (3.3.6) and (3.3.7), we can easily find that σ_0 and $u_{\sigma 0}$ are, respectively, the electrical conductivity and the drift mobility of an impurity-free semiconductor in the absence of degeneracy ($\mu^* \ll 0$). To do this, it is sufficient to use Eq. (2.5.5). We then find:

$$\sigma_0 = \frac{8\pi e^2}{3m^* h^3} (2m^* k_0 T)^{3/2} \tau_{0L} e^{\mu^*}, \qquad (3.3.8)$$

$$u_{\sigma 0} = \frac{4}{3\sqrt{\pi}} \frac{e}{m^*} \tau_{0L}. \qquad (3.3.9)$$

If it is assumed that $\mu^* \gg 0$ in Eqs. (3.3.6) and (3.3.7), expressions are obtained which are valid for the same impurity-free semiconductor but in the strongly degenerate case:

$$\sigma_0 = \frac{8\pi e^2}{3m^* h^3} (2m^* k_0 T)^{3/2} \tau_{0L} \mu^*, \qquad (3.3.10)$$

$$u_{\sigma 0} = \frac{e}{m^*} \tau_{0L} (\mu^*)^{-1/2}. \qquad (3.3.11)$$

In these expressions, we have used the approximations of Eq. (2.5.11) for $F_0(\mu^*)$ and for $F_{1/2}(\mu^*)$.

We shall now consider the opposite case: $a^2 \gg (\varepsilon^*)^2$, which represents the case when only the impurity scattering is important. By dropping the term $(\varepsilon^*)^2$ from the denominator of Eq. (3.2.22), we find that the integral $\Phi_3(\mu^*,\ a)$ transforms into $3 \cdot F_2(\mu^*)/a^2$ and the expressions for the electrical conductivity [Eq. (3.3.3)] and for the drift mobility [Eq. (3.3.4)] transform to

$$\sigma_i = \frac{8\pi\epsilon^2}{m^*\hbar^3}\ (2m^*k_0T)^{3/2}\ \tau_{0i}F_2\,(\mu^*)\,, \tag{3.3.12}$$

$$u_{\text{d}i} = \frac{2e}{m^*}\tau_{0i}\frac{F_2(\mu^*)}{F_{1/2}\,(\mu^*)}. \tag{3.3.13}$$

The subscript "i" is used to denote the quantities referring to the case when only the impurity scattering is of importance.

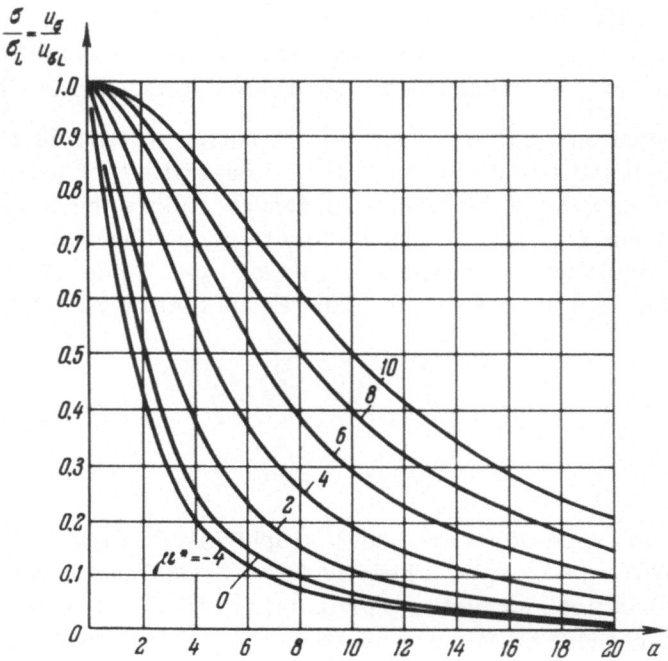

Fig. 3.6. Contribution of the lattice scattering to the electrical conductivity or the drift mobility of electrons.

As in the case of scattering by the acoustical phonons, we shall now obtain expressions for σ_i and $u_{\sigma i}$ in the two limiting cases: $\mu^* \ll 0$ and $\mu^* \gg 0$,

$$\mu^* \ll 0 \left\{ \begin{array}{ll} \sigma_i = \dfrac{16\pi e^2}{m^* h^3} (2m^* k_0 T)^{3/2} e^{\mu^*}, & (3.3.14) \\[3mm] u_{\sigma i} = \dfrac{8e}{m^* \sqrt{\pi}} \tau_{0i}, & (3.3.15) \end{array} \right.$$

$$\mu^* \gg 0 \left\{ \begin{array}{ll} \sigma_i = \dfrac{8\pi e^2}{3m^* h^3} (2m^* k_0 T) \tau_{0i} (\mu^*)^3, & (3.3.16) \\[3mm] u_{\sigma i} = \dfrac{e}{m^*} \tau_{0i} (\mu^*)^{3/2}. & (3.3.17) \end{array} \right.$$

We shall modify Eq. (3.2.23) using Eqs. (3.3.7) and (3.3.13):

$$a^2 = 3 \frac{F_2(\mu^*)}{F_0(\mu^*)} \frac{u_{\sigma L}}{u_{\sigma i}}. \qquad (3.3.18)$$

The parameter a^2 has been used in this form by Mansfield [40] to allow for the influence of the degeneracy on the electrical conductivity. In the Boltzmann statistics case ($\mu^* \ll 0$), Eq. (3.3.18) reduces to $a^2 = 6 u_{\sigma L}/u_{\sigma i}$. This parameter has been used in the theory of transport phenomena of nondegenerate semiconductors [36–39].

Equations (3.3.3) and (3.3.4) are not very convenient to use in any comparison of the experimental data with the theoretical results for an arbitrary degree of degeneracy since the substitution of τ_{0L} from Eq. (3.2.22) makes them fairly cumbersome. Therefore, it would be convenient to eliminate τ_{0L}. To do this, we shall divide Eq. (3.3.3) by Eq. (3.3.8) and Eq. (3.3.4) by Eq. (3.3.9):

$$\frac{\sigma}{\sigma_0} = \Phi_3(\mu^*, a) e^{-\mu^*}, \qquad (3.3.19)$$

$$\frac{u_\sigma}{u_{\sigma 0}} = \frac{\sqrt{\pi}}{2} \frac{\Phi_3(\mu^*, a)}{F_{1/2}(\mu^*)}. \qquad (3.3.20)$$

These two expressions are shown graphically in Figs. 3.4 and 3.5. By measuring σ or u_σ of a heavily doped crystal and employing the known electrical conductivity or mobility of an impurity-free semiconductor, these curves can be used to find a (we are assuming that μ^* is known), i.e., we can determine from Eq. (3.3.18) the true value of the ratio $u_{0L}/u_{\sigma i}$ of the investigated heavily doped semiconductor.

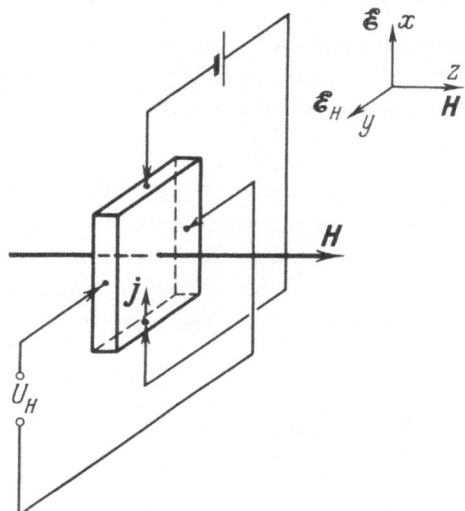

Fig. 3.7. Arrangement of a sample in the electric and magnetic fields in the measurement of
the Hall effect.

This approach applies also to the case of mixed scattering
in a nondegenerate semiconductor ($\mu* \leq -4$), which has already
been considered [36].

We shall now determine the range of validity of the expressions (3.3.6) and (3.3.7), representing the scattering by phonons,
and of the expressions (3.3.12) and (3.3.13), which apply to the scattering on ions alone, i.e., we shall estimate when we can assume
that $a^2 \ll (\varepsilon*)^2$ and when $a^2 \gg (\varepsilon*)^2$. To do this, we shall find the
ratio σ/σ_L (or, which is equivalent, $u_\sigma/u_{\sigma L}$) by dividing the relevant expressions:

$$\frac{\sigma}{\sigma_L} = \frac{u_\sigma}{u_{\sigma L}} = \frac{\Phi_3(\mu*, a)}{F_0(\mu*)} \qquad (3.3.21)$$

It is evident from the curves in Fig. 3.6 that in the absence of degeneracy ($\mu* \leq -4$), the contribution of the scattering by ions is
important in all cases with the exception of very small values of
a. However, the narrow range of such values of a (from 0 to 0.2)
may be misleading, because these values correspond to a very im-

portant range of impurity concentrations (pure and lightly doped semiconductors) and to moderate and high temperatures.

In the degenerate case ($\mu^* > 0$), we can ignore the scattering by ions for much larger values of a (Fig. 3.6). For example, when $\mu^* = 6$ and $a = 3$, the phonon scattering contribution to the carrier mobility is up to 20%.

Hall Effect. We shall now consider a crystal placed in electric and magnetic fields of intensities $\mathcal{E}$ and $\mathbf{H}$, respectively (Fig. 3.7). We shall again assume that the temperature along a sample is constant ($\nabla T = 0$). Using the expression for the force acting on the electrons (Table 3.1), we can write the transport equation in the form

$$\frac{e}{\hbar}\left(\mathcal{E} + \frac{1}{c}\,[vH]\,\right)\nabla_k f = -\frac{C(E)\,\boldsymbol{k}\,\frac{\partial f}{\partial E}}{\tau(\boldsymbol{k})}. \tag{3.3.22}$$

In obtaining this expression, we have assumed that ∇f in Eq. (3.1.3) is equal to zero because $\nabla T = 0$.

We shall calculate separately the second term on the left-hand side of Eq. (3.3.22) by substituting f from Eq. (3.1.14) and using Eq. (3.1.9):

$$-\frac{e}{\hbar c}\,[vH]\,\nabla_k f = -\frac{e}{\hbar c}\left\{[v\,H]\,\frac{\partial f_0}{\partial E}\,\hbar v - \nabla_k\left(\frac{\partial f}{\partial E}\,C(E)\,\boldsymbol{k}\right)[vH]\right\}.$$

The first term in the curly brackets (braces) in the above expression is equal to zero, because $[\mathbf{vH}]\,\mathbf{v} = 0$ and the second term is

$$\nabla_k\left(\frac{\partial f_0}{\partial E}\,C(E)\,\boldsymbol{k}\right)[vH] = [vH]\left\{-\frac{\partial f_0}{\partial E}\,C(E) - \left[\hbar\boldsymbol{k}\,\frac{\partial}{\partial E}\left(\frac{\partial f_0}{\partial E}\right)C(E)\right]\right\}.$$

Neglecting terms of the second order of smallness and using the property of the scalar-vector product, $[\boldsymbol{ab}]\boldsymbol{c} = [\boldsymbol{b}\,\boldsymbol{c}]\boldsymbol{a}$,* we obtain:

$$\frac{e}{\hbar c}\,[vH]\,\nabla_k f = -\frac{e}{\hbar c}\,\frac{\partial f_0}{\partial E}\,[HC]\,v. \tag{3.3.23}$$

*See, for example, V. I. Smirnov, Course of Higher Mathematics (Fizmatgiz, 1962), Vol. II, §105.

Substituting this expression into Eq. (3.3.22), solving it for $\mathbf{C}(E)$, and using $\mathbf{v} = \mathbf{k}h/m^*$, we obtain

$$C(E) = -\frac{e\hbar}{m^*}\tau(k)\left\{\mathcal{E} - \frac{1}{\hbar c}|HC|\right\}. \qquad (3.3.24)$$

To solve this vector equation, we recall that if

$$x = a + |bx|,$$

where

$$bx = ba \ [\text{ since } (b|bx|) \equiv 0], \qquad (3.3.25)$$

the substitution of the above expression for x on the right-hand side yields

$$x = a + |ba| + |b|bx||,$$

but*

$$[b|bx|] = b(bx) - xb^2 \qquad (3.3.26)$$

so that finally

$$x = \frac{a + |ba| + (ab)b}{1 + b^2}.$$

Turning back to Eq. (3.3.24), we can rewrite it thus

$$C(E) = -\frac{e\hbar}{m^*}\tau(k)\left\{\frac{\mathcal{E} + \frac{e\tau}{m^*c}[H\mathcal{E}] + \left(\frac{e\tau}{m^*c}\right)^2(H\mathcal{E})H}{1 + \left(\frac{e\tau H}{m^*c}\right)^2}\right\}. \qquad (3.3.27)$$

Since right from the beginning, we have limited our discussion to weak fields (forces), it follows that the parameter $e\tau H/m^*c$ can

---

*See, for example, V. I. Smirnov, Course of Higher Mathematics (Fizmatgiz, 1962), Vol. II, §105.

be assumed to be small and we can expand Eq. (3.3.27) as a series in terms of this parameter.

Hence, we obtain a criterion for weak magnetic fields:

$$H \ll \frac{m^* c}{e\tau}.$$ (3.3.28)

Thus, expanding Eq. (3.3.27) as a series and including only terms in the first power of H, we obtain

$$C(E) = -\frac{e\hbar}{m^*}\tau(k)\left\{\mathscr{E} + \frac{e\tau}{m^* c}[H\mathscr{E}]\right\}.$$ (3.3.29)

We can now substitute this expression into Eq. (3.1.16) in exactly the same way as in the derivation of the electrical conductivity formula:

$$j_c = -\frac{e^2}{3\pi^2 m^*}\int_0^\infty k^3\tau\,\frac{\partial f}{\partial k}\left\{\mathscr{E} + \frac{e\tau}{m^* c}[H\mathscr{E}]\right\}dk$$ (3.3.30)

or

$$j_c = -\frac{e^2\mathscr{E}}{3\pi^2 m^*}\int_0^\infty k^3\tau\,\frac{\partial f_0}{\partial k} + \frac{e^3[H\mathscr{E}]}{3\pi^2 m^{*2}c}\int_0^\infty k^3\tau^2\frac{\partial f_0}{\partial k}\,dk.$$ (3.3.31)

The first term, representing the ordinary electric current through a crystal, has been analyzed before, and the second term, proportional to $[H\mathscr{E}]$ represents the Lorentz force acting on carriers moving in a magnetic field. Thus, the second term describes a current at right angles to the plane containing $\mathscr{E}$ and H, i.e., the Hall current j_H. It is known that

$$\mathscr{E}_H = RjH,$$ (3.3.32)

where the quantity $\mathscr{E}_H \equiv \mathscr{E}_y$ is the Hall field; R is a coefficient of proportionality known as the Hall coefficient; j is the density of the current flowing through a sample.

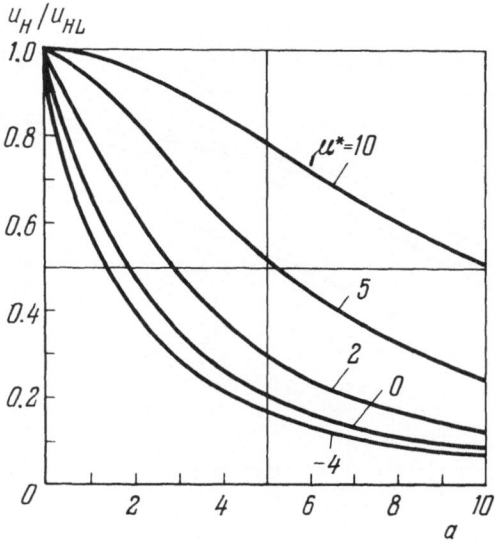

Fig. 3.8. Contribution of the lattice scattering to the Hall mobility of electrons.

We shall now go over from vector to scalar quantities in Eq. (3.3.31):

$$j_x = -\frac{e^2 \mathcal{E}_x}{3\pi^2 m^*} I_1 + \frac{e^3 H \mathcal{E}_y}{3\pi^2 m^{*2}c} I_2, \left.\vphantom{\begin{array}{c} 1 \\ 1 \end{array}}\right\}$$
$$j_y = -\frac{e^2 \mathcal{E}_y}{3\pi^2 m^*} I_1 + \frac{e^3 H \mathcal{E}_x}{3\pi^2 m^{*2}c} I_2, \left.\vphantom{\begin{array}{c} 1 \\ 1 \end{array}}\right\} \qquad (3.3.33)$$

where:

$$I_i = \int_0^\infty k^3 \tau^i \frac{\partial f_0}{\partial k} \, dk. \qquad (3.3.34)$$

Since the current flows through a sample only along the x axis ($j \equiv j_x$) it follows that $j_y = 0$ and the second equation in the system (3.3.33) yields:

$$\mathcal{E}_y = \frac{eH}{c} \mathcal{E}_x \frac{I_2}{I_1}.$$

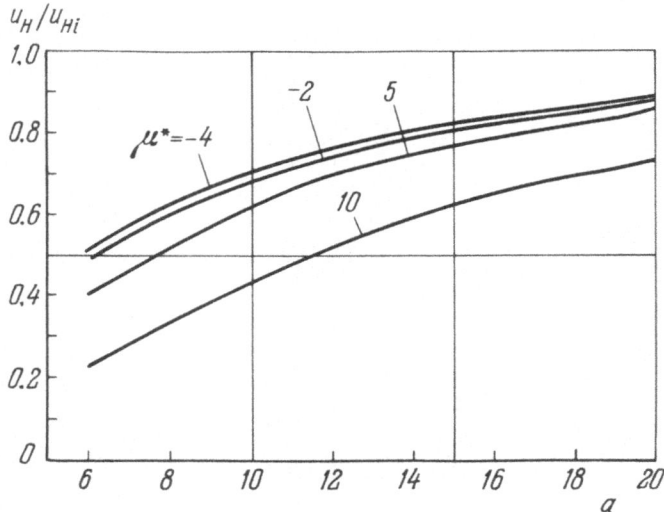

Fig. 3.9. Contribution of the impurity scattering to the Hall
mobility of electrons.

Substituting $\mathscr{E}_x = j/\sigma$, we obtain

$$\mathscr{E}_y \equiv \mathscr{E}_H = \frac{eH}{c} j \frac{1}{\sigma} \frac{I_2}{I_1}. \tag{3.3.35}$$

Comparing this expression with Eq. (3.3.32), we find:

$$R = \frac{e}{c} \frac{1}{\sigma} \frac{I_2}{I_1} \tag{3.3.36}$$

and substituting the general form of σ from Eq. (3.1.18), we obtain:

$$R = \frac{3\pi^2 m^*}{ec} \frac{\displaystyle\int_0^\infty k^3 \tau^2 \frac{\partial f_0}{\partial k} \, dk}{\left[\displaystyle\int_0^\infty k^3 \tau \frac{\partial f_0}{\partial k} \, dk \right]^2}. \tag{3.3.37}$$

It follows from Eq. (3.3.36) that the product $R\sigma$ has the dimensions
of mobility since the ratio I_2/I_1 has the dimensions of time. Therefore, we can introduce the "Hall mobility" u_H, which is defined as

$$u_H = R\sigma = \frac{e}{c} \frac{\displaystyle\int_0^\infty k^3 \tau^2 \frac{\partial f_0}{\partial k} dk}{\displaystyle\int_0^\infty k^3 \tau \frac{\partial f_0}{\partial k} dk}. \qquad (3.3.38)$$

Further analysis of the Hall coefficient and of the Hall mobility will be carried out in the same way as in the case of the electrical conductivity and the drift mobility, i.e., we shall use Eq. (3.2.22) for τ and we shall take f_0 to be the Fermi distribution function. Then, integrating I_1 and I_2 by parts, replacing k with E, and using the dimensionless energy ε^*, we obtain

$$R = \frac{3h^3}{8\pi e \,(2m^*k_0T)^{3/2}} \cdot \frac{\Phi_{9/2}(\mu^*, a)}{[\Phi_3(\mu^*, a)]^2}, \qquad (3.3.39)$$

$$u_H = \frac{e}{m^*} \tau_{0L} \frac{\Phi_{9/2}(\mu^*, a)}{\Phi_3(\mu^*, a)}, \qquad (3.3.40)$$

where the integral $\Phi_{9/2}(\mu^*, a)$ has the form:

$$\Phi_{9/2}(\mu^*, a) = \int_0^\infty \frac{(\varepsilon^*)^{9/2}\exp(\varepsilon^* - \mu^*)\,d\varepsilon^*}{[(\varepsilon^*)^2 + a^2]^2\,[1 + \exp(\varepsilon^* - \mu^*)]^2}. \qquad (3.3.41)$$

We shall analyze first the Hall mobility. Let us assume that $a^2 \ll (\varepsilon^*)^2$ (scattering by the acoustical phonons only). We can easily show [integrating Eq. (3.3.41) by parts] that

$$\Phi_{9/2}(\mu^*, a) \to \frac{1}{2} F_{-1/2}(\mu^*)$$

and, consequently,

$$u_H \to u_{HL} = \frac{e}{2m^*} \tau_{0L} \frac{F_{-1/2}(\mu^*)}{F_0(\mu^*)}. \qquad (3.3.42)$$

Figure 3.8 gives the ratio

$$\frac{u_H}{u_{HL}} = 2 \frac{\Phi_{9/2}(\mu^*, a)}{\Phi_3(\mu^*, a)} \frac{F_0(\mu^*)}{F_{-1/2}(\mu^*)}, \qquad (3.3.43)$$

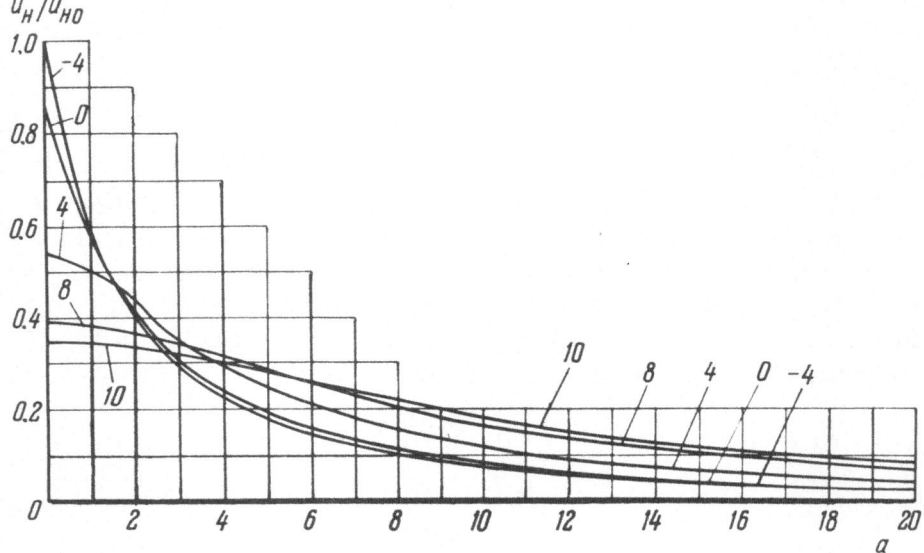

Fig. 3.10. Calculated Hall mobility for the mixed scattering of electrons.

which, like the ratio (3.3.21), shows for which values of μ^* and a we can assume the inequality $a^2 \ll (\varepsilon^*)^2$ to be valid. The curves in Fig. 3.8 are somewhat more convenient than the curves in Fig. 3.6, since in the majority of cases the experimentally determined quantity is the Hall mobility and not the drift mobility. The drift mobility of heavily doped semiconductors is practically impossible to measure because of experimental difficulties. We can also use the measured values of u_H and u_{HL} and the curves in Fig. 3.8 to find the value of a and then we can employ the curves in Fig. 3.6 to estimate the drift mobility in heavily doped samples. Knowing a, the value of u_σ can be found also directly from Eq. (3.3.18).

The Hall mobility can be easily found from Eq. (3.3.42) in two special cases: absence of degeneracy ($\mu^* \ll 0$) and strong degeneracy ($\mu^* \gg 0$). Using the expressions (2.5.6) and (2.5.11), we obtain

$$\mu^* \ll 0: \quad u_{H_0} = \frac{\sqrt{\pi}}{2} \frac{e}{m^*} \tau_{0L}, \tag{3.3.44}$$

$$\mu^* \gg 0: \quad u_{H_0} = \frac{e}{m^*} \tau_{0L} (\mu^*)^{-1/2}. \tag{3.3.45}$$

TABLE 3.4. Hall Mobility of Important Semiconductors

Semiconductor	u_{H0}, cm$^2 \cdot$ V$^{-1} \cdot$ sec^{-1} [64, 65]	
	Electrons	Holes
Ge	$4\,500 \left(\dfrac{T}{300}\right)^{-1.6}$	$3\,500 \left(\dfrac{T}{300}\right)^{-2.33}$
Si	$1\,300 \left(\dfrac{T}{300}\right)^{-2.0}$	$500 \left(\dfrac{T}{300}\right)^{-2.7}$
AlSb	$200 \left(\dfrac{T}{300}\right)^{-?}$	$400 \left(\dfrac{T}{300}\right)^{-1.8}$
GaSb	$4\,000 \left(\dfrac{T}{300}\right)^{-2.0}$	$1\,400 \left(\dfrac{T}{300}\right)^{-0.9}$
InSb	$78,000 \left(\dfrac{T}{300}\right)^{-1.6}$	$750 \left(\dfrac{T}{300}\right)^{-2.1}$
GaAs	$8\,500 \left(\dfrac{T}{300}\right)^{-1.0}$	$420 \left(\dfrac{T}{300}\right)^{-2.1}$
InAs	$33,000 \left(\dfrac{T}{300}\right)^{-1.2}$	$460 \left(\dfrac{T}{300}\right)^{-2.3}$
InP	$4\,600 \left(\dfrac{T}{300}\right)^{-2.0}$	$150 \left(\dfrac{T}{300}\right)^{-2.4}$
GaP	$110 \left(\dfrac{T}{300}\right)^{-1.5}$	$75 \left(\dfrac{T}{300}\right)^{-1.5}$
PbS	$500 \left(\dfrac{T}{300}\right)^{-2.5}$	$500 \left(\dfrac{T}{300}\right)^{-2.5}$
PbTe	$1\,400 \left(\dfrac{T}{300}\right)^{-2.5}$	$1\,400 \left(\dfrac{T}{300}\right)^{-2.5}$
PbSe	$2\,000 \left(\dfrac{T}{300}\right)^{-2.5}$	$2\,000 \left(\dfrac{T}{300}\right)^{-2.5}$

If only the scattering by ions is important, we can assume that $a^2 \gg (\varepsilon^*)^2$ and $\Phi_{9/2} \to (9/2a^4) F_{7/2}(\mu^*)$, so that

$$u_H \to u_{Hi} = \frac{3}{2} \frac{e}{m^*} \tau_{0i} \frac{F_{7/2}(\mu^*)}{F_2(\mu^*)}. \qquad (3.3.46)$$

The mobilities corresponding to the cases $\mu^* \ll 0$ and $\mu^* \gg 0$ have the form

$$\mu^* \ll 0: \quad u_{Hi} = \frac{315\sqrt{\pi}}{64} \frac{e}{m^*} \tau_{0i}, \tag{3.3.47}$$

$$\mu^* \gg 0: \quad u_{Hi} = \frac{e}{m^*} \tau_{0i} (\mu^*)^{3/2}. \tag{3.3.48}$$

For the sake of generality, we shall write a^2 [as in the derivation of Eq. (3.3.18)] as a function of the Hall mobilities:

$$a^2 = 3 \frac{F_0(\mu^*) F_{7/2}(\mu^*)}{F_{-1/2}(\mu^*) F_2(\mu^*)} \frac{u_{HL}}{u_{Hi}}. \tag{3.3.49}$$

It is interesting to estimate under what conditions the inequality $a^2 \gg (\varepsilon^*)^2$ remain valid. Figure 3.9 shows the dependence of u_H/u_{Hi} on a. It is evident from this figure that the phonon scattering in nondegenerate semiconductors is important up to $a \approx 12$. For degenerate ($\mu^* > 0$), we can assume that the mobility is mainly due to ionized impurities only if $a > 15$.

To compare the experimental values of the Hall mobilities with theory, we shall divide Eq. (3.3.40) by u_{H0} of an impurity-free semiconductor [Eq. (3.3.44)]:

$$\frac{u_H}{u_{H0}} = \frac{2}{\sqrt{\pi}} \frac{\Phi_{5/2}(\mu^*, a)}{\Phi_3(\mu^*, a)}. \tag{3.3.50}$$

For convenience, Table 3.4 gives the values of the Hall mobilities of electrons and holes for the most important semiconductors.

Appendix A.7 lists the values of u_H/u_{H0} for various values of μ^* and a. For clarity, some of the results in that Appendix are also given in Fig. 3.10. A similar table for a smaller number of values of a and μ^* (and with some numerical errors) is given also in [41].

In conclusion, we shall consider the possibility of finding the total mobility u from the law

$$\frac{1}{u} = \frac{1}{u_{HL}} + \frac{1}{u_{Hi}}. \tag{3.3.51}$$

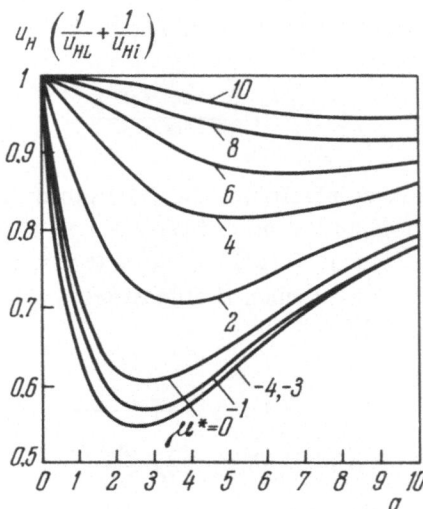

Fig. 3.11. Error in the determination of the Hall mobility of electrons by adding u_{HL} and u_{Hi}.

Because of the different energy dependence of τ_L and τ_i, this relationship is, generally speaking, invalid, although it is used quite widely.

Using the formulas deduced here, we can plot (Fig. 3.11) the dependence of the quantity $u_H(1/u_{HL} + 1/u_{Hi})$ on the parameter a for various values of μ^*. This quantity is the error in the determination of u_H using Eq. (3.3.51). It is evident from Fig. 3.11 that the law represented by Eq. (3.3.51) is valid for nondegenerate semiconductors only for small values of a. In the case of strong degeneracy ($\mu^* \geq 5$), this law is fairly accurate even at large values of a.

We shall now consider the Hall coefficient. First, we shall compare Eq. (3.3.38) with the expression for the electron density given by Eq. (2.3.4). Combining these expressions, we obtain R in the form

$$R = \frac{\sigma_f}{ne}. \qquad (3.3.52)$$

Hall Factor. The quantity $\mathscr{A}$, known as the Hall factor, is the following combination of integrals [51]:

$$\mathscr{A} = \frac{3}{2} \frac{F_{1/_2}(\mu^*) \, \Phi_{\bullet/_2}(\mu^*, a)}{[\Phi_3(\mu^*, a)]^2} .$$

(3.3.53)

Like other transport coefficients, the Hall factor is also a function of the scattering mechanism and of the degree of degeneracy. Assuming, as before, that $a^2 \ll (\varepsilon^*)^2$ or $a^2 \gg (\varepsilon^*)^2$, we can easily find expressions for $\mathscr{A}$, valid when the scattering takes place only on phonons or only on ions:

$$\mathscr{A}_L = \frac{3}{4} F_{1/_2}(\mu^*) \frac{F_{-1/_2}(\mu^*)}{F_0^2(\mu^*)}$$

(3.3.54)

$$\mathscr{A}_i = \frac{3}{4} F_{1/_2}(\mu^*) \frac{F_{7/_2}(\mu^*)}{F_3^2(\mu^*)} .$$

(3.3.55)

Using $\mu^* \ll 0$ and $\mu^* \gg 0$ in the above two expressions, we obtain:

$$\mu_* \ll 0: \quad \begin{cases} \mathscr{A}_L = \dfrac{3\pi}{8} = 1.18, & (3.3.56) \\[2mm] \mathscr{A}_i = \dfrac{315}{512}\pi = 1.93, & (3.3.57) \end{cases}$$

$$\mu^* \gg 0: \quad \mathscr{A}_L = \mathscr{A}_i = 1.$$

Naturally, Eqs. (3.3.56) and (3.3.57) are in good agreement with the well-known values [52] which apply to these special cases.

The values of the Hall factor for the general case of any degree of degeneracy are given in Appendix A.8 and Fig. 3.12. We can easily show that when $\mu^* < -4$, the curves $\mathscr{A}(\mu^*, a)$ are the same as the curve for $\mu^* = -4$. The value of $\mathscr{A}$ for all nondegenerate semiconductors can be found from this curve [51] (the value of a is assumed to be known, for example, from Table A.7).

Determination of the Hall factor for a nondegenerate semiconductor was considered in [36, 37, 39]. These workers obtained a dependence $\mathscr{A}(\rho_i/\rho)$, where ρ_i is the resistivity of a semiconductor in the case of scattering on ions only, and ρ is the resistivity in the case of mixed scattering on ions and phonons. Since the parameter ρ_i/ρ is very indeterminate, the dependences obtained are not of much practical value.

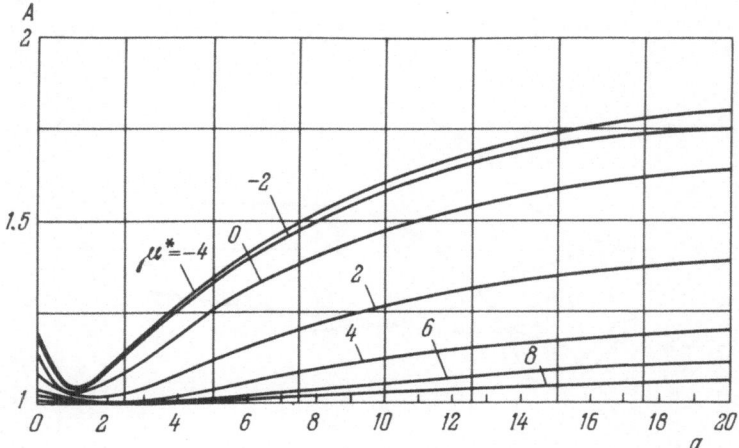

Fig. 3.12. Calculated values of the Hall factor for the mixed scattering of electrons.

Dependence of Transport Phenomena on Carrier Density and Temperature. From Figs. 3.4, 3.6, and 3.10, or from the corresponding formulas (3.3.19), (3.3.20), and (3.3.50), it follows that the quantities σ, u_σ, and u_H depend in a complex manner on the carrier density and temperature.

We shall consider first the dependence on the carrier density. It is manifested mainly by a change in $\mu^*(n)$, given by Eq. (2.3.4) (Fig. 2.9). The numerical values of $\mu^*(n)$, and consequently of σ, u_σ, and u_H, depend on the nature of a semiconductor (through m^*) and on the temperature. Moreover, when the carrier density is altered, the contributions of the individual scattering mechanisms change as well, i.e., the value of a changes. Fistul' et al. [53] have demonstrated that in the limiting cases, when only one type of scattering predominates, the mobilities u_L and u_i can be represented in the form

$$u_L \propto n^{-s/2}; \; u_i \propto n^{\frac{3}{2}s - 1 - \gamma},$$ (3.3.58)

where s and γ are, respectively, the exponents in the dependences

$$\overline{\varepsilon^*} \propto n^{s(\mu^*)}; \quad g(b) \propto n^{\gamma(\mu^*)}.$$ (3.3.59)

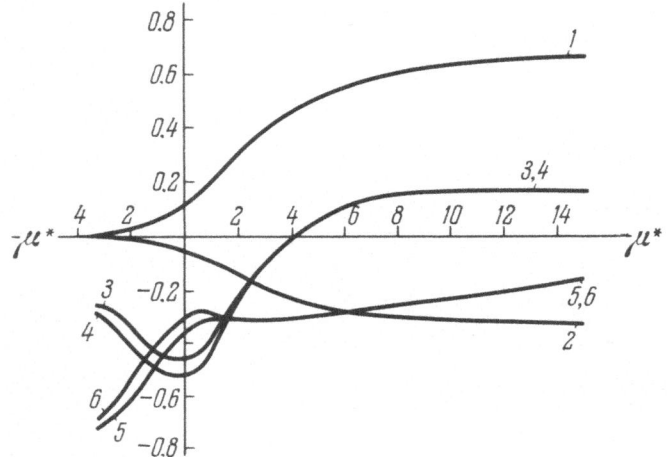

Fig. 3.13. Auxiliary curves for the explanation of the dependence
of the electron mobility on the electron density in heavily doped
silicon (following [53]).

The dependence of the exponent $-s/2$ on the degree of degeneracy
is shown in Fig. 3.13 by curve 2. Curve 1 represents the values of
s. The values of γ are given by curves 3 and 4 for 300 and 100°K,
respectively. The dependence of the exponent $(3s/2 - 1 - \gamma)$ on μ^*
is represented by curves 5 and 6 for the same temperatures of 300
and 100°K. The latter two curves show that the carrier-density dependence of the mobility cannot be represented by a single power
law with a constant exponent.

The curves in Fig. 3.13 refer to silicon. Because of different effective masses, the curves showing the dependence $\gamma(\mu^*)$ for
other semiconductors are slightly different and therefore the final
curves 5 and 6 are also different. It follows that it is sensible to
analyze the expected carrier-density dependences only in special
cases, namely for $a^2 \ll (\varepsilon^*)^2$ and $a^2 \gg (\varepsilon^*)^2$ for $\mu^* \ll 0$ and $\mu^* \gg 0$.
Such an analysis can be carried out easily by the reader, using the
formulas given in the present section.

This applies also to the temperature dependences of the phenomena considered so far.

Need to Allow for Screening. In the theory presented here, the screening factor g(b) in the Brooks−Herring

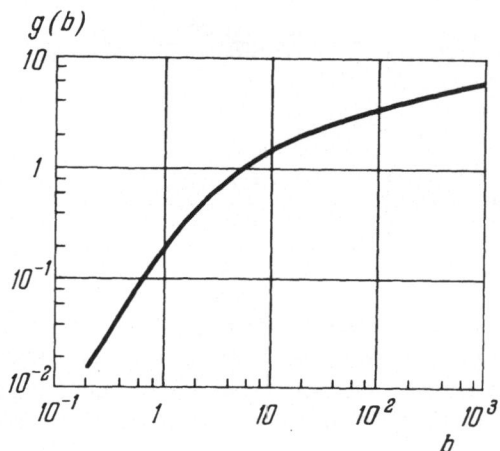

Fig. 3.14. Dependence $g(b) = \ln(1+b) - [b/(b+1)]$.

formula for τ_i has been assumed to be constant, i.e., independent of energy. Therefore, this factor is taken outside the integral sign in the derivation of the transport integrals Φ_m. In many cases, this approximation is unjustified. We shall now consider Fig. 3.14, which shows the dependence $g(b)$. We can easily see that at high values of b ($> 10^2$), we can indeed assume with sufficient accuracy that $g(b)$ is a constant. The maximum carrier densities, at which we can still assume that $g(b) = \text{const}$, depend on the nature of the semiconductor. Thus, for example, a numerical estimate shows that at room temperature the maximum carrier density is approximately $1 \cdot 10^{17}$ cm^{-3} for germanium and $5 \cdot 10^{17}$ cm^{-3} for silicon. When the temperature is lowered to 78°K, these quantities become almost an order of magnitude less. When the concentration of the dopant in a semiconductor is increased, the dependence of $g(b)$ on the energy becomes stronger so that the scattering factor r in the law (3.2.21) is no longer equal to 2 for the scattering by impurity ions.

Allowance for the energy dependence of the screening factor does not alter the form of the final expressions obtained in the present section for the transport phenomena. The only difference is the replacement of the integrals $\Phi_3(\mu^*, a)$ and $\Phi_{9/2}(\mu^*, a)$, respectively, with the integrals

$$\mathscr{F}_3(\mu^*, a) = \int_0^\infty \frac{(\varepsilon^*)^3 \exp(\varepsilon^* - \mu^*)\, d\varepsilon^*}{[(\varepsilon^*)^2 + a^2 g(b)][1 + \exp(\varepsilon^* - \mu^*)]^2} \qquad (3.3.60)$$

and

$$\mathscr{F}_{5/2}(\mu^*, a) = \int_0^\infty \frac{(\varepsilon^*)^{5/2} \exp(\varepsilon^* - \mu^*)\, d\varepsilon^*}{[(\varepsilon^*)^2 + a^2 g(b)]^2 [1 + \exp(\varepsilon^* - \mu^*)]^2}. \qquad (3.3.61)$$

Allowance for the dependence of $g(b)$ on the energy is particularly important in the case of scattering by ions. Therefore, we can assume, approximately, that it is important if $a^2 g(b) > (\varepsilon^*)^2$. Then, the integrals (3.3.60) and (3.3.61) simplify to become

$$\mathscr{F}_{5/2} \to \frac{1}{a^4}\, \varphi_{5/2} \quad \text{and} \quad \mathscr{F}_3 \to \frac{1}{a^2}\, \varphi_3,$$

where

$$\varphi_{5/2} = \int_0^\infty \frac{(\varepsilon^*)^{5/2} \exp(\varepsilon^* - \mu^*)\, d\varepsilon^*}{[g(b)]^2 [1 + \exp(\varepsilon^* - \mu^*)]^2}, \qquad (3.3.62)$$

$$\varphi_3 = \int_0^\infty \frac{(\varepsilon^*)^3 \exp(\varepsilon^* - \mu^*)\, d\varepsilon^*}{g(b)[1 + \exp(\varepsilon^* - \mu^*)]^2}, \qquad (3.3.63)$$

shown graphically in Appendix A.6.

Need to Allow for Anisotropy. The transport phenomena in semiconductors with nonspherical constant-energy surfaces are anisotropic. The anisotropy is of two types: the effective-mass anisotropy, which is characterized by a quantity $K_m = m^*_\| / m^*_\perp$, and the relaxation-time anisotropy, represented by the parameter $K_\tau = \tau_\| / \tau_\perp$.

Glicksman [54] has shown that if the relaxation time tensor has the same symmetry as the effective mass tensor, then the anisotropy parameter is the ratio $K = K_m / K_\tau$. This applies to n-type germanium and silicon. In general, K_m and K_τ are interdependent; the relationship between them was found by Samoilovich et al. and it has been considered in the preceding section of the present chap-

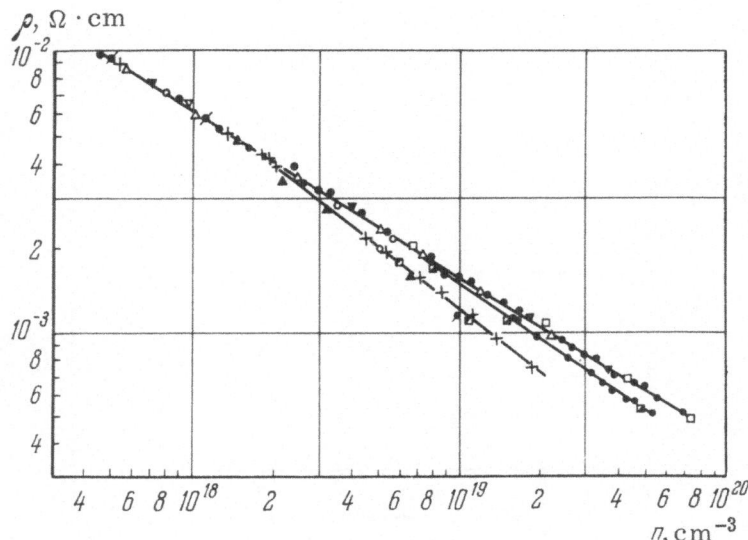

Fig. 3.15. Dependence of the resistivity of heavily doped n-type germanium crystals on the carrier density (T = 300°K). Lower curve, Sb dopant; middle curve, P; upper curve, As.

ter. In the same section (§3.2), we have given the formulas (3.2.25) and (3.2.26) for the calculation of the relaxation times $\tau_i^{||}$ and $\tau_i^{\perp}$ of heavily doped semiconductors. The relaxation times in the case of scattering on the acoustical phonons, $\tau_L^{||}$ and $\tau_L^{\perp}$, in heavily doped crystals are the same as for impurity-free semiconductors. They can be calculated from the formulas given in [56]:

$$\tau_L^{||} = \frac{\tau_0}{s_0} \text{ and } \tau_L^{\perp} = \frac{\tau_0}{s_1},\qquad (3.3.64)$$

where

$$\tau_0 = \frac{\pi C_{11} \hbar^4}{kT E_i^2 \sqrt{2m_{\perp}^2 m_{||} E}}$$

(for notation see Table 3.2),

$$s_0 = 1 + 1.88 \frac{c_2}{c_1} + 1.03 \left(\frac{c_2}{c_1}\right)^2,$$

$$s_1 = 1 + 1.24 \frac{c_2}{c_1} + 0.87 \left(\frac{c_2}{c_1}\right)^2,$$

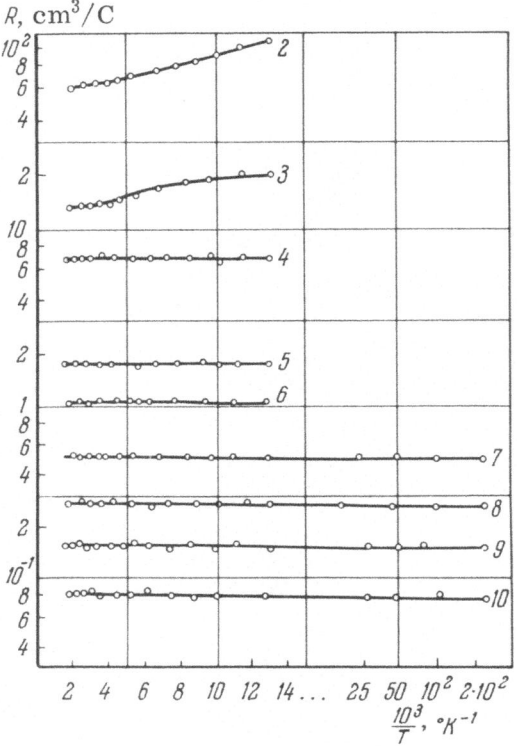

Fig. 3.16. Temperature dependence of the Hall coefficient of germanium heavily doped with arsenic
[50]. The numbers alongside the curves have the
same meaning as in Fig. 3.17.

where c_1 and c_2 are the deformation potential constants, the substitution of which gives the following expressions for germanium [56]:

$$\tau_L^{\parallel} = \frac{3.45 \cdot 10^{17}}{T \sqrt{E}}; \quad \tau_L^{\perp} = \frac{2.94 \cdot 10^{17}}{T \sqrt{E}}. \qquad (3.3.65)$$

The Hall mobility in the anisotropic case should be calculated
from the formula [9]:

$$u_H = \frac{e}{m_{\perp}^*} \cdot \frac{2 \langle \tau_{\parallel} \tau_{\perp} \rangle + \dfrac{m_{\parallel}^*}{m_{\perp}^*} \langle \tau_{\perp}^2 \rangle}{\langle \tau_{\parallel} \rangle + 2 \dfrac{m_{\parallel}^*}{m_{\perp}^*} \langle \tau_{\perp} \rangle}, \qquad (3.3.66)$$

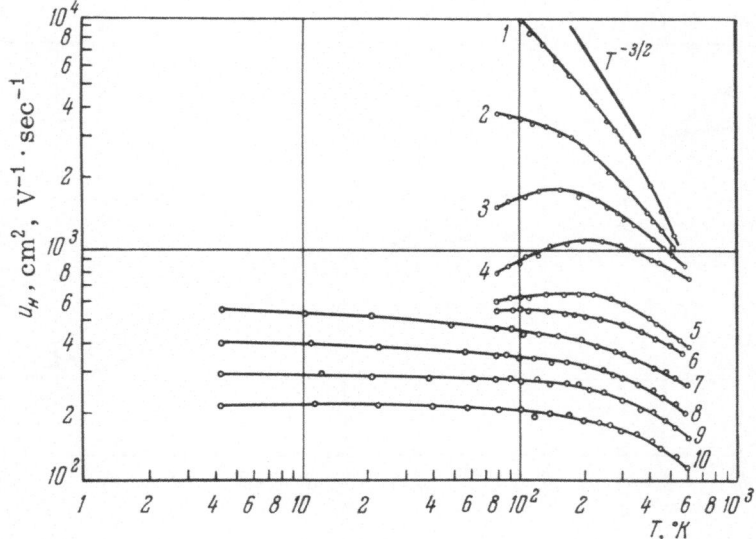

Fig. 3.17. Temperature dependence of the Hall mobility of electrons in arsenic-doped germanium [63]. Different curves represent different degrees of doping (the carrier densities n are given in cm^{-3}): 1) $1.8 \cdot 10^{15}$; 2) $9.6 \cdot 10^{16}$; 3) $4.3 \cdot 10^{17}$; 4) $8.85 \cdot 10^{17}$; 5) $3.5 \cdot 10^{18}$; 6) $6 \cdot 10^{18}$; 7) $1.2 \cdot 10^{19}$; 8) $2.3 \cdot 10^{19}$; 9) $4 \cdot 10^{19}$; 10) $7.7 \cdot 10^{19}$.

where

$$\frac{1}{\tau_{\parallel}} = \frac{1}{\tau_L^{\parallel}} + \frac{1}{\tau_i^{\parallel}}; \qquad \frac{1}{\tau_{\perp}} = \frac{1}{\tau_L^{\perp}} + \frac{1}{\tau_i^{\perp}} \qquad (3.3.67)$$

and

$$\langle y \rangle = \frac{2}{3 F_{1/_2}(\mu^*)} \int_0^\infty (\varepsilon^*)^{3/_2} y \frac{\partial f}{\partial \varepsilon^*} d\varepsilon^*, \qquad (3.3.68)$$

and f in the case of heavily doped semiconductors represents the Fermi−Dirac distribution function.

Experimental Results (n−Type Germanium) [50, 57−61]. The results of different workers on the carrier-density dependence of the resistivity of germanium with As, P, and Sb impurities are in good mutual agreement (results taken from [57−61] and [73] are shown in Fig. 3.15).

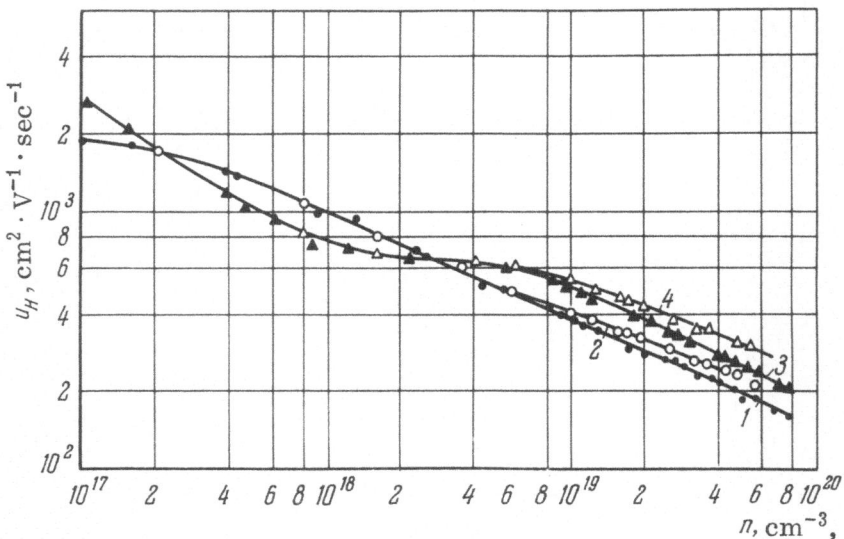

Fig. 3.18. Dependence of the Hall mobility of electrons on the electron density in heavily doped germanium [50]. Arsenic impurity: 1) 300°K; 3) 78°K; phosphorus impurity: 2) 300°K; 4) 78°K.

It is very important to stress that in the case of heavily doped samples the electron scattering depends strongly on the nature of the dopant. The threshold carrier density n_0, at which the "individuality" of an impurity becomes important, is different for samples doped with different impurities and it depends on the temperature. Several workers [30, 58] have drawn attention to the correlation between n_0 and the impurity ionization energy. However, this correlation is of a purely formal nature since at densities of the order of n_0 the ionization energy itself is equal to zero, as demonstrated in Chap. 1.

It is evident from Fig. 3.15 that the results of measurements fit well logarithmic straight lines, which can be described, for example, in the case of arsenic-doped germanium, by an empirical relationship of the type [61]:

$$\rho = 1.67 \cdot 10^8 \cdot n^{-0.58} \quad \text{for} \quad 4 \cdot 10^{17} < n < 7 \cdot 10^{19} \text{ cm}^{-3},$$

which differs considerably from the theoretical dependence of Johnson and Lark-Horovitz [62]

$$\rho = 6.27 \cdot 10^3 \cdot n^{-0.33},$$

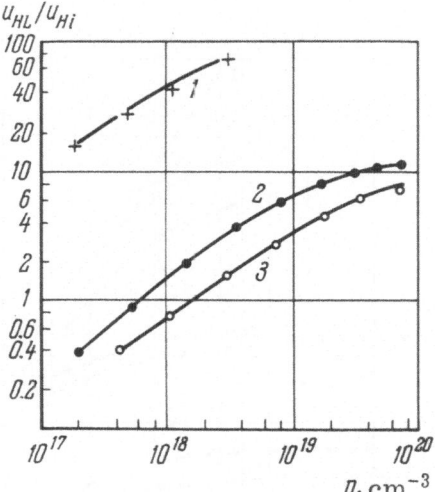

Fig. 3.19. Ratio of the Hall mobilities governed by the lattice and impurity mechanisms
of electron scattering in heavily doped germanium: 1) 100°K; 2) 300°K; 3) 500°K.

obtained by these authors for a degenerate electron gas using a
classical method analogous to the Conwell—Weisskopf method.

The Hall coefficient is practically independent of temperature
(Fig. 3.16). The R(T) dependence is observed only for samples having carrier densities less than $5 \cdot 10^{17}$ cm^{-3}. This dependence is
due to the temperature dependence of the Hall factor $\mathcal{A}$ resulting
from a change in the contributions of the phonon and ion components to the total scattering.

The Hall mobility u_H depends on temperature (Fig. 3.17). It
is evident from this figure that when the degree of doping is increased, the slope of the $u_H(T)$ curves decreases from a value which
is close to the theoretical ($-3/2$ for sample 1) to very low values
for heavily doped samples. Thus, for samples 7-10 the mobility is
practically independent of the temperature, beginning from helium
temperatures. However, when the temperature is increased from
room temperature to 600°K, the slope of the $u_H(T)$ curves becomes
considerable, which may be due to a reduction in the degree of degeneracy and an increase in the contribution of the lattice scattering component.

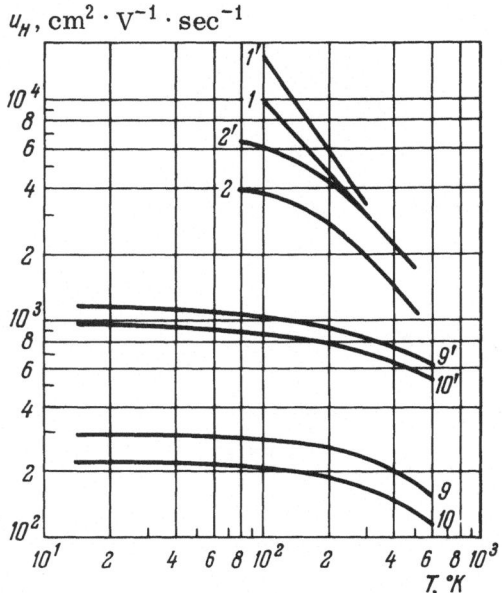

Fig. 3.20. Comparison of the experimental values of
the electron mobility in germanium with the values
calculated allowing for the anisotropy.

The temperature dependences of u_H for Ge samples with other impurities (for example, with phosphorus [50]) (Fig. 3.18) show that the dependence on the nature of the impurity is strongly influenced by temperature. When the temperature is reduced to 78°K, the dependence on the nature of the impurity appears at lower densities than at 300°K.

The experimental results on the Hall mobility make it possible to find a relationship between the scattering on the acoustical vibrations of the lattice atoms and the scattering on impurity ions. This can be done using Table A.7 or the curves in Fig. 3.10 and the following temperature dependence of u_{H_0} [64] (cf. Table 3.4):

$$u_{H_0} = 4500 \left(\frac{T}{300}\right)^{-1.6} \text{cm}^2 \cdot \text{V}^{-1} \cdot \text{sec}^{-1}.$$

The value of μ^*, which is necessary in such a calculation, is found from Eq. (2.3.4). The results are presented in Fig. 3.19,

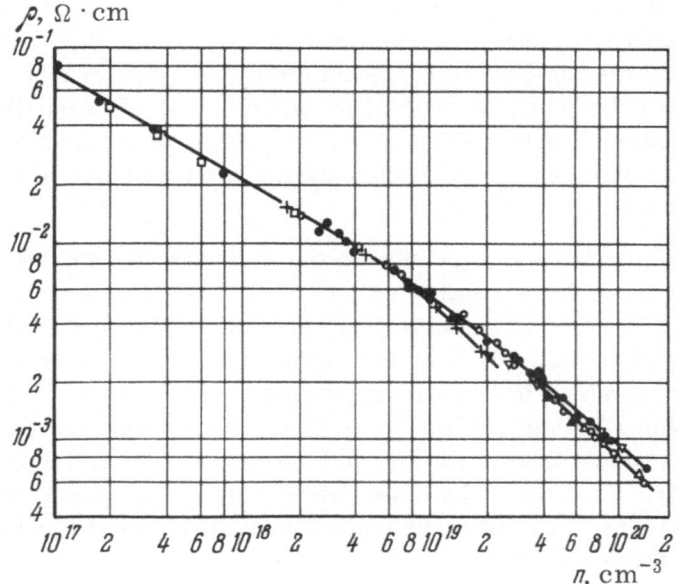

Fig. 3.21. Dependence of the resistivity of heavily doped n-type Si crystals on the electron density (T = 300°K) [53, 69-73]. Lower curve, Sb impurity; middle curve, P; upper curve, As.

which shows that when the impurity concentration is increased, the contribution of the ion scattering component increases rapidly so that, beginning from n > 5 · 10^{18} cm^{-3}, we may assume that the electron mobility at 300°K is governed mainly by the scattering on the impurity ions.

At liquid nitrogen temperatures, the influence of the thermal scattering mechanism is negligibly small (a > 20), beginning at least from 3 · 10^{17} cm^{-3}. The nature of curve 3 in Fig. 3.19 shows that the scattering by the impurity ions also predominates at 500°K, although at that temperature the relative contribution of the thermal scattering mechanism is greater than at 300°K.

The decrease in the slopes of curves 2 and 3 at higher impurity concentrations may be explained by the "heating" of carriers in degenerate semiconductors, as pointed out first by Samoilovich and Korenblit [66]. This "heating" effect is due to an increase in the electron gas degeneracy, i.e., an increase in the average energy of the scattered particles, when the degree of doping is increased. The consequence of this effect may be a reduction in the contribu-

tion of the Rutherford scattering to the total scattering. Extrapo-
lating hypothetically curves 2 and 3 in Fig. 3.19 to still higher impurity concentrations, we find, that in the $n > 10^{20}$ cm^{-3} range (which
cannot be reached because of the limited solubility of impurities in
Ge), the slopes of these curves may decrease to zero and even
change their sign, which would indicate the predominance of the
phonon scattering mechanism.

The dependence of u_{HL}/u_{Hi} on the impurity concentration is
in qualitative agreement with the theory. In fact, if we bear in mind
that in the absence of degeneracy u_{HL} is independent of n, and
$u_{Hi} \propto n^{-1}$ [cf. Eq. (3.3.58)], it follows that u_{HL}/u_{Hi} should be proportional to n, which is in good agreement with the results for concentrations lower than 10^{19} cm^{-3}. In the strongly degenerate case,
we have – in accordance with Eq. (3.3.58) $u_{HL} \propto n^{-1/3}$ and u_{Hi}
$\propto 1/g[b(\mu *)]$. Therefore, the dependence of u_{HL}/u_{Hi} on the impurity concentration may reach saturation at high values of n, which
is in agreement with curves 2 and 3 in Fig. 3.19.

Thus, we may assume that in heavily doped n-type germanium
the scattering by impurity ions is the dominant process. However,
a simple comparison of the experimental data on u_H with the
Brooks–Herring formula for the mobility does not give a satisfactory agreement. True, there is some freedom in the selection
of the value of m*, which is an "adjustable" parameter. The experimental results can be made to fit the Brooks–Herring formula
if we assume a certain dependence of m* on the carrier density,
as indicated by Fistul' in his earlier investigations [67, 68] and by
Zemskov et al. [60]. However, such a dependence has no physical
meaning.

It is more correct to compare the experimental results with
the theory that allows for the scattering anisotropy represented by
means of Eqs. (3.2.25), (3.2.26), (3.3.65), and Eq. (3.3.66). Such a
comparison has been carried out by Fistul' et al. [48, 49]. The
Hall mobility has been calculated for two extreme cases: absence
of degeneracy and strong degeneracy. For the first case, the average energy $\overline{E}$ has been assumed to be 3kT, and for the latter case
$\overline{E} = \mu$, has been used, since exact integration of Eq. (3.3.66) is a
very cumbersome operation. The approximation $\overline{E} = \mu$ has been
used because in a strongly degenerate electron gas most of the
scattered electrons have the Fermi energy μ. The calculations

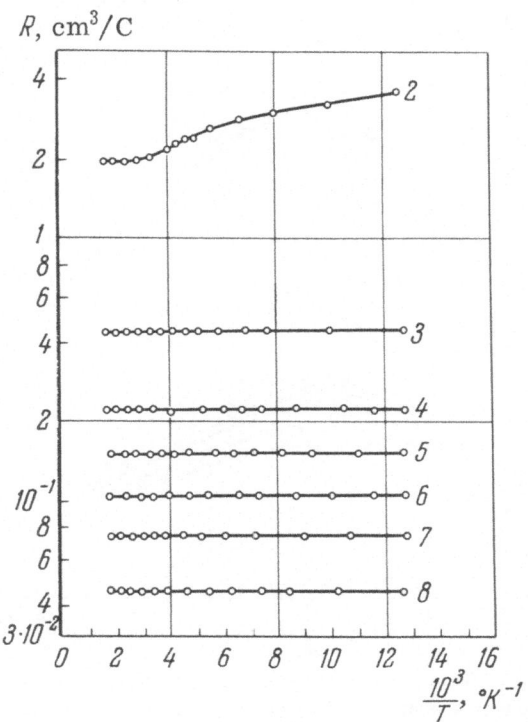

Fig. 3.22. Temperature dependence of the Hall coefficient of silicon [50]. The numbers alongside the
curves have the same meaning as in Fig. 3.23.

reported in [48, 49] have yielded the results presented in Fig. 3.20,
where the experimental curves (corresponding to the similarly num-
bered curves in Fig. 3.17) are denoted by numbers without primes
and the calculated curves are denoted by primes. For lightly doped
samples, the experimental results agree well with the theory. The
difference between the absolute values of u_H does not exceed 50%,
which is due to the use of the approximate expression $\bar{E} = 3kT$. The
agreement for heavily doped samples is only qualitative, i.e., slopes
of the curves agree, but the absolute values of the calculated u_H differ from the experimental ones by a factor of 3-4.

Experimental Results (n-Type Silicon) [53,
69-72]. The experimental results on the electrical conductivity
(Fig. 3.21) and the Hall effect (Fig. 3.22) for n-type Si are very
similar to those for n-type Ge. The nature of the impurity in sili-

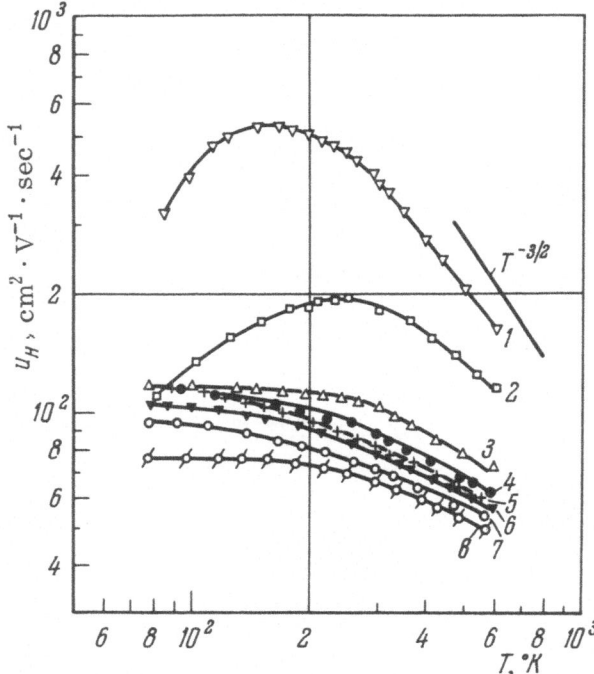

Fig. 3.23. Temperature dependence of the Hall mobility of electrons in arsenic-doped silicon [53]. Different curves represent different carrier densities n (in cm^{-3}): 1) $2.2 \cdot 10^{17}$; 2) $3.1 \cdot 10^{18}$; 3) $1.4 \cdot 10^{19}$; 4) $2.8 \cdot 10^{19}$; 5) $4.2 \cdot 10^{19}$; 6) $6.1 \cdot 10^{19}$; 7) $8.6 \cdot 10^{19}$; 8) $1.4 \cdot 10^{20}$.

con also affects the scattering. However, the impurity concentration threshold n_0, above which the nature of the impurity becomes important, is approximately one and a half orders of magnitude higher than that for germanium.

The temperature dependence of the Hall mobility of As-doped Si samples is also (Fig. 3.23) in qualitative agreement with the corresponding dependence for n-type Ge, but the slopes of the $u_H(T)$ curves for heavily doped silicon samples are somewhat steeper. Bearing in mind that the ordinate scale is enlarged, we can see that electrons in silicon samples Nos. 2–5 have all practically the same mobility at 78°K. On the other hand, at 600°K samples Nos. 5–8 have practically the same mobility. This can be seen more clearly in Fig. 3.24. Here, the temperature dependence of the threshold concentration n_0 is even stronger than that for Ge. It is also worth noting the different nature of the dependence $u_H(n)$ at 78°K for silicon: u_H is constant over a wide range of concentrations from 10^{18} to $3 \cdot 10^{19}$ cm^{-3}.

u_H, cm$^2 \cdot$ V$^{-1} \cdot$ sec^{-1}

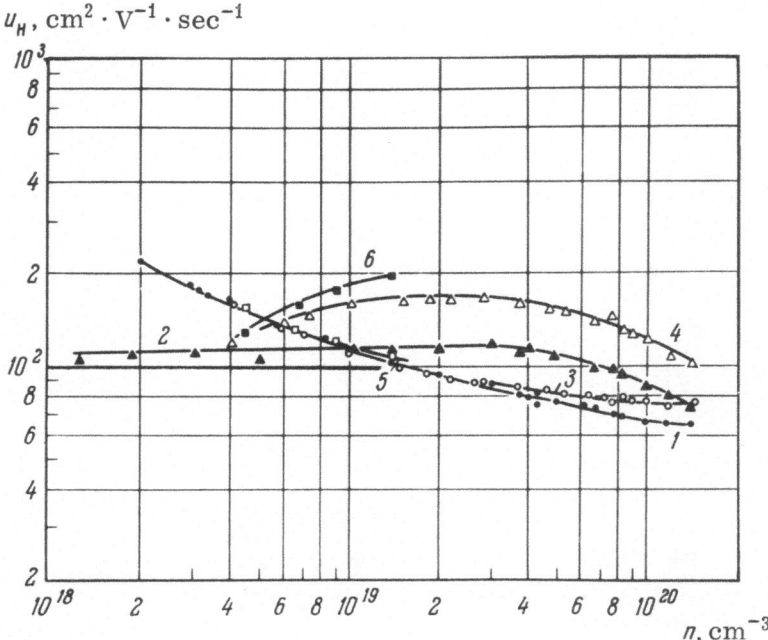

Fig. 3.24. Dependence of the Hall mobility of electrons in heavily doped silicon on the carrier density at two temperatures [53]. As impurity: 1) 300°K; 2) 78°K; P impurity: 3) 300°K; 4) 78°K; Sb impurity: 5) 300°K; 6) 78°K.

The ratio u_{HL}/u_{Hi} was estimated in [42] using the experimental data given in Fig. 3.23 and in Table A.7, as well as expressions for u_{H0}, taken from Table 3.4. The results obtained in this way are given in Fig. 3.25. The curves are very similar to those for germanium. Therefore, the discussion given in the preceding subsection for germanium, applies also to silicon.

We note that in the case of silicon the calculation of u_{HL}/u_{Hi} is less reliable than that for Ge because the experimental data for Si do not satisfy the initial relationships (3.2.21). Nevertheless, in the case of heavily doped silicon we may assume that the dominant type of scattering is the scattering by impurity ions. Comparison of the experimental data with the anisotropic scattering theory [50, 73] shows, that as in the case of germanium, only a qualitative agreement is obtained (Fig. 3.26).

Causes of the Disagreement with the Theory. One of the possible causes of the quantitative disagreement of the

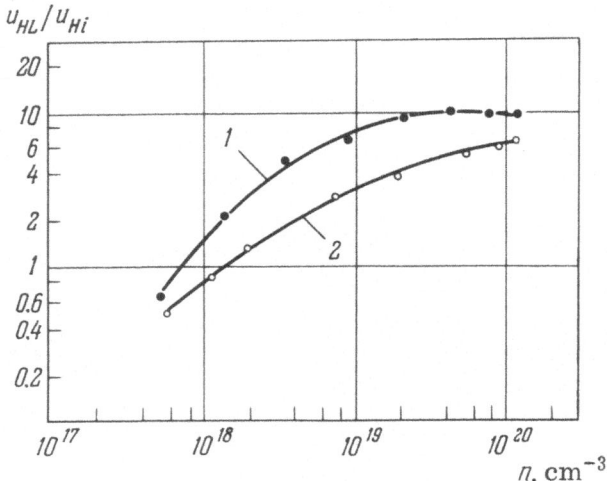

Fig. 3.25. Dependence of the ratio of the Hall mobilities governed by the lattice and impurity electron scattering mechanisms on the carrier density in heavily doped silicon: 1) 300°K; 2) 500°K.

calculated and experimental curves in Fig. 3.26 and 3.20 may be the lack of allowance for the intervalley and electron – electron scattering. If it is assumed that the electron – electron scattering contribution is of the same order as that for lightly doped semiconductors, i.e., about 60% [34], the disagreement reduces to a factor of 1.5-2. The absence of a theory of the electron – electron interaction in heavily doped semiconductors of moderate degeneracy makes it impossible to estimate this correction exactly. The theory developed in [56] is applicable only in the limiting case of a very strong degeneracy, which is not observed in real crystals of the germanium or silicon type.

Another possible cause of the disagreement with the theory may be the lack of allowance for a change in the carrier spectrum, discussed in Chap. 1. In other words, we should not have used the values of the effective masses for pure crystals in our comparison with the experimental data. However, it has been demonstrated in [74] that the corrections to the value of the electrical conductivity (and, consequently, to the value of the mobility), which allow for changes in the electron spectrum, are small.

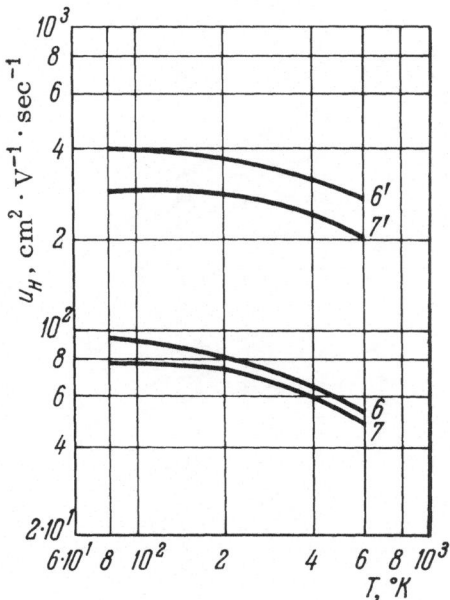

Fig. 3.26 Comparison of the experimental values of the electron mobility in silicon with the values calculated allowing for the scattering anisotropy. The numbers alongside the curves are the same as in Fig. 3.23.

In our opinion, a more important cause of the quantitative disagreement between the theoretical and experimental values of the mobility is the invalidity of the Born approximation in the heavy doping case.

It is known that this approximation is valid if

$$\frac{e^2}{\varkappa \hbar v_F} \ll 1, \tag{3.3.69}$$

where v_F is the velocity of an electron having the Fermi energy. If it is assumed that the values of the permittivity $\varkappa$ of Ge and Si are independent of the impurity concentration and equal to, respectively, 16 and 12, it follows that Eq. (3.3.69) is not obeyed satisfactorily by germanium and silicon. In fact, the validity of the Born approximation is even more doubtful, because the scattering in heavily doped semiconductors takes place over such short distances

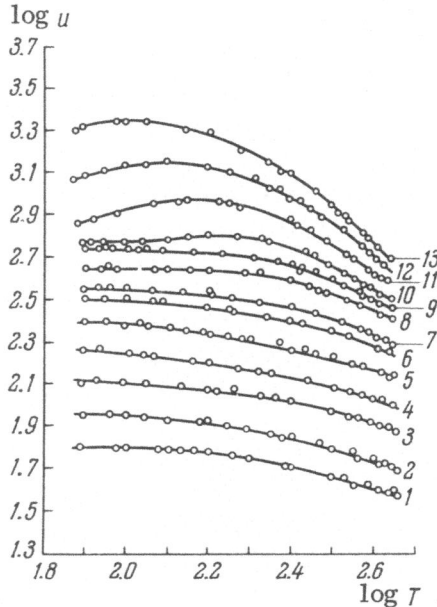

Fig. 3.27. The temperature dependence of the hole mobility in Ge heavily doped with Ga. The numbers alongside the curves correspond to the following hole densities (in cm^{-3}) at $300°K$: 1) $4.2 \cdot 10^{20}$; 2) $2.0 \cdot 10^{20}$; 3) $1.0 \cdot 10^{20}$; 4) $5.8 \cdot 10^{19}$; 5) $2.7 \cdot 10^{19}$; 6) $1.2 \cdot 10^{19}$; 7) $6.9 \cdot 10^{18}$; 8) $4.9 \cdot 10^{18}$; 9) $2.2 \cdot 10^{18}$; 10) $1.1 \cdot 10^{18}$; 11) $6.8 \cdot 10^{17}$; 12) $2.8 \cdot 10^{17}$; 13) $1.2 \cdot 10^{17}$.

that the characteristic scattering length – the Debye radius – becomes equal to the lattice constant of a crystal. In this case the influence of the medium cannot be described by the macroscopic permittivity $\varkappa$. Instead of it we must introduce some effective value $\varkappa_{eff}$, which will be different for different concentrations and which will decrease when the degree of doping is increased, tending in the limit to unity. A somewhat similar situation occurs in the theory of superconductivity [74]. This also explains the difference between the calculated and measured values of the ionization energy of [75]. A theory which allows for this factor should be able to explain the influence of the nature of the impurity on the value of the mobility.

Experimental Results (p-Type Germanium) [76-79]. The Hall mobility of p-type germanium has been in-

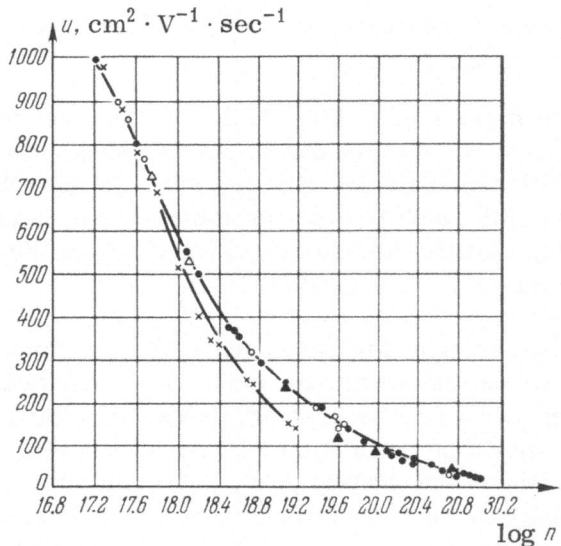

Fig. 3.28. Dependence of the hole mobility in germanium, heavily doped with gallium, aluminum, and indium, on the hole density at 300°K. ●) Al; ○) Ga; ×) In; △) Ga [78].

vestigated in greatest detail [76, 77] for gallium–doped samples with $n = 5 \cdot 10^{13} - 4 \cdot 10^{20}$ cm^{-3}. We shall only quote the experimental results from these papers referring to heavily doped samples. The temperature dependence of the mobility is given in Fig. 3.27 and the dependence on the carrier density at 300°K is shown in Fig. 3.28. The latter figure includes also the results obtained in an earlier paper [78]. We see that the results obtained by different workers are in good agreement.

Figure 3.28 also includes the experimental results obtained for germanium doped with other acceptors: aluminum and indium. Comparison with the theory, carried out in [76, 77], has shown that these samples can be divided into two groups. Samples of the first group comprise those which contain less than $5 \cdot 10^{19}$ cm^{-3} holes. In such crystals, the mobility is in more or less satisfactory agreement with the available theories of scattering. Thus, at 100°K, when the scattering on the optical vibrations is not yet effective, $u_{calc}/u_{exper} \lesssim 1.7$. In this case, u_{calc} is defined as

$$u_{calc} = \frac{u_h + v u_l}{1 + v},$$

where u_h and u_l are, respectively, the mobilities of the heavy and light holes; $\nu = n_l/n_h$.

It has been assumed in [76, 77] that $\nu = 0.04$, as indicated by the cyclotron resonance experiments. Some change in ν should not have affected greatly the value of u_{calc}, since the contribution of the light holes to the mobility is approximately 25% in the case of scattering by the lattice vibrations and about 10% in the case of scattering by the impurity ions [76].

The values of u_h and u_l have been calculated for a mixed scattering on the acoustical vibrations and on the impurity ions, allowing for the degeneracy and screening. Inclusion of the scattering on the optical vibrations and on the neutral impurities, allowance for the scattering of the light by heavy holes and for the interaction between the heavy holes themselves, gives an even better agreement with experiment $(u_{calc}/u_{exper} \lesssim 1.1)$ over the whole temperature range from 78 to 300°K for all the samples in this group [77]. However, this agreement could be accidental.

At high temperatures (300°K < T < 950°K), the agreement between u_{calc} and u_{exper} is no longer observed since the experimentally determined mobility decreases more rapidly with temperature than predicted by the theory.

The situation is even worse in the second group of samples, containing more than $5 \cdot 10^{19}$ cm^{-3} holes. For these samples, the experimental data do not agree with the theory at low or high temperatures. It is concluded in [76, 77] that the disagreement between the experimental and calculated results is most likely to be due to a departure of the hole spectrum from the parabolic law when the band is gradually filled, i.e., when the energy of holes increases.

We shall return to this problem in the discussion of the experimental results on the thermoelectric power of p-type germanium.

In concluding this section, we must draw attention to the fact that the nature of the impurity affects the scattering in p-type germanium (Fig. 3.28) but that this influence is less than in n-type germanium.

§3.4. Thermoelectric Phenomena and Thermal Conductivity

Thermoelectric Power. We shall consider the theory of the thermoelectric power in a covalent semiconductor with a simple energy band. We shall assume that only one external force is acting on the carriers: a temperature gradient ∇T. Since the average energy (and also in lightly doped semiconductors the number of carriers) increases when the temperature is increased, a current of carriers appears along the direction of ∇T. Under steady-state conditions, when the sample is not connected to an external electric circuit, the density of this current is zero at all points in the sample. This means that the carrier current due to ∇T is compensated by another current due to a field $\mathscr{E}$ which is established in the crystal. The emf which then appears at the ends of a semiconductor is known as the thermo-emf.

An additional potential difference $(U_{C1} - U_{C2})$ may be superimposed on the true thermo-emf; here U_{C1} and U_{C2} are the contact potential differences between the metal electrodes and the semiconductor. To exclude the influence of the contact potentials, we recall that under statistical equilibrium conditions the chemical (or electrochemical) potentials in all parts of a system of particles are equal (cf. §2.1). Therefore, we may assume that the quantity $\mu - e\varphi$, where φ is the electrostatic potential, is the same on both sides of a metal—semiconductor contact. We shall assume that the temperatures in the metal and in the semiconductor are equal near the contact. Since μ is the same in the metal electrodes to the left and right of the semiconductor, we can eliminate ∇U_C by calculating $\nabla(\varphi - \mu/e)$.

Using an expression for the force $\mathbf{F}$ when $\nabla T \neq 0$ and substituting it into Eq. (3.1.3), we can write the transport equation (3.1.2) in the form

$$\frac{e}{\hbar} \nabla \left(\varphi - \frac{\mu}{e} \right) \nabla_k f - \boldsymbol{v} \nabla f = - \frac{C(E)\, \boldsymbol{k}\, \frac{\partial f_0}{\partial E}}{\tau(\boldsymbol{k})}. \qquad (3.4.1)$$

Although in this case $\mathbf{F} = \mathscr{E}$, as in the electrical conductivity case, the transport equation differs from Eq. (3.1.8) by the presence of the term $\boldsymbol{v}\nabla f$, because ∇f is now not equal to zero because $\nabla T \neq 0$.

In order to write ∇f in an explicit form, we shall, as before, assume that a correction f_1 to the unperturbed function is small compared with f_0. In this case

$$\nabla f \approx \nabla f_0 = \frac{\partial f_0}{\partial E}\left[\frac{\mu - E}{T}\nabla T - \nabla \mu\right], \tag{3.4.2}$$

and the transport equation assumes the form

$$-\frac{e}{\hbar}\nabla\left(\varphi - \frac{\mu}{e}\right)\nabla_k f - v\frac{\partial f_0}{\partial E}\left[\frac{\mu - E}{T}\nabla T - \nabla\mu\right] = -\frac{C(E)\,k\,\frac{\partial f_0}{\partial E}}{\tau(k)}, \tag{3.4.3}$$

or, solving this equation for $C(E)$ after the preliminary replacement of $\nabla_k f$ in accordance with Eq. (3.1.9) and the replacement of v with $(\hbar m^*)k$, we obtain

$$C(E) = -\frac{\hbar}{m^*}\tau(k)\left[\frac{E-\mu}{T}\nabla T + \nabla(\mu - e\varphi)\right]. \tag{3.4.4}$$

We shall now substitute Eq. (3.4.4) into the expression for the current j_e flowing under the action of a field $\mathcal{E}$, i.e., we shall substitute it into Eq. (3.1.16):

$$j_e = \frac{e}{3\pi^2 m^*}\int_0^\infty k^3\frac{\partial f_0}{\partial k}\tau(k)\left[\frac{E-\mu}{T}\nabla T + \nabla(\mu - e\varphi)\right]dk. \tag{3.4.5}$$

Under steady-state conditions $j_e = 0$, i.e.,

$$0 = \frac{e}{3\pi^2 m^*}\left\{\frac{\nabla T}{T}\int_0^\infty k^3\frac{\partial f_0}{\partial E}\tau E\,dk - \frac{\mu\nabla T}{T}\int_0^\infty k^3\frac{\partial f_0}{\partial E}\tau\,dk + \nabla(\mu - e\varphi)\int_0^\infty k^3\frac{\partial f_0}{\partial E}\tau\,dk\right\}.$$

$$\tag{3.4.6}$$

We shall now introduce the concept of the thermoelectric power α, which is defined as

$$\alpha = \frac{\mathcal{E}_T}{\nabla T} = \frac{\nabla\left(\varphi - \frac{\mu}{e}\right)}{\nabla T}, \tag{3.4.7}$$

i.e., the thermoelectric power is the total thermo-emf divided by the temperature drop across the sample. Then, from Eq. (3.4.6) we obtain:

$$\alpha = \frac{\nabla\left(\varphi - \frac{\mu}{e}\right)}{\nabla T} = -\frac{k_0}{e}\left[\frac{\int_0^\infty k^3 \frac{\partial f_0}{\partial k} \tau(k) E \, dk}{\int_0^\infty k^3 \frac{\partial f_0}{\partial k} \tau(k) \, dk} - \mu^*\right]. \qquad (3.4.8)$$

Integrating both integrals by parts [as in the case of integrals in Eqs. (3.1.18) and (3.3.38)] and using the dimensionless quantity ε^*, we obtain:

$$\alpha = -\frac{k_0}{e}\left[\frac{\int_0^\infty (\varepsilon^*)^{5/2} \tau(\varepsilon^*) \frac{\partial f_0}{\partial \varepsilon^*} d\varepsilon^*}{\int_0^\infty (\varepsilon^*)^{3/2} \tau(\varepsilon^*) \frac{\partial f_0}{\partial \varepsilon^*} d\varepsilon^*} - \mu^*\right]. \qquad (3.4.9)$$

As in the preceding section, we shall assume that f_0 represents the Fermi function and instead of τ we shall substitute the expression which applies in the mixed scattering case [Eq. (3.2.22)], to obtain finally

$$\alpha = -\frac{k_0}{e}\left[\frac{\Phi_4(\mu^*, a)}{\Phi_3(\mu^*, a)} - \mu^*\right], \qquad (3.4.10)$$

where Φ_3 is given by Eq. (3.3.5) and Φ_4 is similar:

$$\Phi_4 = \int_0^\infty \frac{(\varepsilon^*)^4 \exp(\varepsilon^* - \mu^*) \, d\varepsilon^*}{[(\varepsilon^*)^2 + a^2][1 + \exp(\varepsilon^* - \mu^*)]^2}. \qquad (3.4.11)$$

The values of Φ_4/Φ_3 are listed in Appendix A.9.

Considering the same special cases as before: $a^2 \ll (\varepsilon^*)^2$ (scattering by phonons) and $a^2 \gg (\varepsilon^*)^2$ (scattering by ions), we can easily find from Eq. (3.4.10) that

$$a^2 \ll (\varepsilon^*)^2: \quad \alpha_L = -\frac{k_0}{e}\left[\frac{2F_1(\mu^*)}{F_0(\mu^*)} - \mu^*\right], \qquad (3.4.12)$$

$$a^2 \gg (\varepsilon^*)^2: \quad \alpha_i = -\frac{k_0}{e}\left[\frac{4F_3(\mu^*)}{3F_2(\mu^*)} - \mu^*\right]. \qquad (3.4.13)$$

When $\mu^* \ll 0$, Eq. (3.4.12) becomes

$$\alpha_L = -\frac{k_0}{e}(2 - \mu^*), \qquad (3.4.14)$$

which is identical with the well-known Pisarenko formula [80], and Eq. (3.4.13) becomes

$$\alpha_i = -\frac{k_0}{e}(4 - \mu^*). \qquad (3.4.15)$$

[Here, we have used the approximations of Eq. (2.5.6).]

If we use the approximation of Eq. (2.5.11), which is valid in the other extreme case when $\mu^* \gg 0$, we easily find that Eqs. (3.4.12) and (3.4.13) become $\alpha_L = \alpha_i = 0$.

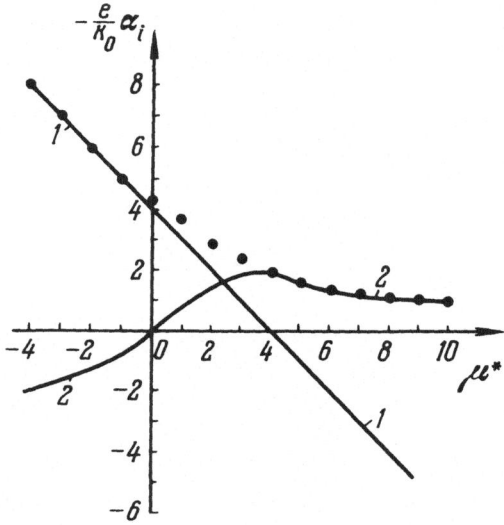

Fig. 3.29. Dependence $\alpha_i(\mu^*)$: 1) approximation represented by Eq. (3.4.15); 2) approximation represented by Eq. (3.4.17); points represent the exact formula (3.4.13).

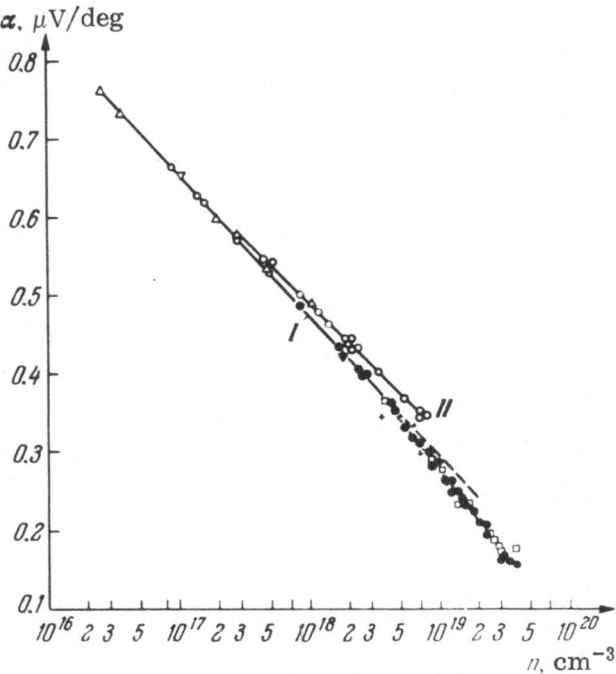

Fig. 3.30. Dependence of the thermoelectric power on the electron density in heavily doped germanium at 300°K. The various
symbols represent the following published results: ▼) [83]; △) [84];
●) As; □) P; ○) Sb. (The last three sets of results in accordance
with [85].)

These zero values of the thermoelectric power are obtained
because the approximation of Eq. (2.5.11) is rough: it includes only
the first term in the expansion of the integrand which occurs in
$F_n(\mu^*)$. If we use a more complex approximation of Eq. (2.5.13),
in which the second term of the expansion is included, we obtain
the formulas

$$\alpha_L = -\frac{k_0}{e}\frac{\pi^2}{3}\frac{1}{\mu^*}, \qquad (3.4.16)$$

$$\alpha_l = -\frac{k_0}{e}\frac{\pi^2\mu^*}{\pi^2 + (\mu^*)^2}, \qquad (3.4.17)$$

which are identical with those derived earlier for metals [81]. From
these formulas, it follows that α decreases considerably when μ^*
is increased but still remains a finite quantity.

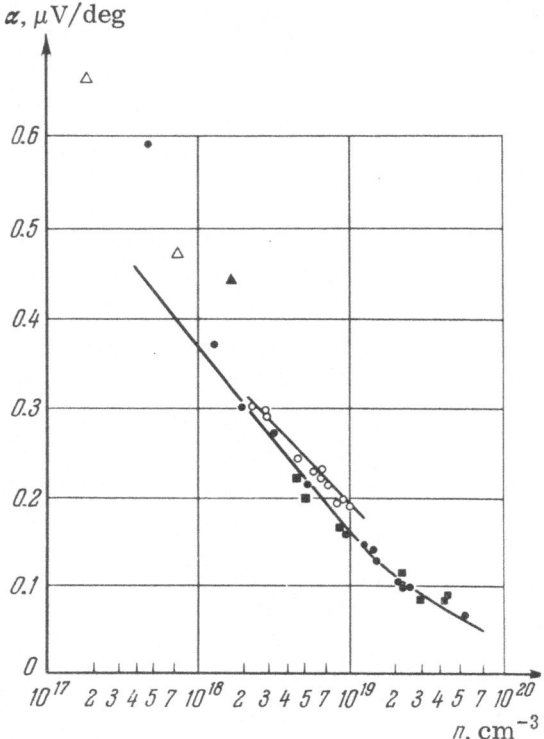

Fig. 3.31. Dependence of the thermoelectric power on
the carrier density in heavily doped germanium at 100°K.
Dopants: ●)As; ○) Sb; ■) P (results reported in [86]); △)
Sb [82]; ▲) As [83].

The range of values of μ^* for which we can use these approximations can be easily obtained from the results in Fig. 3.29.

Experimental Results (Thermoelectric Power
of n-Type Ge and Si). The thermoelectric power of n-type
germanium has been investigated [82-84] using samples with impurity concentrations up to $(1-2) \cdot 10^{18}$ cm^{-3}. Heavily doped samples
were investigated later by Fistul' et al. [85, 86]. All these results
are presented in Fig. 3.30; they indicate that the values of the thermoelectric power of arsenic-doped germanium samples fit the same
straight line plotted in the coordinates α and log n (line I). This
line extends to carrier densities of $5 \cdot 10^{18}$ cm^{-3}. The dashed line
in Fig. 3.30 is an extension of line I. Its slope is 81 μV/deg, in
satisfactory agreement with the theoretical value of the slope

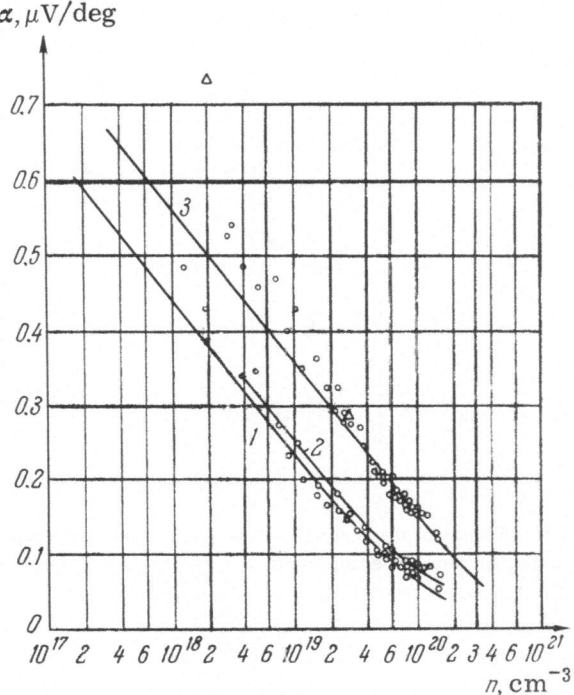

Fig. 3.32. Dependence of the thermoelectric power on the carrier density in heavily doped silicon. 1) As (100°K); 2) P (100°K); 3) As (300°K), P (300°K); Δ) [87].

$k_0/e = 86$ μV/deg. For electron densities exceeding $5 \cdot 10^{18}$ cm^{-3}, the experimental points deviate from this line in the direction of lower values of the thermoelectric power. It is easy to explain this result. In fact, in the absence of degeneracy, $\mu^* \propto \log n$, while in the case of strong degeneracy the dependence of the reduced Fermi level on the carrier density is considerably stronger: $\mu^* \propto n^{2/3}$ and therefore the fall of the value of α at higher carrier densities can be steeper.

The thermoelectric power, like the mobility, is governed by the scattering mechanism and therefore we should expect different values of α for germanium doped with different impurities. In fact, it is evident from Fig. 3.30 that line II, referring to samples doped with antimony lies somewhat higher than line I, which represents samples doped with arsenic. Nevertheless, the quantitative difference is small. The maximum difference between the two lines is

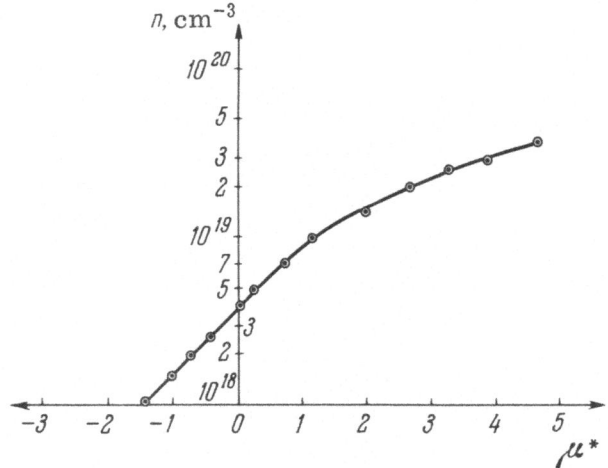

Fig. 3.33. Values of the reduced Fermi level, found from the thermoelectric power measurements, for germanium samples with different electron densities (samples doped with As, $T = 300°K$).

12%, which is about a third of the difference between the mobilities. This can be understood easily using, for the sake of simplicity, Eq. (3.4.16), which is valid in the strongly degenerate case.

Using this expression, we shall assume that $\mu^* \gg 0$ for all the samples. This is not true but the basic nature of the effect can be seen more clearly. We thus obtain:

$$\frac{a_{Sb}}{a_{As}} = \frac{\mu_{As}^*}{\mu_{Sb}^*}.$$

The values of μ^* can be expressed in terms of the mobility, the effective mass, and the scattering cross section Q:

$$u = \frac{e}{m^*} \cdot \frac{1}{n \cdot v \cdot Q}, \qquad (3.4.18)$$

where $v = \sqrt{2\mu^*/m^*}$.

After certain transformations, we can easily show that for the same concentration of arsenic and antimony we have the following expression:

$$\frac{a_{Sb}}{a_{As}} = \left(\frac{u_{Sb}}{u_{As}}\right)^2 \left(\frac{Q_{Sb}}{Q_{As}}\right)^2.$$ (3.4.19)

The values of the effective mass m^*_{Sb} and m^*_{As} are assumed to be the same, although it has been reported in [59] that these masses are, respectively, $0.14m_0$ and $0.15m_0$, which could have been due to experimental errors. The ratio of the scattering cross sections is not known exactly, but in any case it is less than unity [30]. Since the second factor in Eq. (3.4.19) is less than unity, the difference between the thermoelectric powers will be less than the square of the ratio of the mobilities.

It has been reported in [85] that the thermoelectric power of phosphorus-doped samples (Fig. 3.30) does not differ from the thermoelectric power of arsenic-doped samples and the mobilities of these samples are also equal.

The chemical nature of the impurity affects even more strongly the carrier-density dependence of the thermoelectric power at 100°K (Fig. 3.31). This figure shows clearly a characteristic feature of the carrier-density dependence of α at low temperatures: the value of α increases when $n < 10^{18}$ cm^{-3}. Such an increase in the thermoelectric power may be due to the drag on the electrons by the phonons.

The carrier-density dependence of α of arsenic- and phosphorus-doped silicon samples, determined by Fistul' and Cherkas [73], is shown in Fig. 3.32. At room temperature, the values of α of samples doped with As and P are equal, within the limits of the experimental error. At liquid nitrogen temperatures, there is a slight difference. This is in agreement with the carrier density and temperature dependences of the mobilities in these samples (cf. §3.3).

These experimental results allow us to find the degree of degeneracy of the electron gas μ^*. This quantity can be found simply if the carrier scattering mechanism is known. Thus, in the first approximation, assuming that the scattering is solely due to impurity ions, Eq. (3.4.13) yields the curve in Fig. 3.33 for arsenic-doped Ge [85] and the curves of Fig. 3.34 for arsenic- and phosphorus-doped Si [73].

The electron gas in germanium at 300°K is fairly strongly degenerate at carrier densities of the order of $4 \cdot 10^{18}$ cm^{-3} and in

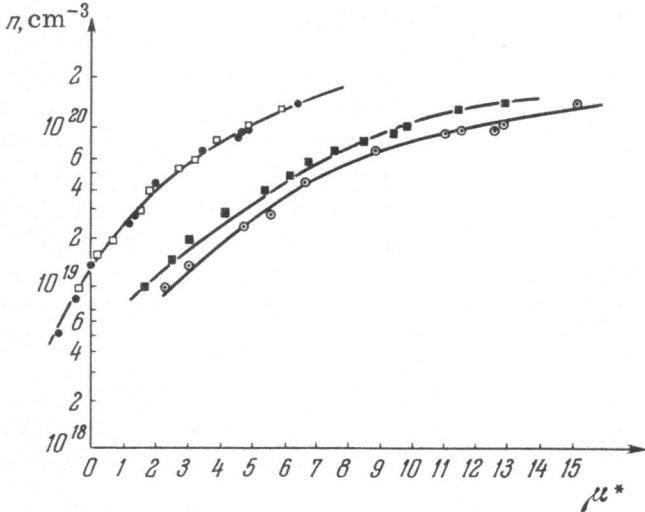

Fig. 3.34. Values of the reduced Fermi level, found from the measurements of the thermoelectric power, for silicon samples with different electron densities. Dopants: ●) As (300°K); □) P (300°K); ⊙) As (100°K); ■) P (100°K).

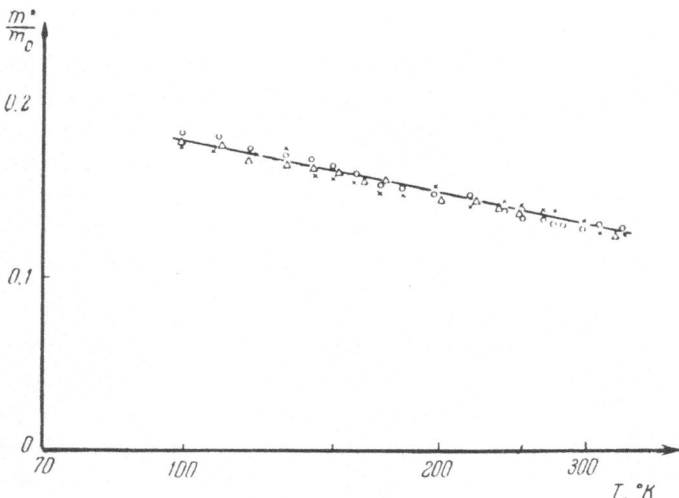

Fig. 3.35. Effective electron mass, found from measurements of the thermoelectric power of heavily doped n-type germanium. Carrier density n (cm^{-3}): ○) $5.3 \cdot 10^{19}$; ×) $1.4 \cdot 10^{19}$; △) $2 \cdot 10^{18}$.

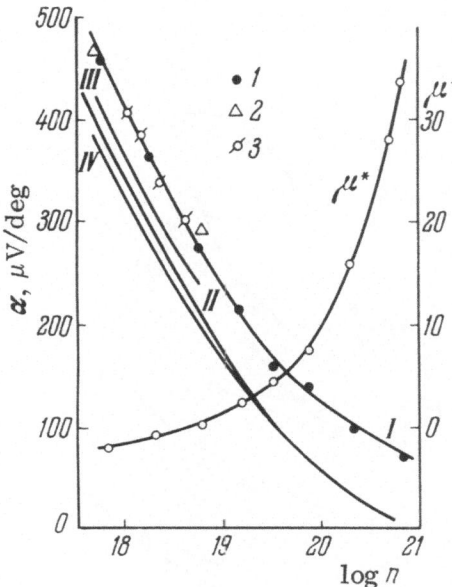

Fig. 3.36. Dependence of the thermoelectric power on the hole density in heavily doped germanium (T = 300°K). I) Experimental curve taken from [90], with experimental points from: 1) [90]; 2) [83]; 3) [89]. II), III), IV) Calculated curves [90]: II) calculated allowing for mixed scattering mechanism (acoustical phonons and ions); III) same as II, but for impurity scattering with allowance for the screening of ions by holes; IV) additional allowance for the scattering of holes on the optical vibrations.

silicon the same happens at 10^{19} cm^{-3}. These quantities are in good agreement with the results of other experiments, for example, with the observation of the tunnel effect in p−n junctions (cf. Chap. 7).

Using these values of μ^* and the dependence of μ^* on the carrier density [Eq. (2.3.4)], we can find the effective density-of-states mass. The values of m_N^* found in this way are independent of the carrier density in germanium and silicon. At room temperature, the values of m_N^* calculated for one of the minima in the many-valley model of the conduction band are, respectively: $0.375m_0$ for Ge and $0.2m_0$ for Si.

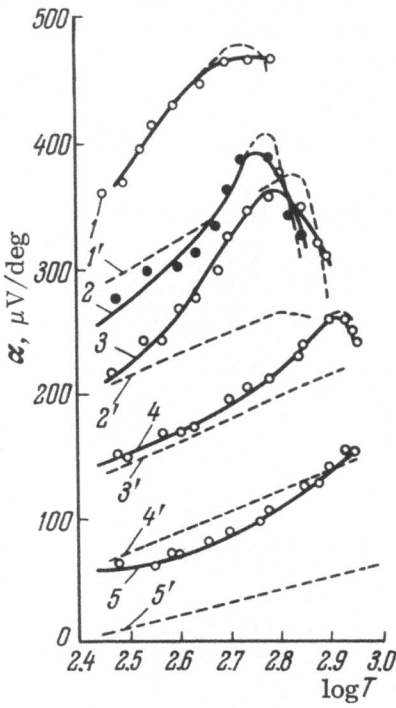

Fig. 3.37. Temperature dependence of the thermoelectric power of heavily doped p-type germanium crystals [90]. 1) $2.2 \cdot 10^{18}$ cm^{-3}; 2) $7 \cdot 10^{18}$ cm^{-3}; 3) $1.2 \cdot 10^{19}$ cm^{-3}; 4) $7.8 \cdot 10^{19}$ cm^{-3}; 5) $7 \cdot 10^{20}$ cm^{-3}; calculated curves are shown dashed.

Analysis of the experimental values of α, followed by the determination of μ^* and m_N^* [86], has led to a discovery of a negative temperature coefficient $\partial m_N^*/\partial T$ (Fig. 3.35). In principle, such a dependence may exist because the effective mass can be written in the form [88]

$$\frac{1}{m^*} \propto \sum \frac{p_\alpha^2}{\Delta E},$$

where p_α is the matrix element of a transition in a unit cell; the subscript α represents the components of the momentum operator.

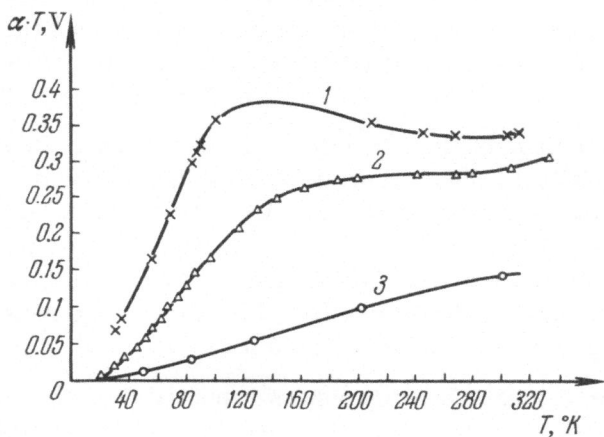

Fig. 3.38. Temperature dependence of the thermoelectric emf
of p-type silicon crystals doped with boron. 1) $2.0 \cdot 10^{17}$ cm^{-3};
2) $1 \cdot 10^{18}$ cm^{-3}; 3) $1.5 \cdot 10^{19}$ cm^{-3}.

The temperature coefficient $\partial m_N^* / \partial T$ should be governed by
the temperature coefficient of the forbidden band width. In fact,
$\partial m_N^* / \partial T$ has the same sign as $\partial E_g / \partial T$ [92].

On the other hand, the dependence $m_N^*(T)$ given in Fig. 3.35
could be an accidental consequence of too much simplification in the
analysis of the experimental data. If we allow for the mixed scattering mechanism, using the values of a given in §3.3., for example,
for arsenic-doped germanium at 300°K, we find that the μ^* (n) curve
changes slightly. The dependence $m_N^*(T)$ in Fig. 3.35 weakens ac-
cordingly, but m_N^* is still not observed to be constant. An allowance for the scattering anisotropy might give even better results,
in the same way as in the discussion of the mobility in §3.3. However, the absence of a theory of the thermoelectric power with an
allowance for the scattering anisotropy prevents us from carrying
out the necessary analysis of the experimental data.

Experimental Results (Thermoelectric Power
of p-Type Ge and Si). There are several papers on the
thermoelectric power of p-type germanium. However, only the
low-temperature range is reported in [83, 89], and [82] deals with
polycrystalline samples. A most thorough investigation was carried out by Vinogradova et al. [90] using single-crystal samples of
germanium doped with gallium in concentrations ranging from $7 \cdot 10^{17}$

to $7 \cdot 10^{20}$ cm^{-3}. Figure 3.36, which gives the carrier-density dependence of α taken from [90] and other papers, shows that results of different workers are in good agreement. The temperature dependence of the thermoelectric power is shown in Fig. 3.37.

A comparison of the experimentally determined dependences $\alpha(n)$ and $\alpha(T)$ with the theory, carried out in [90], showed no agreement although several theoretical variants were tried, allowing for the scattering by the acoustical and optical vibrations and by impurities with and without screening. Corrections for the collisions of carriers with one another have also failed to produce agreement between the experimental and calculated data.

It is said in [90] that the results reported there can be made to agree with the theory if it is assumed that the effective mass depends on temperature in accordance with the law m* $\propto$ T$^{1.7}$, due to a change in the hole spectrum when the hole energy is increased.

Experimental data on the thermoelectric power of p-type silicon are very scarce. A study of boron-doped samples is reported in [83] (Fig. 3.38), but the samples were not of the single-crystal type. There have been no later measurements of the thermoelectric power of p-type silicon.

Electronic Component of the Thermal Conductivity. So far, we have considered the transport of carriers associated with the passage of an electric current through a crystal. However, an electron current is also accompanied by an energy flux, i.e., a heat flux **W**, which can be written in the form

$$W = \int E \boldsymbol{v}\, dn,$$

where dn represents, as in the discussion of the electrical conductivity, the number of carriers in an element of the wave-number space

$$dn = \frac{2}{(2\pi)^3} f(\boldsymbol{k})\, d^3k.$$

Thus

$$W = \frac{\hbar}{m^*} \frac{2}{(2\pi)^3} \int E\boldsymbol{k} f(\boldsymbol{k})\, d^3k.$$

Next, substituting into the above expression f (**k**) in its general form [Eq. (3.1.14)] and integrating in spherical coordinates k, β, φ, we obtain:*

$$W = \frac{1}{3\pi^2\hbar} \int Ek^3 C(E) \frac{\partial f_0}{\partial E} dk. \tag{3.4.20}$$

Since there is a temperature gradient ∇T in the crystal, it follows that $C(E)$ is given by the same expression (3.4.4) as in the derivation of the thermoelectric power. We thus obtain

$$W = -\frac{1}{3\pi^2 m^*}\left[\frac{\nabla T}{T}G_2 - \frac{\mu\nabla T}{T}G_1 + \nabla(\mu - e\varphi)G_1\right]. \tag{3.4.21}$$

where, for the sake of brevity, we have introduced:

$$G_i = \int_0^\infty E^i \tau(k) k^3 \frac{\partial f_0}{\partial k} dk. \tag{3.4.22}$$

The quantity $\nabla(\mu - e\varphi)$ in Eq. (3.4.21) can be found from the expression for the current density $\mathbf{j}_e$ (3.4.6). Then the thermal conductivity, which by definition is

$$\varkappa_e = \frac{W}{\nabla T},$$

becomes:

$$\varkappa_e = \frac{1}{3\pi^2 m^* T}\left(\frac{G_1^2 - G_2 G_0}{G_0}\right). \tag{3.4.23}$$

We note that $G_0 \equiv I_1$, where I_1 is the integral in Eq. (3.3.34), which occurs in all the transport effects discussed so far. Thus, the expression for the electrical conductivity of Eq. (3.1.18) can be writ-

---

*In this derivation, we are assuming that $\int Ekf_0(\mathbf{k})d^3k = 0$, since in the absence of external forces there is no thermal flux in the crystal.

ten in the form

$$\sigma = - \frac{e^2}{3\pi^2 m^*} G_0. \tag{3.4.24}$$

Dividing Eq. (3.4.23) by Eq. (3.4.24), we obtain the relationship

$$\frac{\varkappa_e}{\sigma} = - \frac{1}{e^2 T} \left(\frac{G_1^2 - G_2 G_0}{G_0^2} \right). \tag{3.4.25}$$

On the other hand, it has been established experimentally for many substances that the ratio of the thermal conductivity to the electrical conductivity at a given temperature is constant, i.e.,

$$\frac{\varkappa_e}{\sigma} = LT, \tag{3.4.26}$$

where the constant L is known as the Lorenz number, and Eq. (3.4.26) is known as the Wiedemann−Franz law.

Comparing the last two expressions, we can write:

$$L = \frac{1}{e^2 T^2} \left(\frac{G_2 G_0 - G_1^2}{G_0^2} \right). \tag{3.4.27}$$

To make these expressions more specific, we shall substitute into the integrals G_i the relaxation time $\tau (k)$, valid in the mixed scattering case (3.2.22), and the Fermi distribution function. Then, integrating G_i by parts, we obtain, instead of Eq. (3.4.23),

$$\varkappa_e = \frac{8\pi}{3} \frac{k_0^2 T}{m^*} \frac{(2m^* k_0 T)^{3/2}}{h^3} \tau_{0L} \frac{\Phi_4^2 - \Phi_3 \Phi_5}{\Phi_3}, \tag{3.4.28}$$

instead of Eq. (3.4.25),

$$\frac{\varkappa_e}{\sigma} = - \frac{k_0^2 T}{e^2} \frac{\Phi_4^2 - \Phi_3 \Phi_5}{\Phi_3^2} \tag{3.4.29}$$

and instead of Eq. (3.4.27),

$$L = \frac{k_0^2}{e^2} \frac{\Phi_3 \Phi_5 - \Phi_4^2}{\Phi_3^2}. \tag{3.4.30}$$

We shall analyze in greater detail these relationships. For the sake of brevity, we shall consider only the expressions for the $\varkappa_e$ and L.

In the case of electron scattering by the acoustical vibrations of the lattice atoms [$a^2 \ll (\varepsilon^*)^2$], we obtain

$$\varkappa_{eL} = \frac{8\pi^2}{3} \frac{k_0^2 T}{m^*} \tau_{0L} \frac{4F_1^2(\mu^*) - 3F_0(\mu^*) F_2(\mu^*)}{F_0(\mu^*)}, \qquad (3.4.31)$$

$$L_L = \frac{k_0^2}{e^2} \frac{3F_0(\mu^*) F_2(\mu^*) - 4F_1^2(\mu^*)}{F_0^2(\mu^*)}. \qquad (3.4.32)$$

These formulas simplify further in the two limiting cases: a completely nondegenerate semiconductor ($\mu^* \ll 0$) and a strongly degenerate crystal ($\mu^* \gg 0$). The simplified expressions can be easily obtained using the approximate Fermi integrals given in §2.5, bearing in mind that we must use second-order approximations for $\mu^* \gg 0$).

We thus obtain for $\mu^* \ll 0$

$$\varkappa_{eL} = \frac{16\pi^2}{3} \frac{k_0^2 T}{m^*} \tau_{0L} e^{\mu^*}, \qquad (3.4.33)$$

$$L_L = 2 \left(\frac{k_0}{e} \right)^2, \qquad (3.4.34)$$

and for $\mu^* \gg 0$,

$$\varkappa_{eL} = \frac{8\pi^4}{9} \frac{k_0^2 T}{m^*} \tau_{0L} \left(\frac{\pi^2}{3} \frac{1}{\mu^*} - \mu^* \right), \qquad (3.4.35)$$

$$L_L = \frac{\pi^2}{3} \frac{k_0^2}{e^2} \left(1 - \frac{\pi^2}{3} \frac{1}{(\mu^*)^2} \right). \qquad (3.4.36)$$

The expression (3.4.34) is identical with the well-known formula for impurity-free semiconductors [23] and (3.4.36) is identical with the corresponding formula for metals. In fact, both these formulas confirm the Wiedemann–Franz empirical law.

We shall now consider the case of scattering of electrons by impurity ions [$a^2 \gg (\varepsilon^*)^2$].

Dropping $(\varepsilon^*)^2$ in the denominators of the integrals Φ_n, we obtain from Eqs. (3.4.28) and (3.4.30):

$$\varkappa_{ei} = \frac{8\pi k_0^2 T}{3m^*} \frac{(2m^*k_0 T)^{3/2}}{h^3} \tau_{0i} \frac{16F_3^2(\mu^*) - 15F_2(\mu^*) F_4(\mu^*)}{3F_2(\mu^*)}, \qquad (3.4\ 37)$$

$$L_i = \frac{k_0^2}{e^2} \frac{15F_2 F_4 - 16F_3^2}{9F_2^2}, \qquad (3.4.38)$$

which reduce to the following expressions when $\mu^* \ll 0$:

$$\varkappa_{ei} = \frac{64\pi k_0^2 T}{m^*} \frac{(2m^*k_0 T)^{3/2}}{h^3} \tau_{0i} e^{\mu^*}, \qquad (3.4.39)$$

$$L_i = 4\left(\frac{k_0}{e}\right)^2, \qquad (3.4.40)$$

and when $\mu^* \gg 0$:

$$\varkappa_{ei} = \frac{8\pi k_0^2 T}{3m^*} \frac{(2m^*k_0 T)^{3/2}}{h^3} \tau_{0i} \left[\frac{\mu^*((\mu^*)^3 + 2\pi^2)^2}{(\mu^*)^2 + \pi^2} - (\mu^*)^3 \left((\mu^*)^2 + \frac{10\pi^2}{3}\right)\right], \quad (3.4.41)$$

$$L_i = \frac{k_0^2}{e^2} \frac{1}{(\mu^*)^2 + \pi^2} \left[\left((\mu^*)^2 + \frac{10\pi^2}{3}\right) - \frac{((\mu^*)^3 + 2\pi^2)^2}{(\mu^*)^2 + \pi^2}\right]. \qquad (3.4.42)$$

Lattice Thermal Conductivity. The total thermal conductivity includes the electronic component as well as the thermal conductivity of the crystal lattice. The lattice component is not associated with the transport of heat by moving charges but with the scattering of the phonons themselves.

The theory of the lattice thermal conductivity was developed by Debye [93], who considered the thermal motion as an assembly of all possible vibrations of a body. Each of these vibrations produces elastic waves, which suffer multiple scattering during their propagation in the body. In this theory, the scattering centers are the elastic waves themselves. It is self-evident that an encounter of two ideal harmonic waves cannot result in any scattering and therefore the meeting of such waves makes no contribution to the thermal resistance. However, if the waves are anharmonic, a wave meeting with density fluctuations (compression and rarefaction) in a crystal will change its direction of propagation, i.e., it will be scattered. Such scattering gives rise to a thermal resistance.

Next, Debye introduced the concept of a mean free path $\bar{l}$, a distance traveled by an elastic wave before its intensity decreases by a factor of e compared with its initial value, and he obtained a formula for $\varkappa_1$, similar to the expression which follows from the kinetic theory of gases:*

$$\varkappa_1 = \frac{1}{3} c_v \bar{v} \bar{l} , \qquad (3.4.43)$$

where $\bar{v}$ is the average velocity of propagation of elastic waves in a crystal; c_v is the specific heat per 1 cm^3.

It follows from the Debye theory that when $T > \theta_D$, where θ_D is the Debye temperature (cf. §3.2), the lattice thermal conductivity $\varkappa_1$ should be proportional to 1/T, which has indeed been confirmed by many experimenters [93, 94].

Peierls [95, 96] modified the Debye theory in accordance with the quantum-mechanical ideas by considering the scattering by phonons as the scattering of energy quanta of the vibrational spectrum of the crystal. If there is a temperature gradient in the crystal, the propagation of phonons is not of the equilibrium type. Collisions between phonons may result in the ordering of their motion and a tendency to return to the equilibrium state. The rate of the process of re-establishment of the equilibrium state of the phonons is governed by the thermal conductivity. Here, we have an analogy with the re-establishment of electron equilibrium by the scattering of electrons, represented by the concept of a relaxation time τ.

Peierls has shown that a thermal resistance appears if the direction of energy flow is altered in the phonon collisions, i.e., if the following relationship applies

$$q_1 + q_2 = q_3 + \frac{2\pi}{a} i, \qquad (3.4.44)$$

where q_1 and q_2 are the wave vectors of the phonons before collision; q_3 is the wave vector after collision; a is the crystal lattice parameter; and i is a unit vector whose directions depend on the symmetry of the crystal.

---

*The Debye formula differs only by a numerical factor of 1/4 which replaces 1/3.

Peierls called this type of scattering the U-process (um-klapp process).

The condition (3.4.44) imposes also restrictions on the energy of the phonons taking part in the umklapp process. Only phonons of energies higher than $k_0\theta_D/2$ may suffer umklapp-type collisions [97]. The number of such phonons N_{ph} is

$$N_{ph} \propto \frac{f(T)}{\exp\left(\dfrac{k_0\theta_D}{2k_0T}\right) - 1},$$

which gives $N_{ph} \propto f(T)\exp\left(-\theta_D/2T\right)$ if $T \ll \theta_D$ and the thermal resistance is then

$$\frac{1}{\varkappa_1} \propto f(T)e^{-\frac{\theta_D}{2T}}. \tag{3.4.45}$$

Here, $f(T)$ is a function describing the probability of a three-phonon collision, which increases with temperature approximately as T^n, where n is somewhat greater than unity [98]. It is evident from Eq. (3.4.45) that, as the temperature tends to the absolute zero, the thermal conductivity should increase to infinity. However, at low temperatures, the mean free path of phonons is so large that the scattering of phonons is affected by the physical boundary of the crystal. This case has been considered in [99] and the following expression for $\varkappa_1$ has been obtained for an infinitely long cylindrical crystal with a perfectly rough surface:

$$\varkappa_1 = 2.31 \cdot 10^3 \, rp\,(\bar{v})\, AT^3 \; \text{W} \cdot \text{cm}^{-1} \cdot \text{deg}^{-1}, \tag{3.3.46}$$

where r is the radius of the crystal; $p(\bar{v})$ is a function which depends on the average phonon velocity and is approximately equal to 1.4 for many crystals; A is the constant in the law $c_V = AT^3$.

Thus, the theory predicts two exactly opposite dependences $\varkappa_1(T)$ in the two extreme cases of high and low temperatures. The expected temperature dependences of $\varkappa_1$ are shown in Fig. 3.39.

In the region of a maximum of $\varkappa_1(T)$, several phonon scattering mechanisms are active (umklapp process, scattering by boundaries of a crystal, impurities, point defects):

$$\frac{1}{\varkappa_1} = \sum_i \frac{1}{\varkappa_{1i}}. \tag{3.4.47}$$

The formulas for each of the terms in Eq. (3.4.47) have been deduced by Klemens and are given in [100, 101].

The theory of thermal conductivity for the region to the right of the maximum of the $\varkappa_1(T)$ is much less developed. Therefore, it is usual to employ semiempirical dependences for $\varkappa_1$ when $T > \theta_D$. The formula deduced by A. F. Ioffe is of greatest interest:

$$\varkappa_1 = \frac{v}{3B} \frac{1}{T - \dfrac{\theta_D}{3}}, \tag{3.4.48}$$

where the quantity B represents the degree of anharmonicity and is independent of temperature. The values of B are different for different materials, but they are very similar for substances with the same kind of interatomic binding forces and the same crystal lattice [102]. Ioffe's formula is applicable up to values of l close to the value of the lattice parameter.

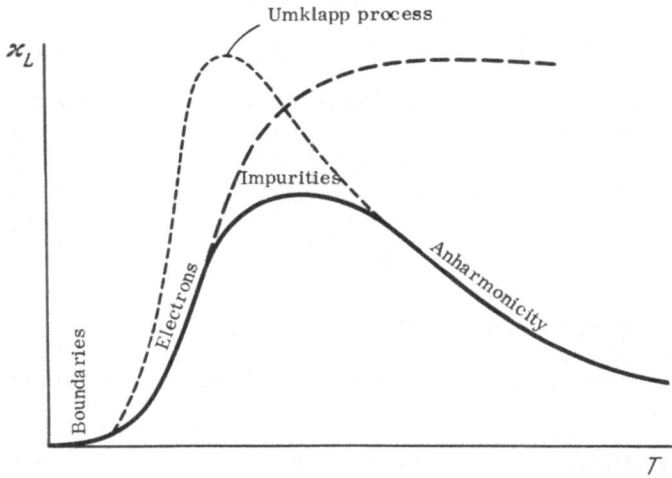

Fig. 3.39. Temperature dependence of the lattice thermal conductivity for various phonon scattering mechanisms.

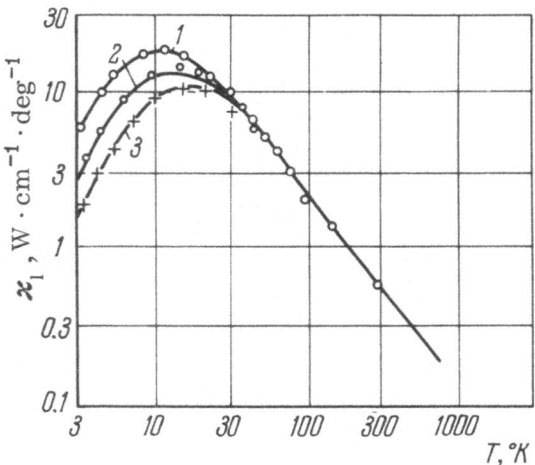

Fig. 3.40. Thermal conductivity of germanium as a
function of the diameter of a sample: 1) 1.06 cm; 2)
0.29 cm; 3) 0.16 cm.

The relationship (3.4.48) can be written out in full by replacing θ_D with an expression which includes the density of the crystal
d, its compressibility χ, and molecular weight A, and we can also
express v as $\sqrt{E/d}$, where E is Young's modulus

$$\varkappa_1 = \frac{1}{3B} \sqrt{\frac{E}{d}} \frac{1}{T - \frac{1.8 \cdot 10^{-3}}{A^{1/3}d^{1/6}\chi^{1/2}}}. \qquad (3.4.49)$$

Another widely used semiempirical formula is due to Leib-
fried and Schlömann [103], which is

$$\varkappa_1 = \frac{1}{\gamma^2}\left(\frac{k_0}{h}\right) Aa\frac{\theta_D^3}{T}, \qquad (3.4.50)$$

where γ is the Grüneisen constant.

The same formula can also be obtained by dimensional analysis [104].

Less widely used are the formulas of Keyes [105], which relates the thermal conductivity of a crystal to its melting point, and
Tavernier [106], which also relates $\varkappa_1$ to T_{mp}.

Zhuze [107] has found a correlation between $\varkappa_1$ and the linear expansion coefficient δ; elsewhere [108, 109], $\varkappa_1$ has been related to the crystal hardness.

A more detailed discussion of the various semi-empirical dependences, relating $\varkappa_1$ to various phenomena, is given in Drabble and Goldsmid's monograph [97].

Influence of Impurities on $\varkappa_1$. A. F. and A. V. Ioffe [110, 111] considered the influence of impurities on $\varkappa_1$ at high temperatures $(T > \theta_D)$. They introduced the concept of a scattering cross section s of each impurity center

$$s = \varphi a^2,$$

where φ is a parameter.

If the lattice of a crystal contains N impurity atoms, the total scattering cross section is $S = sN$. The number, ν, of collisions of phonons with impurities in the crystal with a simple cubic lattice is given by

$$\nu = \frac{N\varphi a^2}{N_0 a^3},$$

where N_0 is the number of atoms in the host lattice per 1 cm^3. Then

$$\frac{\varkappa_0}{\varkappa_{imp}} = 1 + \frac{N}{N_0} \varphi \frac{l_0}{a}, \qquad (3.4.51)$$

where $\varkappa_0$ and l_0 refer to an impurity-free crystal.

This formula includes the quantity φ. A. F. and A. V. Ioffe demonstrated that if impurity atoms form a substitutional solid solution, then $\varphi \leq 1$; if they form an interstitial solid solution, then $\varphi > 1$, but in the case of precipitation of impurities at defects, $\varphi < 0$.

The criterion of the validity of the Ioffe formula is the weakness of the concentration N compared with N_0.

Other Mechanisms of Heat Transport in Crystals. In addition to the mechanisms of phonon scattering considered so far, heat may also be transported in crystals by other mechanisms, for example by the diffusion of electron–hole pairs

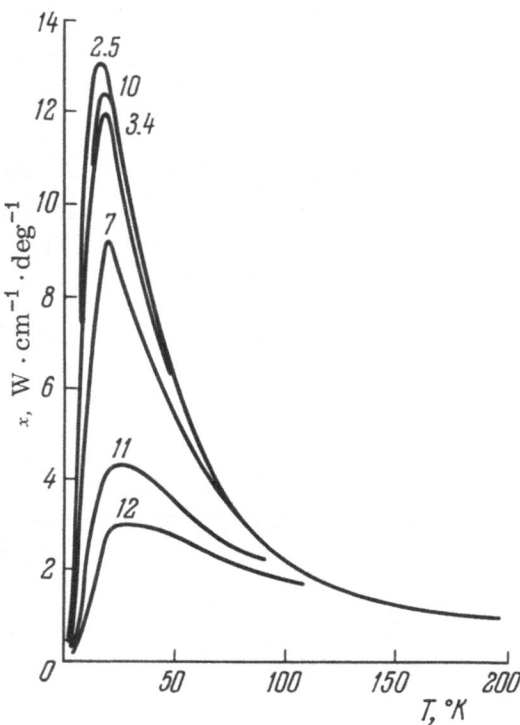

Fig. 3. 41. Temperature dependence of the thermal conductivity of germanium with various impurities [119].

Sample No.	Conduction type	Concentration, cm^{-3}
2; 5	n	$1 \cdot 10^{13}$
10	p	$1 \cdot 10^{15}$
3; 4	p	$1 \cdot 10^{14}$
7	p	$2.3 \cdot 10^{16}$
11	p	$2 \cdot 10^{18}$
12	p	$1 \cdot 10^{19}$

[23, 112], by the diffusion of excitons [113], and by the transport of photons (electromagnetic radiation) [114]. These heat transport mechanisms are unimportant in normal pure crystals and negligible in heavily doped semiconductors.

Experimental Investigations of Lattice Thermal Conductivity. We shall now consider the case of low temperatures, at which the scattering by the boundaries of a crystal should be important and consequently the formula (3.4.46) should be used. This formula has been confirmed for many crystals, including germanium [115] (Fig. 3.40). It is evident from this figure that the lattice thermal conductivity $\varkappa_l$ increases when the diameter of a sample is increased, in agreement with Eq. (3.4.46). However, the experimentally determined temperature dependence is in many cases weaker than T^3. Thus, for diamond $\varkappa_l \propto T^{2.5}$ [116], and for germanium $\varkappa_l \propto T^{2.4}$ [117], etc.

More careful experiments and theoretical calculations have shown that such deviations from the $\varkappa_l \propto T^3$ law are due to a variety of causes. For example, in some experiments these deviations are associated with the mosaic structure of a crystal [118], in others they are due to impurities [119], and in others still they are due to defects [120], or even due to an excess, compared with a natural mixture, of one of the isotopes of germanium [117]. In many experiments, there have been methodological errors (insufficiently rough surface or short samples) [121]. In all cases, the allowance for these factors has made the experimental curve obey again the $\varkappa_l \propto T^3$ law.

Moderately good agreement between the experimental data and theory is also observed at temperatures to the right of a $\varkappa_l(T)$ maximum in Fig. 3.39 but still lower than the Debye temperature. In this region, the exponential law of Eq. (3.4.45) is obeyed, which can be seen quite clearly in Fig. 3.41. The maximum of $\varkappa_l$ of many crystals lies in the range $(0.035-0.045)\theta_D$ [122]. It is evident from Fig. 3.41 that the amplitude of the $\varkappa_l$ maximum depends on the impurity concentration. It is interesting to apply to these results the Ioffe formula of Eq. (3.4.51). According to this formula, the quantity $[(\varkappa_0/\varkappa_{imp}) - 1]$ should be a linear function of the impurity concentration. If $\varkappa_0$ is taken to be the value of $\varkappa_{max}$ for sample No. 10 in Fig. 3.41, the experimental results are found to fit the straight

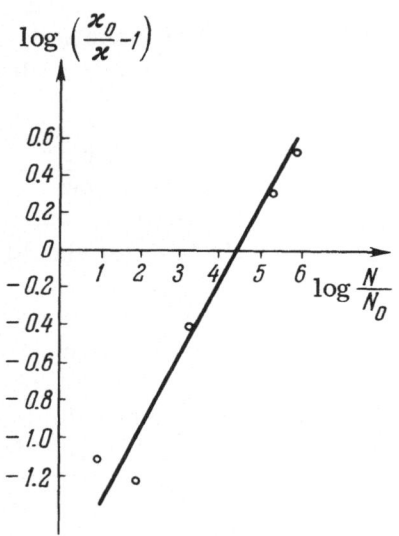

Fig. 3.42. Results of analysis of the data in Fig. 3.41 using Ioffe's formula.

line shown in Fig. 3.42. The point of intersection of this line with the abscissa makes it possible to find φ from the relationship

$$\frac{N}{N_0}\bigg|_{\log\left(\frac{\varkappa_0}{\varkappa_{imp}}-1\right)=0} = \varphi \cdot \frac{l_0}{a}.$$

Substituting $a \approx 5.6 \cdot 10^{-8}$ cm [123] and $l_0 \approx 10^{-1}$ cm (cf. Fig. 3.40 at T $\approx$ 25°K), we obtain $\varphi \approx 10^{-2}$ for Ge, i.e., impurities form a substitutional solid solution, as expected for gallium in germanium.

Germanium heavily doped with donor impurities was investigated in detail by Fistul' et al. [124, 125] (cf. Figs. 3.43 and 3.44). A similar dependence $\varkappa(n)$ for silicon heavily doped with arsenic [126] is given in Fig. 3.45.

The application of the Ioffe formula (3.4.51) to n-type germanium and silicon samples shows that the parameter φ is not constant (in contrast to p-type germanium). Thus, for example, for n-type germanium the value of this parameter ranges from 50 for

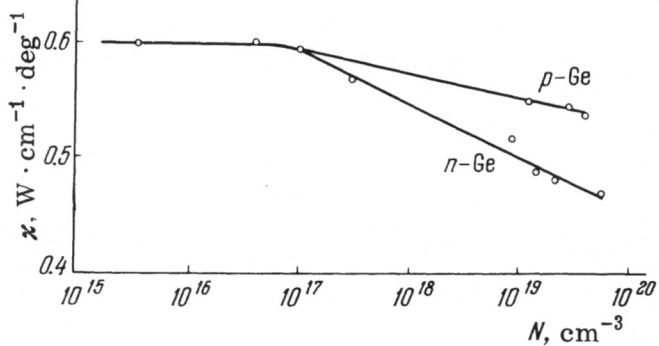

Fig. 3.43. Dependence of the thermal conductivity on the impurity concentration in germanium at 300°K.

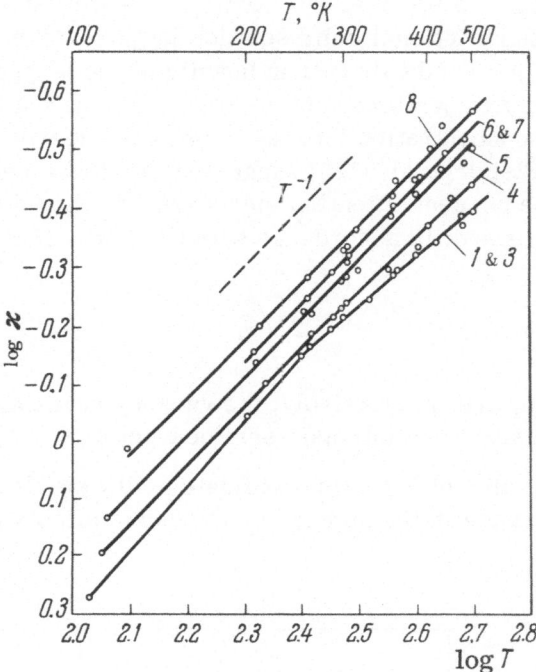

Fig. 3.44. Temperature of the thermal conductivity of germanium doped with the following amounts of arsenic (in cm^{-3}): 1) $3.6 \cdot 10^{15}$; 3) $1 \cdot 10^{17}$; 4) $3 \cdot 10^{17}$; 5) $8.8 \cdot 10^{18}$; 6) $1.3 \cdot 10^{19}$; 7) $2 \cdot 10^{19}$; 8) $5.4 \cdot 10^{19}$.

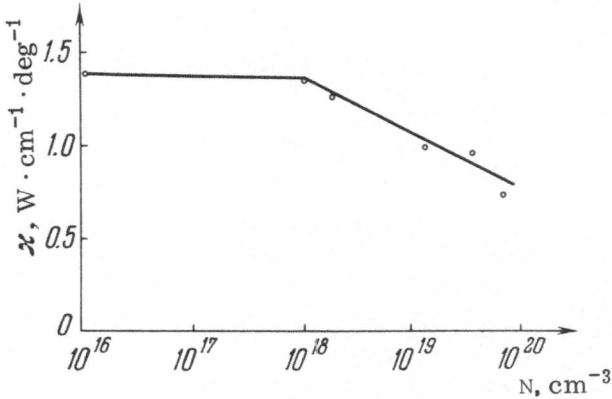

Fig. 3.45. Dependence of the thermal conductivity on the impurity concentration in silicon at 300°K.

$N = 5 \cdot 10^{16}$ cm^{-3} to 3.3 for $N = 5 \cdot 10^{19}$ cm^{-3}. It is well known that arsenic forms a substitutional solid solution in germanium, and therefore the values $\varphi > 1$ indicate that in heavily doped n-type germanium phonons are scattered not on arsenic atoms but on other centers, whose concentration increases when the degree of doping is increased. Fistul' et al. [125] suggested that these centers are vacancies whose concentration N_V increases when the donor doping level is increased, in accordance with the law [127]:

$$N_V = N_{Vi} \frac{N}{n_i}, \qquad (3.4.52)$$

where N_{Vi} and n_i are, respectively, the vacancy concentration and the electron density in an intrinsic semiconductor.

Since the value of N_{Vi} is approximately 10^{17} cm^{-3}, it is evident that an increase in the number of vacancies affects the proper-

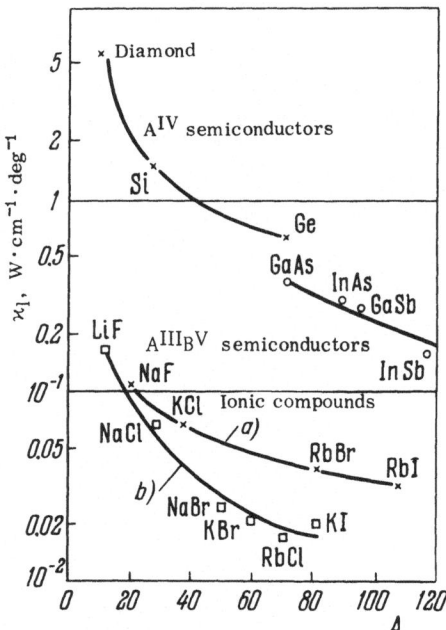

Fig. 3.46. Dependence of the thermal conductivity on the average molecular weight of a semiconductor.

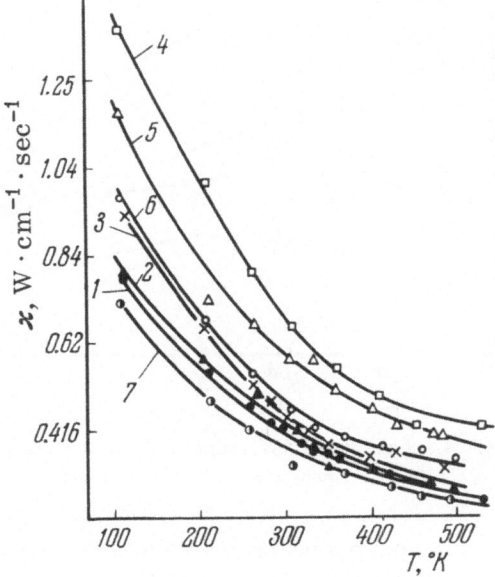

Fig. 3.47. Temperature dependence of the thermal conductivity of n- and p-type gallium arsenide. Carrier density (in cm^{-3}): 1) $n = 2 \cdot 10^{18}$; 2) $n = 6.4 \cdot 10^{17}$; 3) $n = 4.0 \cdot 10^{17}$; 4) $p = 10^{15}$; 5) $p = 10^{19}$-10^{20}; 6) $n = 1 \cdot 10^{17}$; 7) $n = 4 \cdot 10^{18}$.

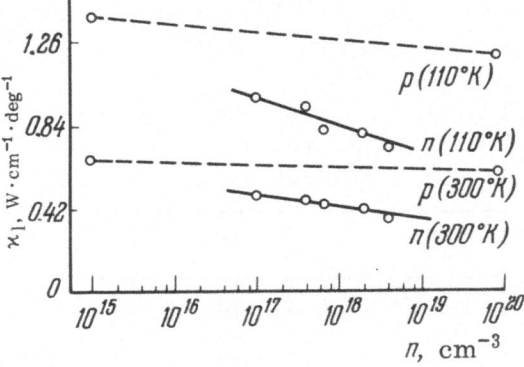

Fig. 3.48. Dependence of the lattice thermal conductivity of gallium arsenide on the carrier density.

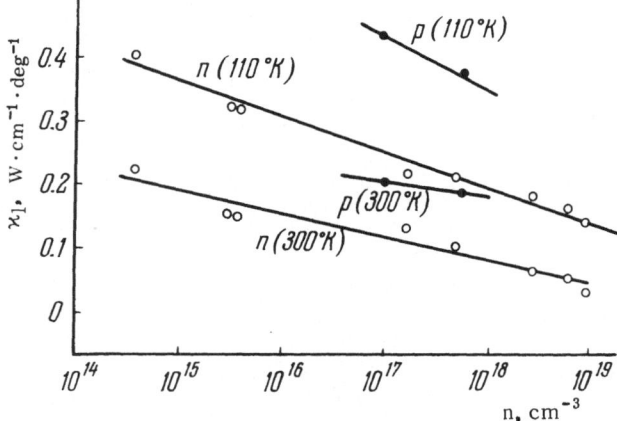

Fig. 3.49. Dependence of the lattice thermal conductivity of
indium antimonide on the carrier density.

ties only if the donor concentration is of the same order, which is
indeed the case (cf. Fig. 3.43). However, it would be desirable to
carry out more detailed investigations in order to confirm the vacancy mechanism of the thermal conductivity in heavily doped germanium and silicon.

Few experimental data are available for the range of temperatures exceeding the Debye temperature. The experimental
curves in Fig. 3.46 support the formula (3.4.49) which predicts, that
at any given temperature log $(1/\varkappa_l)$, should be proportional to $(1/3)$
log A. As the contribution of the ionic binding increases, $\varkappa_l$ decreases in agreement with Eq. (3.4.48) because the Debye temperature of semiconductors with ionic binding is lower than that of crystals with covalent bonds (cf. Table 3.5) [65].

TABLE 3.5. Debye Temperatures
of Ge, Si, and of $A^{III}B^V$
Semiconductors

Semi-conductor	θ_D, °K	Semi-conductor	θ, °K
Ge	406	GaSb	250
Si	689	InP	420
AlSb	370	InAs	280
GaAs	355	InSb	170

TABLE 3.6. Lattice Thermal Conductivity
of Semiconductors at Room Temperature

Semi-conductor	$\varkappa_1$, $W \cdot cm^{-1} \cdot deg^{-1}$	Semi-conductor	$\varkappa_1$, $W \cdot cm^{-1} \cdot deg^{-1}$
Ge	0.64	AlSb	0.6
Si	1.45	GaAs	0.37
Se	0.02	InAs	0.29
Te	0.03	GaSb	0.27
GaP	1.1	InSb	0.15
InP	0.8		

The experimentally determined dependences $\varkappa_1(T)$ and $\varkappa_1(n)$ for some $A^{III}B^{V}$ compounds are shown in Figs. 3.47–3.49. These dependences were obtained by subtracting $\varkappa_e(T)$ from the measured total thermal conductivity $\varkappa(T)$. The values of $\varkappa_e$ were calculated very approximately, assuming that the degeneracy was absent and that carriers were scattered solely on the acoustical vibrations of the lattice, i.e., $\varkappa_e$ was calculated from the Wiedemann–Franz law using the formula (3.4.34) for the Lorentz number.

Allowance for the degeneracy, i.e., the use of the formula (3.4.32), alters somewhat the dependences $\varkappa_1(T)$.

Table 3.6 gives the experimental data on the lattice thermal conductivity $\varkappa_1$, compiled from the results given in [65, 97].

Relationship between the Electronic and Lattice Components of the Thermal Conductivity. This relationship is different for dielectrics and metals, i.e., it is different for materials with different carrier densities. In dielectrics, there are no free electrons and therefore $\varkappa = \varkappa_1$. In metals, there are very many electrons ($\approx 10^{22}$ cm^{-3}) and therefore $\varkappa_e > \varkappa_1$. For example, it has been estimated in [128] that $\varkappa_e/\varkappa_1$ for metals is $3 \cdot 10^3$.

Semiconductors occupy an intermediate position between metals and dielectrics. A rough estimate for the two limiting cases of heavily doped and pure semiconductors can be obtained quite easily. We shall find $\varkappa_e$ for metals using Eq. (3.4.36), which is valid when $\mu^* \gg 0$. Then, the Wiedemann–Franz law gives

$$\varkappa_e = 2.45 \cdot 10^{-8} \sigma T \ W \cdot cm^{-1} \cdot deg^{-1}.$$

Calculating $\varkappa_1$ using Eq. (3.4.51), we find that the ratio $\varkappa_e/\varkappa_1$ is

$$\frac{\varkappa_e}{\varkappa_1} = 2.45 \cdot 10^{-8} \sigma T \frac{1}{\varkappa_0} \left(1 + \varphi \frac{N}{N_0} \frac{l_0'}{a} \right). \qquad (3.4.53)$$

It is clear from this expression that the contribution of the electronic component increases when the impurity concentration is increased.

The ratio $\varkappa_e/\varkappa_1$ depends in a complex manner on temperature because of the temperature dependences $\sigma(T)$ and $\varkappa_0(T)$. In the simplest case when the carrier density and mobility can be assumed to be independent of temperature and $\varkappa_0 \propto 1/T$, the ratio $\varkappa_e/\varkappa_1$ is almost independent of temperature.

In the range $\mu^* < 0$, the Lorenz number is found from Eq. (3.4.34) and then

$$\varkappa_e = 1.47 \cdot 10^{-8} \sigma T \frac{1}{\varkappa_0} \left(1 + \frac{N}{N_0} \frac{l_0}{a} \right). \qquad (3.4.54)$$

The values of $\varkappa_e/\varkappa_1$ for heavily doped germanium and silicon are small but measurable. Thus, for example, for very heavily doped germanium ($N = 5 \cdot 10^{19}$ cm^{-3}) and silicon ($N = 8 \cdot 10^{19}$ cm^{-3}) samples, the values of $\varkappa_e/\varkappa_1$ are, respectively, 0.02 and 0.006. These values have been calculated on the assumption that the scattering by the ion impurities is the dominant process. If an allowance is made for the mixed nature of scattering, we can easily show (using suitable formulas given above), that the values of $\varkappa_e/\varkappa_1$ are hardly affected. This is because the Lorenz number is not very sensitive to the scattering mechanism or to the degree of degeneracy.

For other semiconductors, there are fewer experimental data. Thus, $\varkappa_e$ has been measured for PbTe [129], for bismuth selenide Bi$_2$Se$_3$ [130] and certain other thermoelectric materials. Analysis of the experimental data on semiconductors is given in [97, 98, 131].

§3.5. Thermomagnetic Effects

Nernst − Ettingshausen (NE) Transverse Effect. Let us assume that a temperature gradient ∇T and a magnetic field $\mathbf{H}$ are applied to a semiconductor (Fig. 3.50). The

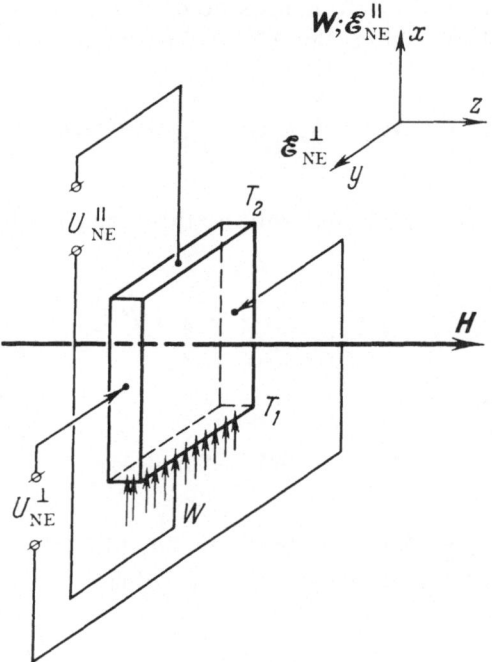

Fig. 3.50. Method of application of a magnetic field
and a temperature gradient to a sample in studies of
thermomagnetic effects.

resultant Lorentz force deflects the carriers as in the ordinary Hall
effect (cf. §3.3) but the motion of free carriers is subject not only
to an electric field $\mathscr{E}$, but also to a field $\nabla(\varphi - \mu/e)$, as in the case
of the thermoelectric power. The additional field $\mathscr{E}_{NE}^{\perp}$ along the
y axis is known as the Nernst—Ettingshausen field (NE field). The
explicit form of this field can be found from an expression for the
current density $\mathbf{j}$, which we shall find—as before—from Eq. (3.1.16)
by substituting $\mathbf{C}(E)$. Thus, the problem reduces to finding $\mathbf{C}(E)$.
To calculate this quantity, we shall use the corresponding expression for F (Table 3.1) and write the transport equation in the form

$$\frac{e}{\hbar}\left\{\nabla\left(\varphi - \frac{\mu}{e}\right) + \frac{1}{\hbar c}\,[\mathbf{v}H]\right\}\nabla_k f - \mathbf{v}\nabla f = -\frac{C(E)\,\mathbf{k}\,\frac{\partial f_0}{\partial E}}{\tau(\mathbf{k})}. \qquad (3.5.1)$$

We shall transform the left-hand side of this equation using Eqs.
(3.3.23) and (3.4.2). Then, Eq. (3.5.1) becomes

$$v \frac{\partial f_0}{\partial E} \left[\frac{\mu - E}{T} \nabla T - \nabla \mu \right] + \frac{\hbar}{m^*} \nabla (e\varphi) + \frac{e}{\hbar c} v \frac{\partial f_0}{\partial E} [HC] = - \frac{C(E) k \frac{\partial f_0}{\partial E}}{\tau(k)}.$$

Replacing v with $(\hbar/m^*)k$ and solving this equation for $C(E)$, we
obtain:

$$C(E) = - \frac{\hbar}{m^*} \left\{ P - \frac{e}{\hbar c} [HC] \right\}, \tag{3.5.2}$$

where P denotes

$$P = \frac{E - \mu}{T} \nabla T + \nabla (\mu - e\varphi). \tag{3.5.3}$$

Equation (3.5.2) has the same form as Eq. (3.3.24). Using the same
method as for the latter equation, we obtain:

$$C(E) = - \frac{\hbar}{\mu^*} \tau(k) \left\{ \frac{P + \frac{e}{m^*c} \tau(k) [HP] + \left(\frac{e}{m^*c}\right)^2 \tau^2(k) (HP) H}{1 + \left[\frac{e}{m^*c} \tau(k) H\right]^2} \right\}. \tag{3.5.4}$$

We shall consider only the NE effect for weak fields H, for which
we obtain, by analogy with Eq. (3.3.29):

$$C(E) = - \frac{\hbar}{m^*} \tau \left\{ P + \frac{e}{m^*c} \tau [H,P] \right\}. \tag{3.5.5}$$

Substituting this expression into Eq. (3.1.16) and replacing the vectors with the corresponding scalar quantities, we obtain the following expressions for the components of j_e along the coordinate axes:

$$j_{ex} = a_1 \nabla_x \left(\frac{\mu}{e} - \varphi \right) + b_1 \nabla_x T - a_2 \nabla_y \left(\frac{\mu}{e} - \varphi \right) - b_2 \nabla_y T,$$

$$j_{ey} = a_2 \nabla_x \left(\frac{\mu}{e} - \varphi \right) + b_2 \nabla_x T - a_1 \nabla_y \left(\frac{\mu}{e} - \varphi \right) + b_1 \nabla_y T,$$

$$i_{ez} = 0, \tag{3.5.6}$$

where the coefficients a_1, a_2, b_1, b_2 are:

$$a_1 = -\frac{e^2}{3\pi^2 m^*} \int_0^\infty k^3 \tau \frac{\partial f_0}{\partial k} \, dk, \tag{3.5.7}$$

$$a_2 = -\frac{e^3}{3\pi^2 (m^*)^2 c} H \int_0^\infty k^3 \tau^2 \frac{\partial f_0}{\partial k} \, dk, \tag{3.5.8}$$

$$b_1 = -\frac{ek_0}{3\pi^2 m^*} \left\{ \int_0^\infty k^3 \tau E \frac{\partial f_0}{\partial k} \, dk - \frac{\mu}{k_0 T} \int_0^\infty k^3 \tau \frac{\partial f_0}{\partial k} \, dk \right\}, \tag{3.5.9}$$

$$b_2 = -\frac{e^2 k_0}{3\pi^2 (m^*)^2 c} \left\{ \int_0^\infty k^3 \tau^2 E \frac{\partial f_0}{\partial k} \, dk - \frac{\mu}{k_0 T} \int_0^\infty k^3 \tau^2 \frac{\partial f_0}{\partial k} \, dk \right\}. \tag{3.5.10}$$

We can easily show that Eq. (3.5.6) includes the transport effects considered earlier. Thus, if we assume that $\nabla T = 0$ (and also that $\nabla \mu = 0$) and that $H = 0$, we easily obtain Eq. (3.1.17) and therefore the value of σ. Substituting $\nabla T = 0$ but $H \neq 0$ into Eq. (3.5.6), we obtain a system of two equations, which are identical with those in Eq. (3.3.33) and represent the Hall effect. If we assume that in Eq. (3.5.6) $H = 0$ but $\nabla T \neq 0$, we can use $j_{ex} = 0$ to find easily $\nabla_x (\mu/e - \varphi)$, and therefore the thermoelectric power α, i.e., Eq. (3.4.7).

In the same way, we obtain the NE coefficient $Q^\perp$, which we shall define using the relationship

$$\mathscr{E}_{NE} \equiv \mathscr{E}_y = -Q^\perp H \nabla_x T, \tag{3.5.11}$$

where $\mathscr{E}_y$ is a field shown in Fig. 3.54, and the superscript "$\perp$" shows that the magnetic field vector H is perpendicular to the temperature gradient ∇T.

The NE effect is isothermal, i.e., $\nabla_y T = 0$. Therefore,

$$\nabla_y \left(\frac{\mu}{e} - \varphi \right) = -\nabla_y \varphi = \mathscr{E}_y.$$

From Eq. (3.5.6), we can easily find that

$$\nabla_y \left(\frac{\mu}{e} - \varphi \right) = \mathscr{E}_y = -\frac{a_1 b_2 - a_2 b_1}{a_1^2 + a_2^2} \nabla_x T. \tag{3.5.12}$$

Comparing this expression with Eq. (3.5.11) after the substitution of the coefficients given by Eqs. (3.5.7)-(3.5.10), we obtain:

$$Q^\perp = \frac{k_0}{m^*c} \frac{I_1 (I_{2E} - \mu^*/_2) + I_2 (I_{2E} - \mu^*/_1)}{I_1^2 + \frac{e^2}{(m^*)^2 c^2} I_2^2},$$
(3.5.13)

where

$$I_i = \int_0^\infty k^3 \tau^i \frac{\partial f_0}{\partial k} dk,$$
(3.5.14)

$$I_{iE} = \int_0^\infty k^3 \tau^i E \frac{\partial f_0}{\partial k} dk.$$
(3.5.15)

In the subsequent analysis of Eq. (3.5.13), we shall proceed as in the studies of other transport coefficients: we shall substitute the Fermi distribution function for f_0 in the integrals and we shall replace τ with the expression for the mixed scattering case [Eq. (3.2.22)].

Integrating Eq. (3.5.13) by parts and using the dimensionless energy ε^*, we obtain:

$$Q^\perp = \frac{k_0}{m^*c} \tau_{0L} \frac{\Phi_{9/_2} (\mu^*, a) \Phi_4 (\mu^*, a) - \Phi_3 (\mu^*, a) \Phi_{11/_2} (\mu^*, a)}{\Phi_3^2 (\mu^*, a)},$$
(3.5.16)

where

$$\Phi_{11/_2} (\mu^*, a) = \int_0^\infty \frac{(\varepsilon^*)^{11/_2} \exp (\varepsilon^* - \mu^*) d\varepsilon^*}{[(\varepsilon^*)^2 + a^2]^2 [1 + \exp (\varepsilon^* - \mu^*)]^2};$$
(3.5.17)

the integrals Φ_3, Φ_4, and $\Phi_{9/2}$ have been discussed in preceding sections.

When $a^2 \ll (\varepsilon^*)^2$, i.e., in the absence of impurity scattering, Eq. (3.5.16) becomes

$$Q_L^\perp = \frac{k_0}{m^*c} \tau_{0L} \frac{F_1 (\mu^*) F_{-1/_2} (\mu^*) - \frac{3}{2} F_0 (\mu^*) F_{1/_2} (\mu^*)}{F_0^2 (\mu^*)},$$
(3.5.18)

which is identical with the formulas obtained in [11, 66].

We can also assume that $\mu^* \ll 0$ to obtain a well-known expression valid for pure semiconductors:

$$Q_{L0}^{\perp} = \frac{\sqrt{\pi}}{4} \frac{k_0}{m^* c} \tau_{0L},$$

(3.5.19)

which can be rewritten using Eq. (3.3.44) as follows

$$Q_{L0}^{\perp} = \frac{1}{2} \frac{k_0}{ec} u_{H0},$$

(3.5.20)

or, using Eq. (3.3.9), we can rewrite it to obtain

$$Q_{L0}^{\perp} = \frac{3\pi}{16} \frac{k_0}{ec} u_{\sigma 0},$$

(3.5.21)

where u_{H0} and $u_{\sigma 0}$ are the Hall and drift mobilities in an impurity-free semiconductor.

The above formula is usually quoted without specifying which mobility is meant [11]. Therefore, the mobility calculated from Eq. (3.5.21) is frequently identified as the Hall mobility, which, of course, is incorrect.

In the other limiting case, when the carriers are scattered only by impurity ions, $a^2 \gg (\varepsilon^*)^2$, Eq. (3.5.16) becomes

$$Q_i^{\perp} = \frac{k_0}{m^* c} \tau_{0i} \frac{18 F_3(\mu^*) F_{7/2}(\mu^*) - \frac{33}{2} F_2(\mu^*) F_{9/2}(\mu^*)}{9 F_3^2(\mu^*)},$$

(3.5.22)

which gives the following expression for $\mu^* \ll 0$:

$$Q_i^{\perp} = -\frac{945}{128} \sqrt{\pi} \frac{k_0}{m^* c} \tau_{0i},$$

(3.5.23)

or, using Eqs. (3.3.47) and (3.3.15), we can obtain

$$Q_i^{\perp} = -\frac{3}{2} \frac{k_0}{ec} u_{Hi},$$

(3.5.24)

$$Q_i^{\perp} = -\frac{945}{974} \pi \frac{k_0}{ec} u_{\sigma i}.$$

(3.5.25)

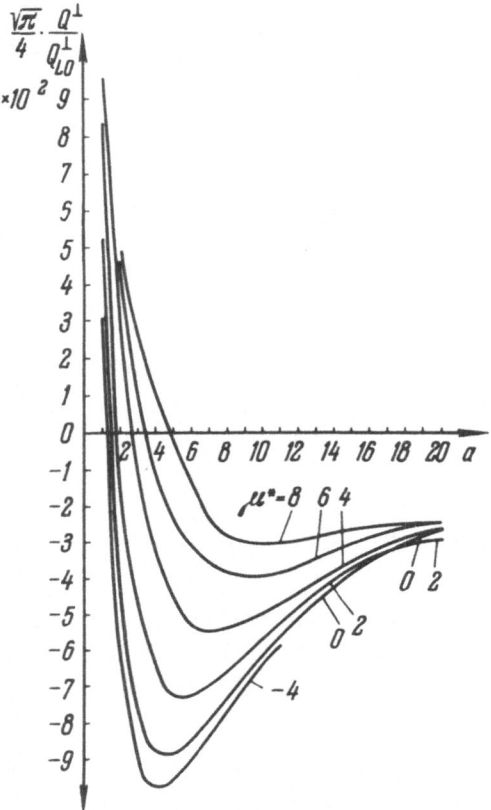

Fig. 3.51. Calculated transverse NE coefficient for
mixed electron scattering.

If the first-order approximation of Eq. (2.5.11) is applied to
heavily doped semiconductors ($\mu^* \gg 0$), it is found that $Q_L^\perp$ and $Q_i^\perp$
are both equal to zero. If we use a second-order approximation of
Eq. (2.5.13), we then obtain

$$Q_L^\perp = \frac{\pi^2}{6} \frac{k_0}{m^*c} \tau_{0L} (\mu^*)^{-3/2}, \qquad (3.5.26)$$

$$Q_i^\perp = -\frac{\pi^2}{2} \frac{k_0}{m^*c} \tau_{0i} (\mu^*)^{1/2}. \qquad (3.5.27)$$

In general, for any degree of degeneracy and any value of a,
the experimental results should be compared with the theory using

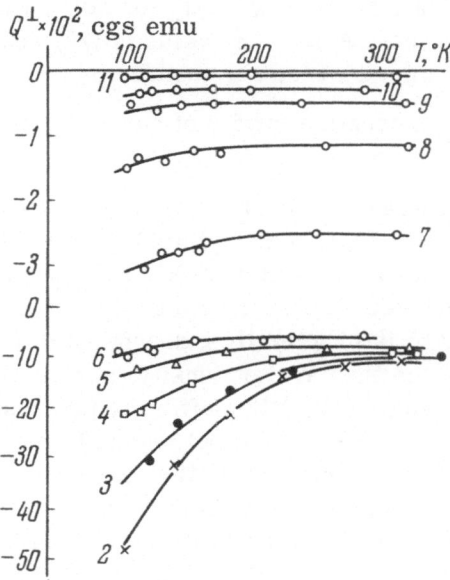

Fig. 3.52. Temperature dependence of the NE coefficient for germanium heavily doped with arsenic.

Sample No.	2	3	4	5	6
Electron density, cm^{-3} (T = 300°K)	$6.5 \cdot 10^{16}$	$1.1 \cdot 10^{17}$	$1.4 \cdot 10^{17}$	$3.2 \cdot 10^{17}$	$4.3 \cdot 10^{17}$
Sample No.	7	8	9	10	11
Electron density, cm^{-3} (T = 300°K)	$3.2 \cdot 10^{18}$	$5.9 \cdot 10^{18}$	$1.1 \cdot 10^{19}$	$2.5 \cdot 10^{19}$	$3.9 \cdot 10^{19}$

Eq. (3.5.28), which is easily obtained by dividing Eq. (3.5.16) by Eq. (3.5.19):

$$\frac{Q^{\perp}}{Q_{t\,0}^{\perp}} = \frac{4}{\sqrt{\pi}} \frac{\Phi_{9/2}\Phi_4 - \Phi_3\Phi_{11/2}}{\Phi_3^2}. \qquad (3.5.28)$$

The numerical values of $Q^\perp/Q^\perp_{L_0}$ for some values of μ^* and a are given in Appendix A.10. The value of $Q^\perp_{L_0}$ represents the NE coefficient of an impurity-free semiconductor.

Figure 3.51 presents a family of curves plotted on the basis of Eq. (3.5.28).

Experimental Results. From the results reported in [132], and shown in Figs. 3.52 and 3.53, it is evident that when the impurity concentration is increased, the dependence $Q^\perp(T)$ becomes weaker, and when the carrier density $n > 10^{18}$ cm^{-3} the value of the coefficient $Q^\perp$ is practically independent of temperature. The dependence of $Q^\perp$ on the carrier density, measured at 120°K, shows that the absolute value of the NE coefficient increases at first when the degree of doping is increased, but begins to decrease rapidly when $n > 10^{17}$ cm^{-3}. It is interesting that the NE effect of all the substances investigated so far has been found to be negative. These results have been accounted for qualitatively in [132].*

In general, the expression for the Nernst — Ettingshausen effect can be written in the form

$$Q^\perp = \frac{k_0}{ec} \mathcal{K}(\mu^*) \frac{u_H}{\mathcal{A}},\qquad(3.5.29)$$

where $\mathcal{K}(\mu^*)$ is a combination of Fermi integrals with various indices.

The function $\mathcal{K}(\mu^*)$ is different for different values of r in the law $\tau = \tau_0(\varepsilon^*)^{r-1/2}$ (Fig. 3.54). The same figure includes an experimental $\mathcal{K}(\mu^*)$, curve for germanium, i.e., a curve calculated from the relationship (3.5.29) substituting in it the mobility u_H found experimentally at 120°K. The Hall factor $\mathcal{A}$ can be assumed to be equal to unity. The values of μ^* are calculated from Eq. (2.3.4), using the experimental value of n and $m^* = 0.57m_0$ for germanium.

At low impurity concentrations, the function $\mathcal{K}(\mu^*)$, calculated from the experimental data is close to the theoretical function for the case of scattering by the acoustical phonons. This

_____

*The results for a sample with $n = 1.7 \cdot 10^{16}$ cm^{-3} were reported incorrectly in [132]. In fact, this sample had a negative sign of the NE effect.

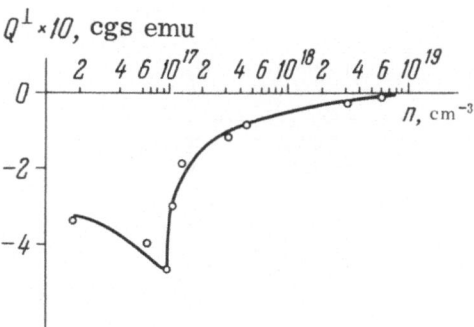

Fig. 3.53. Dependence of the NE coefficient at 120°K on the carrier density in arsenic-doped germanium.

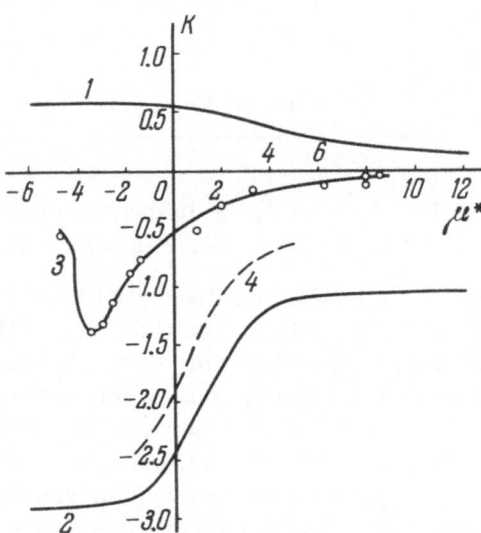

Fig. 3.54. Dependence of the transport coefficient for the transverse NE effect in n-type germanium on the degree of degeneracy of the electron gas: 1) scattering on acoustical phonons; 2) scattering on impurity ions; 3) experimental results [132]; 4) calculation with allowance for screening.

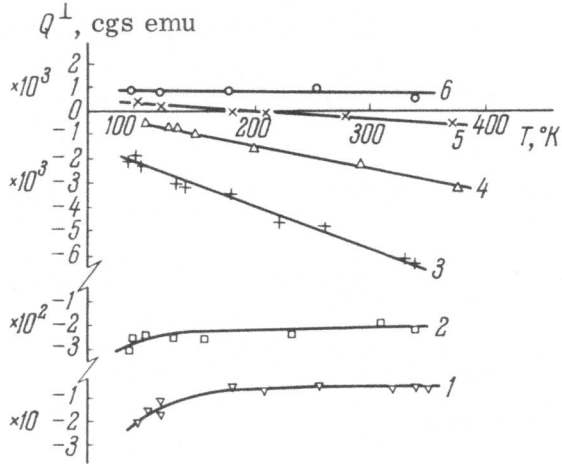

Fig. 3.55. Temperature dependence of the transverse NE
coefficient for silicon heavily doped with arsenic.

Sample No.	1	2	3	4	5	6
Carrier density, cm^{-3} (77°K)	$1.2 \cdot 10^{16}$	$3.9 \cdot 10^{17}$	$7.7 \cdot 10^{17}$	$1.9 \cdot 10^{19}$	$3.9 \cdot 10^{19}$	$9.5 \cdot 10^{19}$

means that when $n \leq 1.7 \cdot 10^{16}$ cm^{-3}, the interaction between the electrons and thermal vibrations of the lattice is important.

Nevertheless, at such impurity concentrations, the impurity scattering is dominant. This follows from the fact that $\mathscr{H}(\mu^*)$ is negative, which represents the scattering by the impurity ions.* However, at high impurity concentrations, when electrons are scattered mainly by ions, the screening factor g(b) in the formula for the relaxation time begins to depend on the electron energy. Consequently (cf. §3.3), the factor r decreases and curves 3 and 2 in Figs. 3.54 diverge when the degree of degeneracy increases.

For a quantitative comparison of the theory with experiment, we shall use integrals of the type given in Eqs. (3.3.62) and (3.3.63), which are shown graphically in Appendix A. 6.

*Outside the Soviet Union, the notation is frequently opposite: $Q^{\perp} < 0$ is used to repre-
sent the scattering of the electrons by the acoustical phonons, and $Q^{\perp} > 0$ is used for
the scattering by the impurity ions.

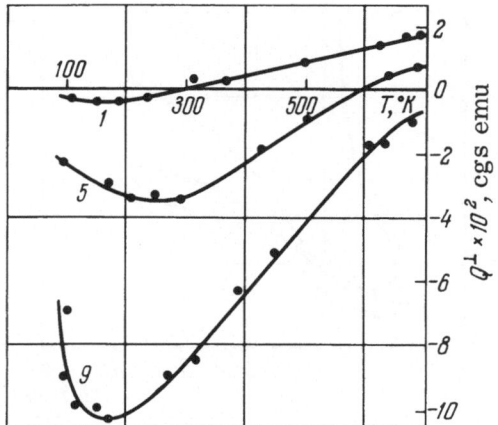

Fig. 3.56. Temperature dependence of the NE coefficient of gallium arsenide [134]. 1) $n = 1 \cdot 10^{19}$
cm^{-3}; 5) $n = 3 \cdot 10^{18}$ cm^{-3}; 9) $n = 2 \cdot 10^{16}$ cm^{-3}.

Then the expression for $Q^{\perp}$ becomes:

$$Q^{\perp} = \frac{2k_0}{\sqrt{\pi}\, m^* c}\, u_{H_0} \frac{\varphi_{9/_2}\varphi_4 - \varphi_{11/_2}\varphi_3}{\varphi_3^2}. \qquad (3.5.30)$$

The calculation of $\mathscr{K}(\mu^*)$ now gives curve 4 in Fig. 3.54, which agrees somewhat better with the experimental data at high concentrations but there is still no complete match. Satisfactory analysis of the experimental data on the NE effect requires a theory of this effect which allows for the scattering anisotropy.

The dependences $Q^{\perp}(n, T)$ for heavily doped silicon samples (Fig. 3.55), reported in [133], are similar. A characteristic feature of these results is the change of the sign of $Q^{\perp}(\mu^*)$ from negative to positive at $\mu^* = 1$ (300°K) and at $\mu^* = 3$ (100°K). These points correspond to impurity concentrations of $4.5 \cdot 10^{19}$ cm^{-3} and $1 \cdot 10^{19}$ cm^{-3}. An estimate of the ratio u_{HL}/u_{Hi} (cf. Fig. 3.25) shows that this reversal of the sign of $Q^{\perp}$ cannot be explained by the dominant role of the lattice scattering. A qualitative explanation of this observation, proposed in [133], is based on the assumption that the electron scattering does not take place in the Coulomb field of impurity ions but in some other short-range field. In other words, in a heavily doped material a scattered electron is affected by the core

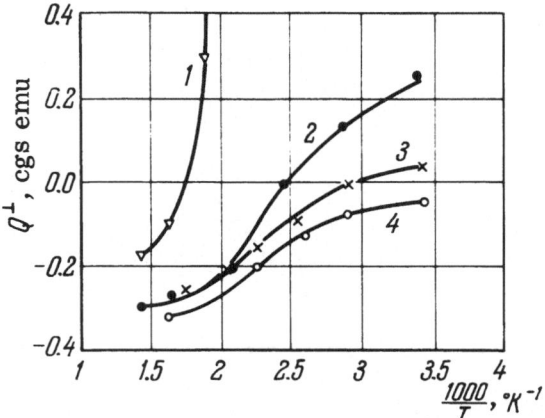

Fig. 3.57   Temperature dependence of the NE coefficient of indium antimonide [154]. 1) $p = 5 \cdot 10^{17}$ cm^{-3};
2) $n = 3.5 \cdot 10^{15}$ cm^{-3}; 3) $n = 2.4 \cdot 10^{16}$ cm^{-3}; 4) $n = 8 \cdot 10^{17}$
cm^{-3}.

of an ion. This approximate theory [133] gives only an order-of-
magnitude agreement between the calculated and experimental
values of $Q^{\perp}$.

The results of the experimental investigations of the NE effect
in gallium arsenide and in indium antimonide are presented in Figs.
3.56 and 3.57. These results are discussed in §3.7. Among the
other thermomagnetic effects, it is worth mentioning the longitudinal Nernst—Ettingshausen effect; the mechanism of appearance of
this effect is shown schematically in Fig. 3.50. There are as yet
no experimental data on this effect in heavily doped semiconductors.
Experimental data on lightly doped crystals are reported in [11].
Attempts to measure the longitudinal Nernst—Ettingshausen effect
in heavily doped germanium have shown that this effect is almost
an order of magnitude weaker than the transverse NE effect. Therefore, the experimental data on the longitudinal effect are not sufficiently reliable to use them to obtain information on the electron
scattering mechanism.

§3.6.   Magnetoresistance of Semiconductors

In the preceding sections, we have considered transport phenomena (the Hall and Nernst—Ettingshausen effects) governed by
the first power of the magnetic field intensity. They are known as

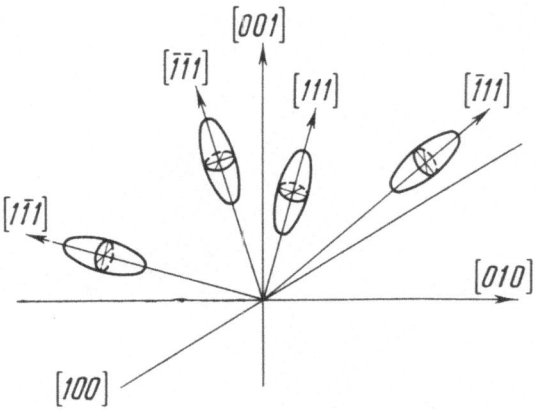

Fig. 3.58. Orientation of energy ellipsoids
in n-type germanium.

the odd effects (with respect to the magnetic field). There are also
other effects, known as the even transport effects, because they are
governed by the square of the magnetic field intensity. Of these,
we shall consider only the relative change in the resistivity of a
semiconductor $\Delta\rho/\rho_0$ in a magnetic field, which is known as the
magnetoresistance.

General Theory of Magnetoresistance. We
shall consider a semiconductor with a many-valley electron energy
surface, e.g., n-type germanium or silicon.

We shall make the following assumptions.

a) The electron energy near each minimum may be approximated (using suitably selected coordinates) by a quadratic function
of the crystal momentum $\hbar\mathbf{k}$, i.e.,

$$E = \frac{\hbar^2}{2}\left(\frac{k_1^2}{m_1} + \frac{k_2^2}{m_2} + \frac{k_3^2}{m_3}\right),$$ (3.6.1)

where m_i are the components of the effective mass tensor along the
i-th axis and k_i are the components of the wave vector along the
same axis.

b) The constant-energy surface for electrons in germanium
is assumed to consist of four crystallographically equivalent ellipsoids of revolution oriented along the [111] directions in the

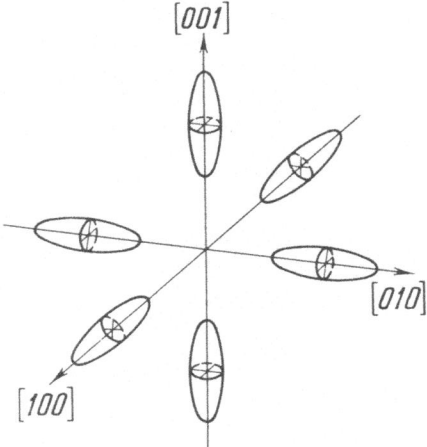

Fig. 3.59. Orientation of energy ellipsoids
in n-type silicon.

Brillouin zone (Fig. 3.58) and the corresponding surface of silicon
consists of six ellipsoids oriented along the [100] directions (Fig.
3.59). Thus, $m_1 = m_2 = m_\perp$ and $m_3 = m_\parallel$.

c) In relaxation processes, transitions take place only between states within the limits of one set of ellipsoids, i.e., between ellipsoids describing the same minimum. This assumption
is justified at not too high temperatures.

d) The relaxation time is anisotropic and, in terms of the
principal axes of the energy ellipsoids of Eq. (3.6.1), it can be described by a diagonal tensor $\{\tau_{ii}\}$ with the following components:

$$\tau_{11} = \tau_{22} = \tau_\perp, \qquad \tau_{33} = \tau_\parallel. \qquad (3.6.2)$$

This assumption is justified in those cases when relaxation processes either produce a random distribution of electron velocities
or do not change the electron energy [135].

e) At a fixed temperature, the components of the relaxation
tensor given in Eq. (3.6.2) depend only on the electron energy.

f) There is no temperature gradient across the sample.

When all these assumptions are made, the transport equation becomes:

$$\frac{e}{\hbar}\left(\mathscr{E} + \frac{1}{c}[vH]\right)\nabla_k f = -\frac{C(E) k \frac{\partial f_0}{\partial E}}{\tau_{ii}}.$$ (3.6.3)

Then, we obtain the following algebraic equation for the determination of $C(E)$:

$$\frac{\{m_{ii}\}}{\{\tau_{ii}\}} C(E) - \frac{e}{c}[CH] = \mathscr{E}$$

or

$$\{u_0\}^{-1} C(E) - \frac{e}{c}[CH] = \mathscr{E},$$ (3.6.4)

where $\{u_0\}$ is the mobility tensor in the absence of a magnetic field. The components of this tensor are given by:

$$u_{0ik} = m_{il}^{-1}\tau_{lk}.$$ (3.6.5)

The solution of Eq. (3.6.4) for an arbitrary magnetic field is [136]

$$C(E) = \frac{\{u_0\}\,\mathscr{E} + \|u_0\|\frac{e}{c}[\mathscr{E}\{u_0\}^{-1}H] + \frac{e^2}{c}\|u_0\|(\mathscr{E}H)H}{1 + \frac{e^2}{c^2}\|u_0\|(\{u_0\}^{-1}HH)}.$$ (3.6.6)

Here, $\|u_0\|$ is the determinant of the tensor (3.6.5).

For weak magnetic fields, Eq. (3.6.6) simplifies and becomes

$$C(E) = \{u_0\}\,\mathscr{E} + \|u_0\|\frac{e}{c}[\mathscr{E}\{u_0\}^{-1}H] + \frac{e^2}{c}\|u_0\|(\mathscr{E}H)H.$$ (3.6.7)

Substituting Eq. (3.6.7) into the expression for the current density j_e of Eq. (3.1.16), which is valid for one set of ellipsoids, we obtain the i-th component

$$j_i = \sigma_{ik}\mathscr{E}_k + \sigma_{ikl}\mathscr{E}_k H_l + \sigma_{iklm}\mathscr{E}_k H_l H_m$$ (3.6.8)

and we find expressions for the components of the tensors $\{\sigma\}$, which were first given in [135]:

$$\sigma_{ik} = -\frac{aB_i}{m_i^2}\langle\tau_i\rangle\,\delta_{ik},\tag{3.6.9}$$

$$\sigma_{ikl} = \frac{abB_i}{m_i^2 m_k}\langle\tau_i\tau_k\rangle\,\delta_{ikl},\tag{3.6.10}$$

$$\sigma_{iklm} = -\frac{ab^2B_i}{m_i^2 m_k m_l}\langle\tau_i\tau_k\tau_l\rangle\,\frac{\delta_{ilk}\delta_{lim}+\delta_{lmk}\delta_{lil}}{2}.\tag{3.6.11}$$

In these expressions, we have used the following notation:

$$\left.\begin{aligned}a &= \frac{e^2}{4\pi^3\hbar^3};\qquad b = \frac{e}{c},\\ B_1 &= B_2 = 2^{7/2}\,3^{-1}\pi m_\perp^2\,m_\parallel^{1/2},\\ B_3 &= 2^{7/2}\,3^{-1}\pi m_\perp m_\parallel^{3/2}\end{aligned}\right\}\tag{3.6.12}$$

and the following tensor components:

$$\left.\begin{aligned}\delta_{ik} &= \begin{cases}1 \text{ for } i = k,\\ 0 \text{ for } i \neq k,\end{cases}\\ \delta_{123} &= \delta_{231} = \delta_{312} = +1,\\ \delta_{213} &= \delta_{132} = \delta_{321} = -1.\end{aligned}\right\}\tag{3.6.13}$$

All the other components of the tensor δ_{ikl} are equal to zero. The symbol $\langle x\rangle$ represents averaging of the type

$$\langle x\rangle = \int_0^\infty xE^{3/2}\frac{\partial f_0}{\partial E}\,dE.\tag{3.6.14}$$

Finally, the components of the tensors $\{\sigma\}$ become:

$$\left.\begin{aligned}\sigma_{11} &= \sigma_{22} = -aB_1\frac{\langle\tau_\perp\rangle}{m_\perp^2},\\ \sigma_{33} &= -aB_3\frac{\langle\tau_\parallel\rangle}{m_\parallel^2},\end{aligned}\right\}\tag{3.6.15}$$

$$\sigma_{123} = -\sigma_{213} = abB_1 \frac{\langle \tau_\perp^3 \rangle}{m_\perp^3},$$

$$\sigma_{231} = \sigma_{312} = -\sigma_{132} = -\sigma_{321} = abB_1 \frac{\langle \tau_\perp \tau_\parallel \rangle}{m_\perp^2 m_\parallel}, \tag{3.6.16}$$

$$\sigma_{1122} = \sigma_{2211} = ab^2 B_1 \frac{\langle \tau_\perp^2 \tau_\parallel \rangle}{m_\perp^3 m_\parallel},$$

$$\sigma_{1133} = \sigma_{2233} = ab^2 B_1 \frac{\langle \tau_\perp^3 \rangle}{m_\perp^4},$$

$$\sigma_{3311} = \sigma_{3322} = ab^2 B_3 \frac{\langle \tau_\perp \tau_\parallel^2 \rangle}{m_\perp m_\parallel^3}, \tag{3.6.17}$$

$$\sigma_{1212} = \sigma_{2121} = \sigma_{2323} = \sigma_{3232} =$$

$$= \sigma_{3131} = \sigma_{1313} = -ab^2 B_1 \frac{\langle \tau_\perp^2 \tau_\parallel \rangle}{m_\perp^3 m_\parallel}.$$

All the other components of the tensors $\{\sigma\}$ are equal to zero.

Further calculations will be carried out in the same way as in [137].

Magnetoresistance of Germanium. In accordance with assumption (b), the tensors (3.6.15)-(3.6.17) are obtained in a coordinate system in which one of the axes coincides with the [111] direction in germanium. However, it is more convenient to transform these tensors to a new system of coordinates with the axes directed along the edges of a cubic cell. The matrix for such a transformation (Fig. 3.60) is:

$$A_{i'}^i = \frac{1}{2\sqrt{3}} \begin{vmatrix} 1+\sqrt{3} & 1-\sqrt{3} & -2 \\ 1-\sqrt{3} & 1+\sqrt{3} & -2 \\ 2 & 2 & 2 \end{vmatrix}.$$

Here, the lower index with a prime represents the number of the column, and the upper index without a prime represents the number of the row. We shall transform the tensors (3.6.15)-(3.6.17) in accordance with the well-known rule

$$\sigma_{i'k'} = \sum_{i,k} \sigma_{ik} A_{i'}^i A_{k'}^k, \quad \sigma_{i'k'l'} = \sum_{i,k,l} \sigma_{ikl} A_{i'}^i A_{k'}^k A_{l'}^l,$$

$$\sigma_{i'k'l'm'} = \sum_{i,k,l,m} \sigma_{iklm} A_{i'}^i A_{k'}^k A_{l'}^l A_{m'}^m. \tag{3.6.18}$$

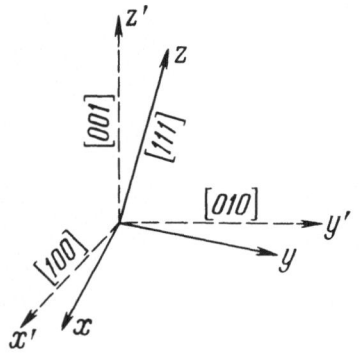

Fig. 3.60. Transformation of the coordinate system.

The new components of the tensors $\{\sigma\}$ are:

$$\sigma_{i'k'} = \frac{1}{3}(2\sigma_{11} + \sigma_{33})\,\delta_{i'k'},$$

$$\sigma_{i'k'l'} = \frac{1}{3}(\sigma_{123} + \sigma_{231} + \sigma_{312})\,\delta_{i'k'l'},$$

$$\sigma_{i'k'l'm'} = \frac{2}{9}(\sigma_{1133} + \sigma_{3311} + 2\sigma_{1313})$$
$$\text{for } i' = k' = l' = m',$$

$$\sigma_{i'k'l'm'} = \frac{1}{18}(7\sigma_{1122} + 4\sigma_{1133} + 4\sigma_{3311} - 3\sigma_{1212})$$
$$\text{for } i' = k' \neq l' = m',$$

$$\sigma_{i'k'l'm'} = \frac{1}{18}(\sigma_{1122} - 2\sigma_{1133} - 2\sigma_{3311} + 15\sigma_{1212})$$
$$\text{for } i' = l' \neq k' = m',$$

$$\sigma_{i'k'l'm'} = \frac{1}{18}(\sigma_{1122} - 2\sigma_{1133} - 2\sigma_{3311} - 3\sigma_{1212})$$
$$\text{for } i' = m' \neq k' = l'.$$

$$(3.6.19)$$

Tensors for the other three ellipsoids can be obtained by the rotation of the coordinate system around one of the cubic axes.

Generally speaking, there are also other components of the tensors (3.6.19) which are not equal to zero but are not quoted because they cancel out in the summation over all four ellipsoids.

We can now find the total density of the current **j** passing through a crystal by summing the densities **j**$_s$ for each set of ellipsoids. The i-th component of the total density of the current is then:

$$j_i = \Sigma_{ik}\mathscr{E}_k + \Sigma_{ikl}\mathscr{E}_k H_l + \Sigma_{iklm}\mathscr{E}_k H_l H_m, \qquad (3.6.20)$$

where the components of the tensors $\{\Sigma\}$ are expressed in terms of the components of the tensors (3.6.19).

Summation over all four sets of ellipsoids gives:

$$
\begin{aligned}
\Sigma_{ik} &= -\frac{4aB_1}{3}\left(2\,\frac{\langle \tau_\perp \rangle}{m_\perp^2} + \frac{\langle \tau_\parallel \rangle}{m_\perp m_\parallel}\right)\delta_{ik}, \\[4pt]
\Sigma_{ikl} &= \frac{4abB_1}{3}\left(\frac{\langle \tau_\perp^2 \rangle}{m_\perp^3} + 2\,\frac{\langle \tau_\perp \tau_\parallel \rangle}{m_\perp^2 m_\parallel}\right)\delta_{ikl}, \\[4pt]
\Sigma_{iklm} &= \frac{8}{9}\,ab^2 B_1\left(\frac{\langle \tau_\perp^3 \rangle}{m_\perp^4} - 2\,\frac{\langle \tau_\perp^2 \tau_\parallel \rangle}{m_\perp^3 m_\parallel} + \frac{\langle \tau_\perp \tau_\parallel^2 \rangle}{m_\perp^2 m_\parallel^2}\right) \\[2pt]
&\hspace{3cm}\text{for } i=k=l=m, \\[6pt]
\Sigma_{iklm} &= \frac{4}{9}\,ab^2 B_1\left(2\,\frac{\langle \tau_\perp^3 \rangle}{m_\perp^4} + 5\,\frac{\langle \tau_\perp^2 \tau_\parallel \rangle}{m_\perp^3 m_\parallel} + \right. \\[2pt]
&\left. \hspace{1cm} + 2\,\frac{\langle \tau_\perp \tau_\parallel^2 \rangle}{m_\perp^2 m_\parallel^2}\right)\text{ for } i=k\neq l=m, \\[6pt]
\Sigma_{iklm} &= \frac{4}{9}\,ab^2 B_1\left(\frac{\langle \tau_\perp^3 \rangle}{m_\perp^4} + 7\,\frac{\langle \tau_\perp^2 \tau_\parallel \rangle}{m_\perp^3 m_\parallel} + \right. \\[2pt]
&\left. \hspace{1cm} + \frac{\langle \tau_\perp \tau_\parallel^2 \rangle}{m_\perp^2 m_\parallel^2}\right)\text{ for } i=l\neq k=m, \\[6pt]
\Sigma_{iklm} &= -\frac{4}{9}\,ab^2 B_1\left(\frac{\langle \tau_\perp^3 \rangle}{m_\perp^4} - 2\,\frac{\langle \tau_\perp^2 \tau_\parallel \rangle}{m_\perp^3 m_\parallel} + \right. \\[2pt]
&\left. \hspace{1cm} + \frac{\langle \tau_\perp \tau_\parallel^2 \rangle}{m_\perp^2 m_\parallel^2}\right)\text{ for } i=m\neq k=l.
\end{aligned}
\tag{3.6.21}
$$

All the other components of the tensors (3.6.21) are equal to zero.

To find the galvanomagnetic transport coefficients, we shall write an equation which is the reciprocal of Eq. (3.6.20):

$$
\mathscr{E}_k = \Lambda_{kr} j_r + \Lambda_{krs} j_r H_s + \Lambda_{krst} j_r H_s H_t. \tag{3.6.22}
$$

Substituting this expression into Eq. (3.6.20), we obtain equations for the determination of the tensors $\{\Lambda\}$

$$
\begin{aligned}
&\Sigma_{ik}\Lambda_{kr} j_r = j_i, \\
&\Sigma_{ik}\Lambda_{krs} j_r H_s + \Sigma_{ikl}\Lambda_{kr} j_r H_l = 0, \\
&\Sigma_{ik}\Lambda_{krst} j_r H_s H_t + \Sigma_{ikl}\Lambda_{krs} j_k H_l H_s + \Sigma_{iklm}\Lambda_{kr} j_r H_l H_m = 0,
\end{aligned}
$$

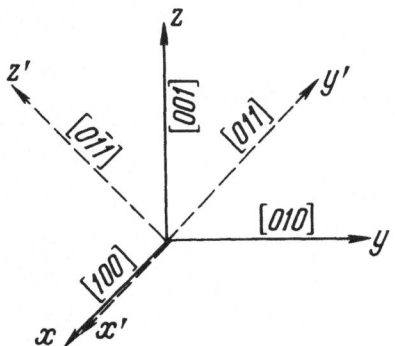

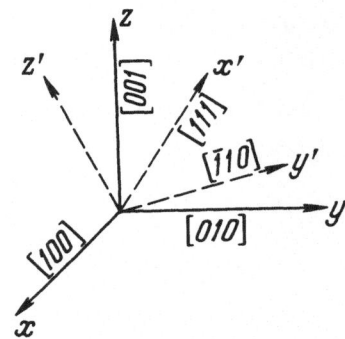

Fig. 3.61. Transformation of the
coordinate system.

Fig. 3.62. Transformation of the
coordinate system.

from which we find:

$$\left.\begin{aligned}
\Lambda_{ik} &= \frac{1}{\sigma_0}\,\delta_{ik}, \\
\Lambda_{ikl} &= -\frac{1}{\sigma_0^2}\,\Sigma_{ikl} = -R\delta_{ikl}, \\
\Lambda_{iklm} &= \frac{1}{\sigma_0^2}\Big[-\Sigma_{iklm} + \frac{1}{\sigma_0}\,\Sigma_{iam}\,\Sigma_{akl}\Big].
\end{aligned}\right\} \qquad (3.6.23)$$

In these equations, R denotes the Hall coefficient and σ_0 denotes the conductivity of the crystal in the absence of a magnetic field, defined by

$$\Sigma_{ik} = \sigma_0\delta_{ik}. \qquad (3.6.24)$$

The components Λ_{1111} and Λ_{1122} determine the longitudinal and transverse magnetoresistance. In fact, it follows from Eq. (3.6.24):

$$\left.\begin{aligned}
\mathscr{H}_{100}^{100} &= \frac{\Lambda_{1111}}{R^2\sigma}, \\
\mathscr{H}_{100}^{010} &= \frac{\Lambda_{1122}}{R^2\sigma},
\end{aligned}\right\} \qquad (3.6.25)$$

where $\mathscr{H}_{ikl}^{mnp}$ are the transport coefficients of the magnetoresistance defined as [138]

$$\left(\frac{\Delta\rho}{\rho_0}\right)_{ikl}^{mnp} = \mathscr{K}_{ikl}^{mnp}\, u_H^2\, H^2. \tag{3.6.26}$$

From Eqs. (3.6.21), (3.6.23), and (3.6.25), we obtain:

$$\left. \begin{aligned}
\mathscr{K}_{100}^{100} &= \frac{2B}{3}\frac{(K_m^2\langle\tau_\perp^3\rangle - 2K_m\langle\tau_\perp^2\tau_\parallel\rangle + \langle\tau_\perp\tau_\parallel^2\rangle)}{K_m\,(K_m\langle\tau_\perp^2\rangle + 2\langle\tau_\perp\tau_\parallel\rangle)^2}, \\
\mathscr{K}_{100}^{010} &= \frac{B}{3}\frac{(2K_m^2\langle\tau_\perp^3\rangle + 5K_m\langle\tau_\perp^2\tau_\parallel\rangle + 2\langle\tau_\perp\tau_\parallel^2\rangle)}{K_m\,(K_m\langle\tau_\perp^2\rangle + 2\langle\tau_\perp\tau_\parallel\rangle)^2} - 1,
\end{aligned} \right\} \tag{3.6.27}$$

where for the sake of simplicity we use the notation $2K_m\langle\tau_\perp\rangle + \langle\tau_\parallel\rangle = B$.

To determine $\mathscr{K}_{110}^{110}$, $\mathscr{K}_{110}^{001}$ and $\mathscr{K}_{110}^{1\bar{1}0}$ it is necessary to transform the tensors (3.6.21) to a new system of coordinates (Fig. 3.61). The matrix for this transformation is

$$A_{i'}^i = \frac{\sqrt{2}}{2}\begin{vmatrix} 2 & 0 & 0 \\ 0 & 1 & -1 \\ 0 & 1 & 1 \end{vmatrix}.$$

Using the rule (3.6.18), we obtain the following expressions in the new system of coordinates

$$\left. \begin{aligned}
\mathscr{K}_{110}^{110} &= \frac{\Lambda_{2'2'2'2'}}{R^2\sigma_0}, \\
\mathscr{K}_{100}^{001} &= \frac{\Lambda_{2'2'1'1'}}{R^2\sigma_0}, \\
\mathscr{K}_{110}^{\bar{1}10} &= \frac{\Lambda_{2'2'3'3'}}{R^2\sigma_0}.
\end{aligned} \right\} \tag{3.6.28}$$

We can easily show that

$$\left. \begin{aligned}
\mathscr{K}_{110}^{110} &= \frac{1}{2}\mathscr{K}_{100}^{100}, \\
\mathscr{K}_{110}^{001} &= \mathscr{K}_{100}^{010}, \\
\mathscr{K}_{110}^{\bar{1}10} &= \frac{1}{2}\mathscr{K}_{100}^{100} + \mathscr{K}_{100}^{010}.
\end{aligned} \right\} \tag{3.6.29}$$

These equations are usually known as the symmetry relationships.

In exactly the same way, we shall now transform Eq. (3.6.21) to a new system of coordinates in accordance with Fig. 3.62. The matrix for this transformation is

$$A^i_{i'} = \frac{1}{\sqrt{3}} \begin{vmatrix} 1 & -\dfrac{\sqrt{6}}{2} & -\dfrac{\sqrt{2}}{2} \\ 1 & \dfrac{\sqrt{6}}{2} & -\dfrac{\sqrt{2}}{2} \\ 1 & 0 & \sqrt{2} \end{vmatrix}.$$

We thus obtain the transport coefficients

$$\left. \begin{aligned} \mathscr{K}^{111}_{111} &= \frac{\Lambda_{1'1'1'1'}}{R^2\sigma_0}, \\ \mathscr{K}^{\bar{1}10}_{110} &= \frac{\Lambda_{1'1'2'2'}}{R^2\sigma_0}, \end{aligned} \right\} \tag{3.6.30}$$

and hence:

$$\left. \begin{aligned} \mathscr{K}^{111}_{111} &= \frac{1}{3} \mathscr{K}^{100}_{100}, \\ \mathscr{K}^{\bar{1}10}_{111} &= \frac{B}{9} \frac{(8K^2_m \langle \tau^3_{\perp} \rangle + 11K_m \langle \tau^2_{\perp}\tau_{\parallel} \rangle + 8 \langle \tau_{\perp}\tau^2_{\parallel} \rangle)}{K_m (K_m \langle \tau^2_{\perp} \rangle + 2 \langle \tau_{\perp}\tau_{\parallel} \rangle)^2} - 1. \end{aligned} \right\} \tag{3.6.31}$$

The formulas for the transport coefficients (3.6.27) and (3.6.31) simplify considerably if we assume that K_T is independent of the energy, i.e., if $\tau_{\parallel}$ and $\tau_{\perp}$ depend in the same way on the energy. The expressions for $\mathscr{K}^{mnp}_{ikl}$, corresponding to this case, are given in Table 3.7.

In the case of the mixed anisotropic scattering, it is necessary to add the reciprocals of like components of the relaxation tensors in accordance with Eq. (3.3.67). The values of $\tau^{\parallel}_i$ and $\tau^{\perp}_i$ are taken from Eqs. (3.2.25) and (3.2.26), while the values of $\tau^{\parallel}_L$ and $\tau^{\perp}_L$ are taken from Eq. (3.3.64). Allowance for the anisotropy in the mixed scattering case gives two parameters a_1 and a_2, which replace Eq. (3.2.23)

$$a^2_1 = \frac{\tau^{\parallel}_{0L}}{\tau^{\parallel}_{0i}} \quad \text{and} \quad a^2_2 = \frac{\tau^{\perp}_{0L}}{\tau^{\perp}_{0i}}. \tag{3.6.32}$$

TABLE 3.7. Magnetoresistance Transport Coefficients of n-Type Germanium

Direction of vector j	Direction of vector H	$\mathcal{H}^{mnp}_{ikl}$ notation	Magnetoresistance transport coefficients
[100]	[100]	$\mathcal{H}^{100}_{100}$	$\dfrac{2}{3} \cdot \dfrac{(K-1)^2 \, (2K+1)}{K \, (K+2)^2} \dfrac{\langle \tau_\perp \tau^2_\parallel \rangle \langle \tau_\parallel \rangle}{\langle \tau_\perp \tau_\parallel \rangle^2}$
[100]	[010]	$\mathcal{H}^{010}_{100}$	$\dfrac{1}{3} \cdot \dfrac{(2K^2+5K+2)\,(2K+1)}{K\,(K+2)^2} \times$ $\times \dfrac{\langle \tau_\perp \tau^2_\parallel \rangle \langle \tau_\parallel \rangle}{\langle \tau_\perp \tau_\parallel \rangle^2} - 1$
[110]	[110]	$\mathcal{H}^{110}_{110}$	$\mathcal{H}^{110}_{110} = \dfrac{1}{2} \mathcal{H}^{100}_{100}$
[110]	[001]	$\mathcal{H}^{001}_{110}$	$\mathcal{H}^{001}_{110} = \mathcal{H}^{010}_{100}$
[110]	[1$\bar{1}$0]	$\mathcal{H}^{1\bar{1}0}_{110}$	$\mathcal{H}^{1\bar{1}0}_{110} = \dfrac{1}{2} \mathcal{H}^{100}_{100} + \mathcal{H}^{010}_{100}$
[111]	[111]	$\mathcal{H}^{111}_{111}$	$\mathcal{H}^{111}_{111} = \dfrac{1}{3} \mathcal{H}^{100}_{100}$
[111]	[110]	$\mathcal{H}^{\bar{1}10}_{111}$	$\dfrac{1}{9} \cdot \dfrac{(8K^2+11K+8)\,(2K+1)}{K\,(K+2)^2} \times$ $\times \dfrac{\langle \tau_\perp \tau^2_\parallel \rangle \langle \tau_\parallel \rangle}{\langle \tau_\perp \tau_\parallel \rangle} - 1$

Consequently, we have

$$\tau_\parallel = \frac{\tau^\parallel_{0L} \, (\varepsilon^*)^{3/2}}{(\varepsilon^*)^2 + a^2_1} \ \text{ and } \ \tau_\perp = \frac{\tau^\perp_{0L} \, (\varepsilon^*)^{3/2}}{(\varepsilon^*)^2 + a^2_3}. \tag{3.6.33}$$

However, if we assume approximately that $a^2_1 = a^2_2$, the substitution of the relationships (3.6.33) and of the Fermi function for f_0 in the expressions of Table 3.7 gives the magnetoresistance formulas listed in Table 3.8. Naturally, these formulas give the best results when $a < 1$ or $a > 1$, while in the region $a \approx 1$ we must use Eq. (3.6.32) and very complex integrals for $\langle \tau^i \rangle$.

If we now assume that $a^2 \ll (\varepsilon^*)^2$ or $a^2 \gg (\varepsilon^*)^2$ in the expressions in Table 3.8, we obtain $\mathcal{H}^{mnp}_{ikl}$, which are valid for the phonon scattering and the ion scattering separately (Table 3.9) [138].

TABLE 3.8. Magnetoresistance Transport Coefficients for Mixed Scattering of Carriers by Ionized Impurities and Acoustical Phonons

$\mathcal{H}_{ikl}^{mpn}$	Magnetoresistance transport coefficients	
	n-type germanium	
$\mathcal{H}_{100}^{100}$	$\dfrac{2}{3} \cdot \dfrac{(K_m - 1)^2 (2K_m + 1)}{K_m (K_m + 2)^2} \dfrac{\Phi_1(\mu^*, a)\, \Phi_3(\mu^*, a)}{\Phi_2^2(\mu^*, a)}$	
$\mathcal{H}_{100}^{100}$	$\dfrac{1}{3} \cdot \dfrac{(2K_m^2 + 5K_m + 2)(2K_m + 1)}{K_m(K_m + 2)^2} \dfrac{\Phi_1(\mu^*, a)\, \Phi_3(\mu^*, a)}{\Phi_2^2(\mu^*, a)} - 1$	
$\mathcal{H}_{111}^{111}$	$\dfrac{1}{9} \cdot \dfrac{(8K_m^2 + 11K_m + 8)(2K_m + 1)}{K_m(K_m + 2)^2} \dfrac{\Phi_1(\mu^*, a)\, \Phi_3(\mu^*, a)}{\Phi_2^2(\mu^*, a)} - 1$	
	n-type silicon	
$\mathcal{H}_{100}^{100}$	0	
$\mathcal{H}_{100}^{010}$	$\dfrac{(K_m^2 + K_m + 1)(2K_m + 1)}{K_m(K_m + 2)^2} \dfrac{\Phi_1(\mu^*, a)\, \Phi_3(\mu^*, a)}{\Phi_2^2(\mu^*, a)} - 1$	
$\mathcal{H}_{110}^{110}$	$\dfrac{1}{2} \cdot \dfrac{(K_m - 1)^2 (2K_m + 1)}{K_m(K_m + 2)^2} \dfrac{\Phi_1(\mu^*, a)\, \Phi_3(\mu^*, a)}{\Phi_2^2(\mu^*, a)}$	
$\mathcal{H}_{111}^{110}$	$\dfrac{1}{3} \cdot \dfrac{(2K_m^2 + 5K_m + 2)(2K_m + 1)}{K_m(K_m + 2)^2} \dfrac{\Phi_1(\mu^*, a)\, \Phi_3(\mu^*, a)}{\Phi_2^2(\mu^*, a)} - 1$	

In each of these cases, we may assume that $\mu^* \gg 0$ or $\mu^* \ll 0$, so as to obtain $\mathcal{H}_{ikl}^{mnp}$, valid for heavily doped [138] and impurity-free [137] semiconductors. For this purpose, we shall use the approximations for the Fermi integrals given in §2.5.

The expressions in Table 3.8 apply also to semiconductors with isotropic scattering ($K_T = 1$) and with isotropic dispersion laws ($K_m = K_T = 1$).

Magnetoresistance of Silicon. Silicon differs from germanium because the tensors (3.6.15)-(3.6.17) are defined in a coordinate system with axes parallel to the edges of a cubic cell. This is the consequence of the orientation of energy ellipsoids in Si along the [100] axes.

TABLE 3.9. Magnetoresistance Transport Coefficients for Scattering by Impurity Ions and Acoustical Phonons

$\mathscr{K}_{ikl}^{mnp}$	Scattering by impurity ions	Scattering by acoustical phonons
	n-type germanium	
$\mathscr{K}_{100}^{100}$	$\dfrac{16}{27}\cdot\dfrac{(K_m-1)^2\,(2K_m+1)}{K_m\,(K_m+2)^2}\times$ $\times\dfrac{F_2\,(\mu^*)\,F_5\,(\mu^*)}{F_{7/2}^2\,(\mu^*)}$	$\dfrac{8}{3}\cdot\dfrac{(K_m-1)^2\,(2K_m+1)}{K_m\,(K_m+2)^2}\times$ $\times\dfrac{f\,(\mu^*)\,F_0\,(\mu^*)}{F_{-1/2}^2\,(\mu^*)}$
$\mathscr{K}_{100}^{010}$	$\dfrac{8}{27}\cdot\dfrac{(2K_m^2+5K_m+2)\,(2K_m+1)}{K_m\,(K_m+2)^2}\times$ $\times\dfrac{F_2\,(\mu^*)\,F_5\,(\mu^*)}{F_{7/2}^2\,(\mu^*)}-1$	$\dfrac{4}{3}\cdot\dfrac{(2K_m^2+5K_m+2)\,(2K_m+1)}{K_m\,(K_m+2)^2}\times$ $\times\dfrac{f\,(\mu^*)\,F_0\,(\mu^*)}{F_{-1/2}^2\,(\mu^*)}-1$
$\mathscr{K}_{111}^{1\bar{1}0}$	$\dfrac{8}{81}\cdot\dfrac{(8K_m^2+11K_m+8)\,(2K_m+1)}{K_m\,(K_m+2)^2}\times$ $\times\dfrac{F_2\,(\mu^*)\,F_5\,(\mu^*)}{F_{7/2}^2\,(\mu^*)}-1$	$\dfrac{4}{9}\cdot\dfrac{(8K_m^2+11K_m+8)\,(2K_m+1)}{K_m\,(K_m+2)^2}\times$ $\times\dfrac{f\,(\mu^*)\,F_0\,(\mu^*)}{F_{-1/2}^2\,(\mu^*)}-1$
	n-type silicon	
$\mathscr{K}_{100}^{100}$	0	0
$\mathscr{K}_{100}^{010}$	$-\dfrac{8}{9}\cdot\dfrac{(K_m^2+K_m+1)\,(2K_m+1)}{K_m\,(K_m+2)^2}\times$ $\times\dfrac{F_2\,(\mu^*)\,F_5\,(\mu^*)}{F_{7/2}^2\,(\mu^*)}-1$	$4\cdot\dfrac{(K_m^2+K_m+1)\,(2K_m+1)}{K_m\,(K_m+2)^2}\times$ $\times\dfrac{f\,(\mu^*)\,F_0\,(\mu^*)}{F_{-1/2}^2\,(\mu^*)}-1$
$\mathscr{K}_{110}^{110}$	$\dfrac{4}{9}\cdot\dfrac{(K_m-1)^2\,(2K_m+1)}{K_m\,(K_m+2)^2}\times$ $\times\dfrac{F_2\,(\mu^*)\,F_5\,(\mu^*)}{F_{7/2}^2\,(\mu^*)}$	$2\cdot\dfrac{(K_m+1)^2\,(2K_m+1)}{K_m\,(K_m+2)^2}\times$ $\times\dfrac{f\,(\mu^*)\,F_0\,(\mu^*)}{F_{-1/2}^2\,(\mu^*)}$
$\mathscr{K}_{111}^{1\bar{1}0}$	$\dfrac{8}{27}\cdot\dfrac{(2K_m^2+5K_m+2)\,(2K_m+1)}{K_m\,(K_m+2)^2}\times$ $\times\dfrac{F_2\,(\mu^*)\,F_5\,(\mu^*)}{F_{7/2}^2\,(\mu^*)}-1$	$\dfrac{4}{3}\cdot\dfrac{(2K_m^2+5K_m+2)\,(2K_m+1)}{K_m\,(K_m+2)^2}\times$ $\times\dfrac{f\,(\mu^*)\,F_0\,(\mu^*)}{F_{-1/2}^2\,(\mu^*)}-1$

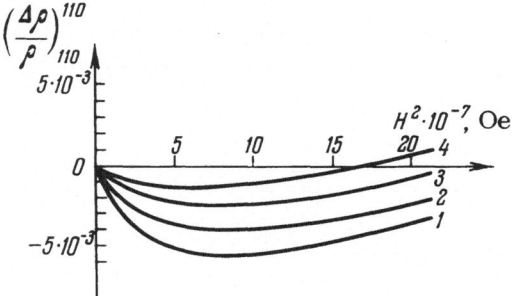

Fig. 3.63. Negative magnetoresistance of germanium heavily doped with arsenic. 1) $T = 1.88°K$; 2) $T = 2.5°K$; 3) $T = 3.3°K$; 4) $T = 4.2°K$.

Summation over all six ellipsoids gives the following expressions for the tensors $\{\Sigma\}$:

$$
\begin{aligned}
\Sigma_{ik} &= 2\,(\sigma_{11} + \sigma_{22} + \sigma_{33})\,\delta_{ik} = \\
&= -2aB_1\Big(2 - \frac{\langle \tau_\perp \rangle}{m_\perp^2} + \frac{\langle \tau_\parallel \rangle}{m_\perp m_\parallel}\Big)\,\delta_{ik}, \\[2mm]
\Sigma_{ikl} &= 2\,(\sigma_{123} + \sigma_{231} + \sigma_{312})\,\delta_{ikl} = \\
&= 2abB_1\Big(\frac{\langle \tau_\perp^2 \rangle}{m_\perp^3} + 2\,\frac{\langle \tau_\perp \tau_\parallel \rangle}{m_\perp^2 m_\parallel}\Big)\,\delta_{ikl}, \\[2mm]
\Sigma_{iklm} &= 2\,(\sigma_{1122} + \sigma_{1133} + \sigma_{3322}) = \\
&= 2ab^2B_1\Big(\frac{\langle \tau_\perp^3 \rangle}{m_\perp^4} + \frac{\langle \tau_\perp^2 \tau_\parallel \rangle}{m_\perp^3 m_\parallel} + \frac{\langle \tau_\perp \tau_\parallel^2 \rangle}{m_\perp^2 m_\parallel^2}\Big) \\
&\qquad\qquad\qquad\text{for } i = k \neq l = m, \\[2mm]
\Sigma_{iklm} &= 2\,(\sigma_{1212} + \sigma_{1313} + \sigma_{3232}) = \\
&= -6ab^2B_1\,\frac{\langle \tau_\perp^2 \tau_\parallel \rangle}{m_\perp^3 m_\parallel}\quad\text{for } i = l \neq k = m.
\end{aligned}
\right\} \quad (3.6.34)
$$

All the other components of the tensors $\{\Sigma\}$ are equal to zero.

The calculations of $\mathscr{K}^{100}_{100}$ and $\mathscr{K}^{010}_{100}$, carried out using Eqs. (3.6.23) and (3.6.25), give

$$
\left.
\begin{aligned}
\mathscr{K}^{100}_{100} &= 0. \\
\mathscr{K}^{010}_{100} &= -\frac{B(K_m^2 \langle \tau_\perp^3 \rangle + K_m \langle \tau_\perp^2 \tau_\parallel \rangle + \langle \tau_\perp \tau_\parallel^2 \rangle)}{K_m\,(K_m \langle \tau_\perp^2 \rangle + 2 \langle \tau_\perp \tau_\parallel \rangle)^2} - 1.
\end{aligned}
\right\} \quad (3.6.35)
$$

TABLE 3.10. Magnetoresistance Transport
Coefficients of n–Type Silicon

Direction of vector j	Direction of vector $\mathbf{H}$	$\mathscr{H}^{mnp}_{ikl}$ notation	Magnetoresistance transport coefficients
[100]	[100]	$\mathscr{H}^{100}_{100}$	0
[100]	[010]	$\mathscr{H}^{010}_{100}$	$\dfrac{(K^2+K+1)(2K+1)}{K(K+2)^2} \times$ $\times \dfrac{\langle \tau_\perp \tau_\parallel^2 \rangle \langle \tau_\parallel \rangle}{\langle \tau_\perp \tau_\parallel \rangle^2} - 1$
[110]	[110]	$\mathscr{H}^{110}_{110}$	$\dfrac{1}{2} \cdot \dfrac{(K-1)^2(2K+1)}{K(K+2)^2} \dfrac{\langle \tau_\perp \tau_\parallel^2 \rangle \langle \tau_\parallel \rangle}{\langle \tau_\perp \tau_\parallel \rangle^2}$
[110]	[001]	$\mathscr{H}^{001}_{110}$	$\mathscr{H}^{001}_{110} = \mathscr{H}^{010}_{100}$
[110]	[1$\bar{1}$0]	$\mathscr{H}^{1\bar{1}0}_{110}$	$\mathscr{H}^{1\bar{1}0}_{110} = \mathscr{H}^{010}_{100} - \mathscr{H}^{110}_{110}$
[111]	[111]	$\mathscr{H}^{111}_{111}$	$\mathscr{H}^{111}_{111} = \dfrac{4}{3} \mathscr{H}^{110}_{110}$
[111]	[1$\bar{1}$0]	$\mathscr{H}^{1\bar{1}0}_{111}$	$\dfrac{1}{3} \cdot \dfrac{(2K^2+5K+2)(2K+1)}{K(K+2)^2} \times$ $\times \dfrac{\langle \tau_\perp \tau_\parallel^2 \rangle \langle \tau_\parallel \rangle}{\langle \tau_\perp \tau_\parallel \rangle^2} - 1$

We can also show that $\mathscr{H}^{100}_{100} = 0$, in the following way. Since the principal directions of the mobility tensor (3.6.5) for each ellipsoid in silicon coincide with the axes [100], [010], and [001], the electron velocity $\mathbf{v}$ should be collinear with the vector $\mathscr{E}$ for the longitudinal magnetoresistance along any of the edges of a cubic cell. Consequently, $\mathbf{v}$ is collinear also with the vector $\mathbf{H}$. This means that the Lorentz force is equal to zero and therefore the magnetoresistance in the longitudinal case is also equal to zero.

Transforming the tensors (3.6.34) to a new system of coordinates (Fig. 3.60) and using Eq. (3.6.28), we obtain:

$$\left.\begin{aligned}
\mathscr{H}^{110}_{110} &= \frac{B}{2} \cdot \frac{(K_m^2 \langle \tau_\perp^3 \rangle - 2K_m \langle \tau_\perp^2 \tau_\parallel \rangle + \langle \tau_\perp \tau_\parallel^2 \rangle)}{K_m (K_m \langle \tau_\perp^2 \rangle + 2 \langle \tau_\perp \tau_\parallel \rangle)^2}, \\
\mathscr{H}^{001}_{110} &= \mathscr{H}^{010}_{110}, \\
\mathscr{H}^{1\bar{1}0}_{110} &= \mathscr{H}^{010}_{100} - \mathscr{H}^{110}_{110}.
\end{aligned}\right\} \qquad (3.6.36)$$

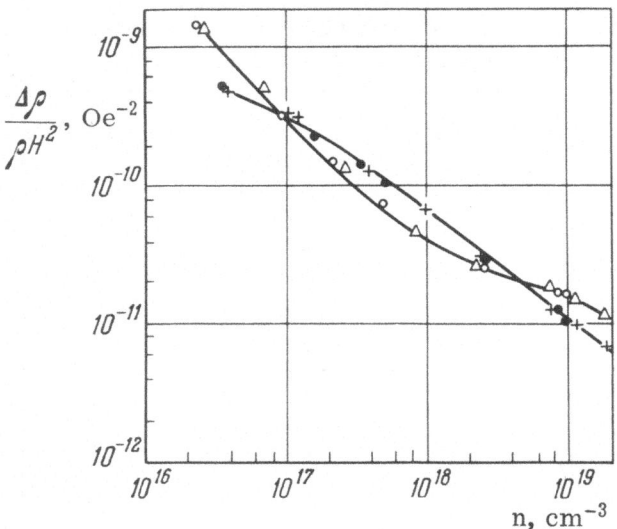

Fig. 3.64. Dependence of $\frac{\Delta\rho}{\rho H^2}\Big|_{100}^{100}$ and $\frac{\Delta\rho}{\rho H^2}\Big|_{110}^{110}$ on the electron density in heavily doped germanium. ●) $-\frac{\Delta\rho}{\rho H^2}\Big|_{100}^{100}$; +) $-2\frac{\Delta\rho}{\rho H^2}\Big|_{110}^{110}$ at 300°K; O) $-\frac{\Delta\rho}{\rho H^2}\Big|_{100}^{100}$, Δ) $-2\frac{\Delta\rho}{\rho H^2}\Big|_{110}^{110}$ at 77°K.

In order to calculate $\mathscr{H}_{111}^{111}$ and $\mathscr{H}_{111}^{\overline{1}10}$, we shall transform the tensors (3.6.34) in accordance with Fig. 3.61 and, using Eq. (3.6.30), we shall find:

$$\left.\begin{aligned}\mathscr{H}_{111}^{111} &= \frac{4}{3}\,\mathscr{H}_{110}^{110}, \\ \mathscr{H}_{111}^{110} &= \frac{B}{3}\cdot\frac{(2K_m^2\langle\tau_{\perp}^3\rangle+5K_m\langle\tau_{\perp}^2\tau_{\parallel}\rangle+2\langle\tau_{\perp}\tau_{\parallel}\rangle)}{K_m(K_m\langle\tau_{\perp}^2\rangle+2\langle\tau_{\perp}\tau_{\parallel}\rangle)^2}-1.\end{aligned}\right\} \quad (3.6.37)$$

The expressions for $\mathscr{H}_{ikl}^{mnp}$ Si, obtained on the assumption that K_T is independent of the energy, are given in Table 3.10 and the explicit forms of these expressions are listed in Tables 3.8 and 3.9.

Experimental Investigations of Magnetore- sistance of Heavily Doped Semiconductors. Investigations of the magnetoresistance of heavily doped semiconductors can be conveniently considered separately for low (helium) and higher (nitrogen and room) temperatures.

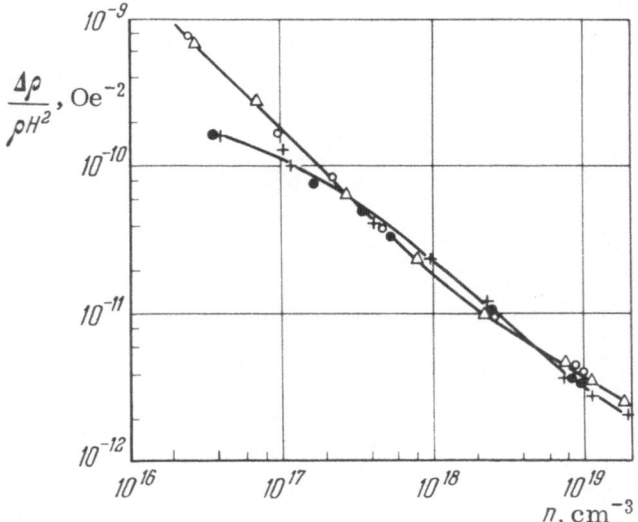

Fig. 3.65. Dependence of $\dfrac{\Delta\rho}{\rho H^2}\Big|_{100}^{010}$ and $\dfrac{\Delta\rho}{\rho H^2}\Big|_{110}^{001}$ on the electron density in heavily doped germanium. ●) $\dfrac{\Delta\rho}{\rho H^2}\Big|_{100}^{010}$, +) $\dfrac{\Delta\rho}{\rho H^2}\Big|_{110}^{001}$ at 300°K; O) $\dfrac{\Delta\rho}{\rho H^2}\Big|_{100}^{010}$, Δ) $\dfrac{\Delta\rho}{\rho H^2}\Big|_{110}^{001}$ at 77°K.

At 4.2°K, semiconductors have a negative magnetoresistance, i.e., the resistance decreases in a magnetic field (Fig. 3.63). It has been reported in [139-142] that the magnetoresistance of arsenic- and antimony-doped germanium consists of two components: a normal component representing an increase in the resistivity of a crystal in a magnetic field and an anomalous component representing $\Delta\rho < 0$. Such a negative magnetoresistance does not obey the theory given earlier in the present section.

The phenomenon of negative magnetoresistance can be considered on the assumption of the additional scattering of carriers by localized electron spins of partly isolated impurity atoms [143]. The upper limits of the weak-field range are quite different for the two components of magnetoresistance: the anomalous component reaches saturation while the normal component still shows the dependence $(\Delta\rho/\rho_0) \propto H^2$. Using this property, the two components have been separated by graphical differentiation of the experimental curves in Fig. 3.63 [144-147].

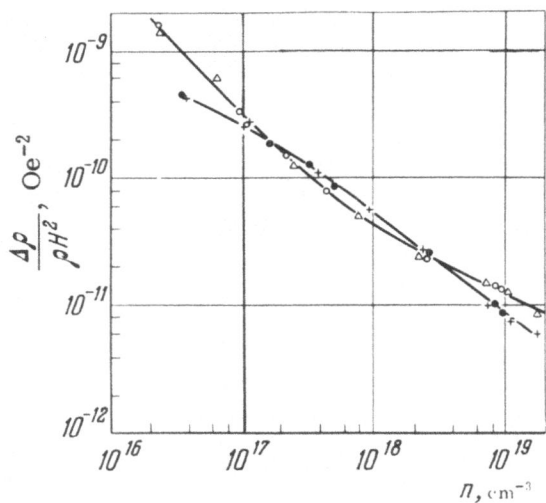

Fig. 3.66. Dependence of $\dfrac{\Delta\rho}{\rho H^2}\Big|_{110}^{1\bar{1}0}$ and $\left(\dfrac{1}{2}\dfrac{\Delta\rho}{\rho H^2}\Big|_{100}^{100}+\dfrac{\Delta\rho}{\rho H^2}\Big|_{100}^{01\,\prime}\right)$ on the electron density in heavily doped germanium. $\bullet$) $\left(\dfrac{1}{2}\dfrac{\Delta\rho}{\rho H^2}\Big|_{100}^{100}+\dfrac{\Delta\rho}{\rho H^2}\Big|_{100}^{010}\right)$; $+$) $\dfrac{\Delta\rho}{\rho H^2}\Big|_{110}^{110}$ at 300°K; $\bigcirc$) $\left(\dfrac{1}{2}\dfrac{\Delta\rho}{\rho H^2}\Big|_{100}^{100}+\dfrac{\Delta\rho}{\rho H^2}\Big|_{100}^{010}\right)$; $\triangle$) $\dfrac{\Delta\rho}{\rho H^2}\Big|_{110}^{110}$ at 77°K.

The dependence of the negative magnetoresistance of germanium on the magnetic field intensity has been investigated in the temperature range from 1.88 to 4.2°K [139].

The negative resistance has been observed also for other semiconductors: n–type InSb [144], p–type InSb [145], n–type GaAs [146–149], n–type InAs [150].

It is mentioned in [142] that the negative magnetoresistance of germanium is observed at such impurity concentrations and such temperatures that the main contribution to the conduction is made by the impurity band.

Analysis of the fairly extensive experimental data on the negative magnetoresistance of semiconductors shows that a reliable interpretation is still lacking.

At higher temperatures (77 and 300°K), the most detailed data are available for the magnetoresistance of arsenic–doped ger-

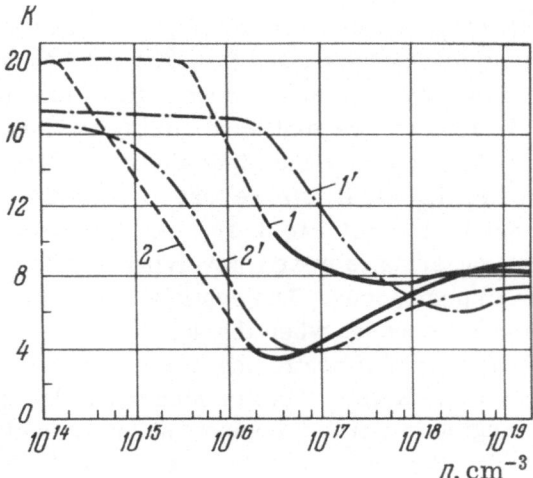

Fig. 3.67. Dependence of the effective anisotropy parameter on the impurity concentration. 1, 1') Experimental and theoretical curves for 300°K; 2, 2') experimental and theoretical curves for 77°K.

manium [140, 141, 151, 152]. Analysis of the experimental data has shown good agreement with the theory described in the present section. All the samples exhibit a quadratic dependence of the magnetoresistance on the magnetic field intensity in the range of fields up to $1 \cdot 10^4$ G.

The main experimental results are shown in Figs. 3.64–3.66. These results show that the symmetry relationships (3.6.29) and (3.6.31) are satisfied quite well at high impurity concentrations. At low impurity concentrations, their validity was established earlier [54].

This allows us to assume that the model of four ellipsoids of revolution, with their principal axes along the [111] directions, describes well the conduction band of germanium even at high impurity concentrations. Therefore, Fistul' and Andrianov [151] have calculated the anisotropy factor K from expressions given in Table 3.7. This can be done easily by eliminating from these expressions the transport integrals (Fig. 3.67). The latter figure also includes the theoretical curves calculated using the formulas (3.3.67), (3.2.25), (3.2.26), and (3.3.64). In the calculation of these theoretical curves, the value of K_m has been assumed to be equal to 20, i.e., it has been assumed that K_m is the same for impurity-free and for heavily doped germanium. Curves 1 and 2 in Fig. 3.67 show clearly four regions. In the first region of curve 1 (free-carrier density $n \leq 4 \cdot 10^{15}$ cm^{-3}) carriers are scattered mainly by the acoustical vibrations of the lattice atoms and since the relaxation time is almost isotropic, i.e., $K_\tau \approx 1$, it follows that $K = K_m = 20$. In the carrier density range from $4 \cdot 10^{15}$ to $3 \cdot 10^{16}$ cm^{-3} (the second region in curve 1), the scattering is mixed, involving both the acoustical phonons and the impurity ions. The contribution of the scattering by charged centers increases when the dopant concentration is increased. Since this form of scattering is strongly anisotropic, it follows that $K_\tau > 1$ and $K < 20$. This is supported also by the fact that the second region in curve 2 appears at lower carrier densities $(2 \cdot 10^{14}$ cm$^{-3} < n < 2 \cdot 10^{16}$ cm$^{-3})$ because at liquid nitrogen temperature the contribution of the scattering by ionized impurities is greater. At still higher dopant levels (the third region in curve 1), the contribution of the ion scattering undoubtedly increases but the influence of the screening of ions by free electrons prevents any further decrease of K. This can be seen more clearly in Fig. 3.2 which shows that when 1/b increases (i.e., when the carrier density n increases) the anisotropy of the scattering by the impurity ions decreases.

It is evident from Fig. 3.67 that the difference between the theoretical and experimental curves for heavily doped samples $(n > 10^{17}$ cm$^{-3})$ does not exceed 25%. Since $\tau_i^{\parallel}$, $\tau_i^{\perp}$ $\tau_L^{\parallel}$, $\tau_L^{\perp}$, K_m, and μ^* have been calculated using values of $m_{\parallel}$ and $m_{\perp}$ determined by the cyclotron resonance method at liquid hydrogen temperature using undoped germanium samples, it is concluded in [48, 49] that the anisotropy of the effective masses in n-type germanium is not affected by an increase in the impurity concentration.

§3.7. Characteristic Features of Transport Phenomena in $A^{III}B^V$ Semiconductors

Semiconducting compounds of the $A^{III}B^V$ type differ from elemental semiconductors mainly because of a certain proportion of ionic bonds in their lattices. Therefore, the scattering of electrons by the optical phonons may play an important role in the transport phenomena in these compounds. This property makes the inter-

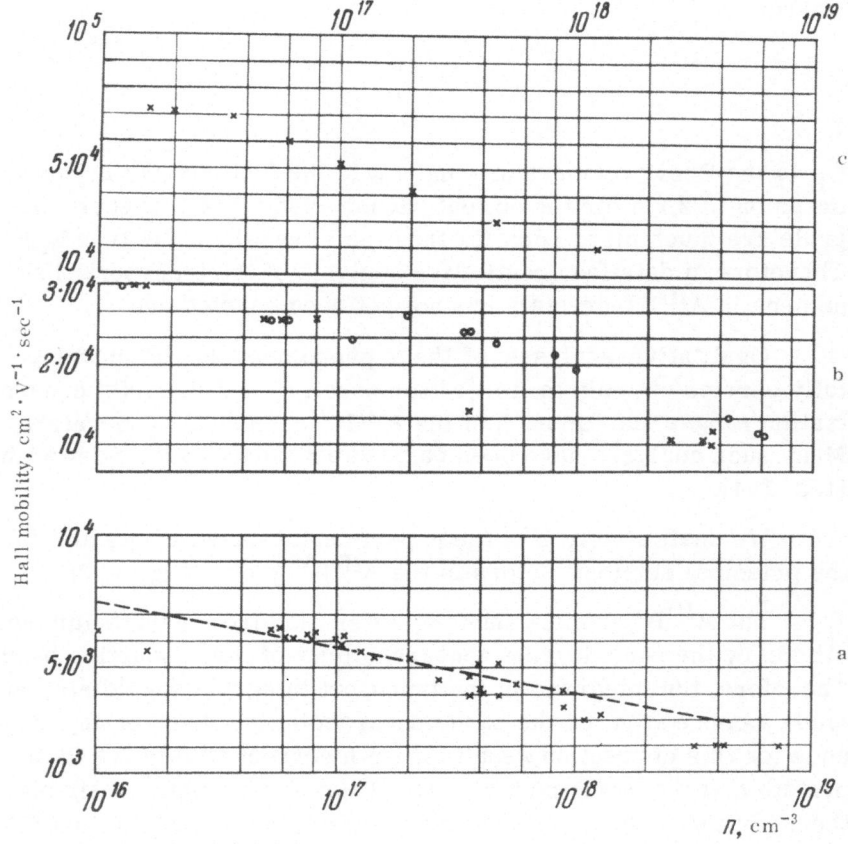

Fig. 3.68. Dependence of the Hall mobility of electrons in InSb, InAs, and GaAs on the carrier density. O) Samples subjected to heat treatment; x) samples not subjected to heat treatment. a) n-type GaAs [153]; b) n-type InAs [153]; c) n-type InSb [155].

pretation of the transport phenomena in A^{III}B^{V} semiconductors extremely difficult. A quantitative analysis of the transport phenomena can be carried out only using the transport integrals, in which the integrand contains a relaxation time which allows for all three scattering mechanisms, i.e., expressions of the type

$$\tau = \frac{\tau_{0L}\,(\varepsilon^*)^{3/2}}{(\varepsilon^*)^2 + b^2\,(\varepsilon^*)^{3/2} + a^2} \quad \text{for } T < \theta_D,$$

$$\tau = \frac{\tau_{0L}\,(\varepsilon^*)^{3/2}}{(\varepsilon^*)^2 + b^2\varepsilon^* + a^2} \quad \text{for } T > \theta_D,$$

$$(3.7.1)$$

where

$$b^2 = \frac{\tau_{00}}{\tau_{0i}},$$

θ_D is the Debye temperature and τ_{00} is the factor $\tau_0(T)$ in the expression (3.2.21) for the optical phonon case. The transport integrals are much more complex than the integrals of the type $\Phi_n(\mu^*, a)$. Therefore, a detailed quantitative analysis of the transport phenomena in $A^{III}B^V$ crystals has not yet been carried out.

Qualitative analyses of these phenomena can be meaningfully carried out only in limited temperature and impurity concentration ranges and for each of the $A^{III}B^V$ compounds separately. Many such analyses have been carried out and they are reviewed in [153, 154].

We shall simply give the experimental dependences u(n) for the principal semiconductors in the $A^{III}B^V$ group (Fig. 3.68).

The $A^{III}B^V$ semiconductors differ also from germanium and silicon by the considerable nonparabolicity of the conduction band. Therefore, the integrand in the transport integrals should also include a correction for the band nonparabolicity. Lack of any allowance for this correction results in an incorrect qualitative interpretation of the experimental data. Thus, for example, a simple theory of the Nernst–Ettingshausen effect (§3.5) predicts that $Q^\perp$ is positive for the scattering of carriers by acoustical vibrations of the lattice atoms and negative for the scattering by impurity ions. Initially, the experimental data on $Q^\perp(n, T)$ of $A^{III}B^V$ compounds (Figs. 3.56 and 3.57) have been interpreted in this way.

However, when the problem is solved allowing for the conduction-band nonparabolicity [156-159] it is found that the Nernst–Ettingshausen coefficient in the strongly degenerate case is defined as

$$Q \propto \frac{1}{2} - r - \left(1 - \frac{m_2}{m_3}\right), \tag{3.7.2}$$

where the following notation is used:

$$m_2 = \left| \hbar^2 k \left(\frac{dE}{dk}\right)^{-1} \right|_{E=\mu}; \quad m_3 = \left| \hbar^2 \left(\frac{d^2E}{dk^2}\right)^{-1} \right|_{E=\mu}.$$

If the conduction band is parabolic, then $m_2 = m_3$ and $Q^{\perp} \propto (1/2 - r)$, i.e., $Q^{\perp}$ is positive only when $r = 0$.

For a nonparabolic band, $m_2 \neq m_3$, and therefore the coefficient $Q^{\perp}$ may be positive even at values of r not equal to zero, i.e., the sign of $Q^{\perp}$ does not give unambiguous information on the scattering mechanism.

Chapter 4

Optical Properties of Heavily Doped Semiconductors

§4.1. Absorption of Light

Absorption Coefficient. The absorption of light by
a solid is represented by an absorption coefficient α, which is defined by the expression

$$\frac{J}{J_0} = \frac{(1 - R)^2}{e^{\alpha d} - R^2 e^{-\alpha d}}, \qquad (4.1.1)$$

where the left-hand side represents the ratio of the intensities of
the transmitted J and incident J_0 light, while R on the right-hand
side represents the coefficient for a single reflection from the surface of a sample of thickness d.

Spectral Dependence of the Absorption Coefficient. The spectral dependence of the absorption coefficient of such materials as Ge or Si is shown in Fig. 4.1. To explain it, we must recall the energy band diagram of, for example,
germanium (Fig. 1.4). When the energy of the incident light quanta
reaches a value sufficient to knock out an electron from the valence
band and to transfer it to the conduction band ($h\nu \gg E_g$), the absorption coefficient α rises rapidly. Since the absolute minimum of the
conduction band on the k axis does not lie opposite the valence band
maximum, the transfer of an electron involves a change in its initial crystal momentum. Such a change requires the participation,
in addition to a photon and an electron, of some third particle which

207

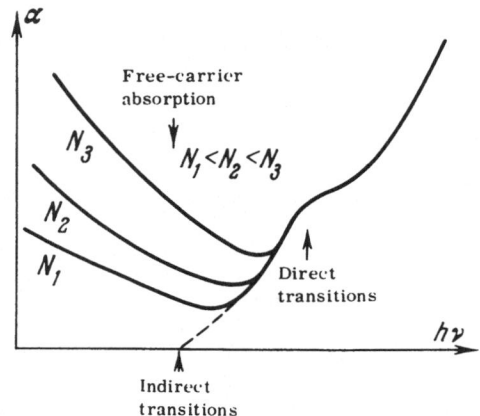

Fig. 4.1. Spectral dependence of the absorption coefficient of light in Ge-type crystals (N represents the impurity concentration).

can carry away some of the crystal momentum. This third particle may be a phonon or an impurity ion. Transitions in which a third particle participates are known as indirect.

A further increase in the energy of the incident quanta to a value $h\nu = E_0$ (cf. Fig. 1.4) gives rise to direct transitions, i.e., the direct transfer of electrons from the top of the valence band to levels in the conduction band having the same value of crystal momentum $\mathbf{k}$. There, the $\alpha(h\nu)$ curve in Fig. 4.1 has an inflection followed by a rapid rise of α.

If the energy of the incident quanta is so small that these quanta cannot transfer electrons from the valence to the conduction band, electrons may undergo transitions within the allowed band, and these transitions also require the absorption of light. This is known as the absorption by free carriers. This type of absorption is more important at high free carrier densities (cf. the left-hand branches of the curve in Fig. 4.1).

In heavily doped semiconductors, the free-carrier absorption is so strong that it effectively masks the fundamental absorption edge, which is usually employed to find the forbidden band width.

Absorption by Free Carriers. According to the classical theory of Drude [1], the real component of the high-frequency conductivity may be written in the form

$$\sigma = \frac{Ne^2}{m} \frac{\tau}{1 + \omega^2\tau^2} = \frac{\sigma_0}{1 + \omega^2\tau^2} , \qquad (4.1.2)$$

where τ is the relaxation time of electrons interacting with scattering centers; σ_0 is the electrical conductivity measured under dc conditions.

The absorption coefficient is related to the conductivity by the expression

$$\alpha = \frac{4\pi\sigma}{cn} , \qquad (4.1.3)$$

where n is the refractive index of light and c is the velocity of light in vacuum.

We have mentioned in Chap. 3 that the room-temperature values of τ for electrons and holes are 10^{-12}-10^{-13} sec. Consequently, at frequencies higher than 10^{12} cps the relationship $\omega\tau \gg 1$ is satisfied and the expression for the absorption coefficient becomes:

$$\alpha = \frac{4\pi}{cn} \frac{\sigma_0}{\omega^2\tau^2} . \qquad (4.1.4)$$

Comparison with experiment shows that the classical theory approximation is too rough for the description of the spectral dependence of the absorption coefficient in the free-carrier absorption region.

A quantum-mechanical treatment of such absorption is given in [2, 3]. Fan, Spitzer, and Collins [2] considered a crystal with a spherical conduction band, which does not represent germanium or silicon. They assumed that phonons are the chief scattering centers. However, in heavily doped semiconductors, scattering takes place mainly on the impurity ions.

Rosenberg and Lax [3] allowed for the nonsphericity of the conduction band, and considered scattering by the impurity ions. They employed a screened Coulomb potential for ions.

A deficiency of Rosenberg and Lax's treatment is the use of the Born approximation, which is not satisfied by the majority of semiconductors [4]. Meyer [5] deduced expressions for the absorption coefficient avoiding the Born approximation but he ignored the screening the Coulomb potential, although such screening is important in the case of the scattering by the impurity ions. The generalization to the screened potential has been described in [6], where the following dependence is given for σ_i:

$$\sigma_i(x) = 7 \cdot 10^{-28} \cdot \frac{NN_1}{v^3 (kT)^{3/2}} (1 - e^{-2x}) I(x), \qquad (4.1.5)$$

where $x = h\nu/2kT$; N_i is the number of ions; N is the number of electrons. For $x \gg 1$

$$I(x) = \frac{2m^* kT}{h^2} \left(1 + \frac{1}{6x} - \frac{1}{15x^2} + \frac{1}{35x^3} - \ldots \right). \qquad (4.1.6)$$

Substituting Eq. (4.1.5) into Eq. (4.1.3), we can easily obtain an expression for the absorption coefficient

$$\alpha \propto NN_i v^{-3}. \qquad (4.1.7)$$

A similar calculation carried out avoiding the Born approximation but with no allowance for the screening [4], gives a stronger frequency dependence for α:

$$\alpha \propto NN_i v^{-3.5}. \qquad (4.1.8)$$

If the scattering takes place not on impurity ions but on phonons, the expressions for α have the following form [7]:

for the acoustical phonons

$$\alpha \propto N v^{-3/2} \qquad (4.1.9)$$

and for the optical phonons

$$\alpha \propto N v^{-3}. \qquad (4.1.10)$$

These expressions show that the frequency dependence of the absorption coefficient is different for different scattering mechanisms. Moreover, the dependence of α on the carrier density or impurity concentration is also different for different scattering mechanisms. In the impurity scattering case, the absorption coefficient is proportional to the product of the impurity ion concentration and the electron density, i.e., it is proportional to the square of the carrier density. In the phonon scattering case, the absorption coefficient depends on the first power of the electron density.

Fundamental Absorption. The nature of the spectral dependence of the absorption coefficient is governed by two factors: the energy dependence of the density of states $\rho(E)$ in the allowed bands and the dependence of the probability of a transition on the energy of the incident light quanta.

Some workers have calculated the $\alpha = f(h\nu)$ dependence for Ge and Si [8-12] making the simplifying assumption that both bands participating in the process of absorption of light are spherical. Naturally, in the vicinity of the absorption edge the dependence $\alpha \propto \nu^2$ calculated in this way does not agree with the experimentally determined $\alpha = f(h\nu)$ curve.

Investigations of the $\alpha = f(h\nu)$ dependence near the absorption edge are particularly important in the presence of degeneracy, i.e., in the case of heavily doped semiconductors. In this case, the energy states lying near an extremum of the band in which the Fermi level is located are inactive, i.e., the inactive region is precisely that one for which the simplifying assumptions made in [8-12] can have any meaning.

An attempt has been made to analyze qualitatively the spectral dependence of the absorption coefficient near the edge without any simplifying assumptions about the dispersion law of electrons or the energy dependence of the transition probability [13, 14]. We shall now use the results obtained in these papers.

Direct Transitions. Let us assume that an electron, which is initially in the valence band (where its wave function is ψ_{ik_i}), absorbs a quantum of light and is transferred to a state with a wave function ψ_{fk_f} in the conduction band. The wave functions of the initial and final states are in the form of Bloch functions:

$$\left. \begin{array}{l} \psi_{ik_i}(r) = U_{ik_i}(r)\exp(ik_i,\ r), \\ \psi_{fk_f}(r) = U_{fk_f}(r)\exp(ik_f,\ r), \end{array} \right\} \qquad (4.1.11)$$

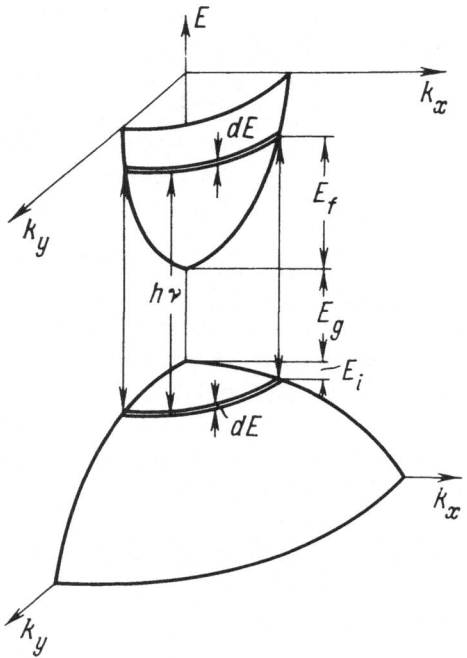

Fig. 4.2. Constant-energy surfaces in the k − E
space, illustrating direct transitions (a section
cut by the $k_z = 0$ plane is shown). The upper
parts represent states taking part in the direct
transitions at a given photon energy.

where $U_{ik_i}(\mathbf{r})$ and $U_{fk_f}(\mathbf{r})$ are potentials with a period equal to the
lattice parameter.

We shall assume that the photon momentum k_{ph} is considerably less than the electron momentum in the initial and final states

$$k_{ph} \ll \boldsymbol{k}_i, \ \boldsymbol{k}_f. \qquad (4.1.12)$$

This assumption is not justified only in a certain region in the immediate vicinity of a band maximum. Thus, for example, for Si
when the photon energy is 1 eV, the relationship (4.1.12) is valid
for states in the valence band which are separated only by 10^{-5} eV
from the band maxima. Then, the law of conservation of momen-

tum for direct transitions assumes the following very simple form:

$$k_i = k_f.$$

The probability of a transition of an electron with a wave function ψ_{ik_i} in the valence band to a state with a wave function ψ_{fk_f} in the conduction band will be represented by a matrix element of the direct-transition probability M_d. The explicit form of this element is not required in our treatment. Nevertheless we can mention that it includes the potentials U_{ik_i} and U_{fk_f} whose exact forms are not known. The total number of direct transitions is found by summing them over all states compatible with the law of conservation of energy, making an allowance for the energy dependence of the transition probability. For any dependence of E on k in the initial and final bands, each state in the valence band with a given value of the vector k_i corresponds only to one state in the conduction band with the same value of the vector k_i. Thus, the electron energy in the initial band E_i and in the final band E_f are defined exactly by the photon energy. In the simplest special case of isotropic bands with a quadratic dispersion law, at a given photon energy $h\nu$, direct transitions take place only between states lying on spheres with energies

$$E_i = \left(\frac{m_f}{m_i + m_f} \right) E \qquad (4.1.13)$$

and

$$E_f = \left(\frac{m_i}{m_i + m_f} \right) E,$$

where m_i and m_f are the effective density-of-states masses in the initial and final bands, and

$$E = h\nu - E_g. \qquad (4.1.14)$$

The energy level scheme in the $k - E$ space, illustrating direct transitions, is given in Fig. 4.2. The density of states in the valence band in the energy range from E_i to $E_i + dE_i$ is

$$\rho(E_i) \, dE_i = \frac{2\pi}{h^3}(m_i)^{3/2} \cdot E_i^{1/2} dE_i, \tag{4.1.15}$$

and the density of states participating in direct transitions at photon energies in the range $h\nu$, $h(\nu + d\nu)$, is

$$N_i(E) \, dE = \frac{2\pi}{h^3} \, m_i^{3/2} \left(\frac{m_f}{m_i + m_f} \right)^{3/2} E^{1/2} dE. \tag{4.1.16}$$

If the transition probability is independent of the energy, the dependence $\alpha(E)$ is given by the simple relationship:

$$\alpha(E) = \frac{n}{N_c} \, |M_d|^2 N_i(E) = AE^{1/2}. \tag{4.1.17}$$

Deviation from the parabolicity of the initial and final bands can be allowed for by introducing energy-dependent effective masses in Eqs. (4.1.13) and (4.1.16).

We must draw attention to the fact that the effective electron mass in all semiconductors is considerably smaller than the effective hole mass. Bearing this in mind, we find from Eq. (4.1.13) that the energy of holes represents only a small fraction of the energy defined by Eq. (4.1.14). Thus, for Ge, for which $m_n^* = 0.036m_0$ and $m_p^* = 0.34m_0$ near $k = 0$, the energy of the holes represents only 10% of $(h\nu - E_g)$, i.e., in this case, we can assume that we are within the limits of the parabolic part of the top of the valence band in considering the edge corresponding to direct transitions.

Thus, only the nonparabolicity of the conduction band may cause a deviation of the experimentally determined $\alpha(h\nu)$ curve for direct transitions from the curve described by Eq. (4.1.17). On the other hand, an energy dependence of the direct-transition probability may also result in a deviation from Eq. (4.1.17) near the absorption edge. If the transition at $k = 0$ is allowed, its probability is independent of the energy. It has been demonstrated in [12, 15] that the transition probability in the case of "forbidden" transitions increases proportionally to k^2. For parabolic bands, this gives an additional factor E in Eq. (4.1.17). For nonparabolic bands, the energy dependence of the transition probability is governed by the

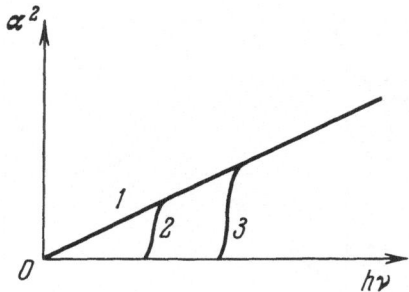

Fig. 4.3. Dependence $\alpha^2 = f(E)$ for direct
transitions between parabolic bands for
energy-independent transition probabilities.
1) Nondegenerate case; 2, 3) degenerate
cases ($\mu_3 > \mu_2$).

nature of the dispersion of electrons in a nonparabolic band and is
more complex. Thus, the band nonparabolicity is manifested in the
spectral dependence in two ways: by an increase in the density of
states compared with the density of states corresponding to a constant effective mass, and by an increase in the transition probability when the energy is increased. These two effects alter the absorption coefficient in the same way and, therefore, they cannot be
distinguished experimentally using only the optical data.

An increase in the transition probability when there is an increase in the energy is formally equivalent to an even greater increase in the density of states at a constant probability. This fact
has been used by Dubrovskii [13, 14] to describe the effective density of states, allowing for the transition probability by introducing
a "virtual" electron mass $m_{gi,f}$. Then Eq. (4.1.15) is replaced with:

$$\rho_{\text{eff}}(E_{i,f})\,dE_{i,f} = |M_d(E_{i,f})|^2\,\rho(E_{i,f})\,dE_{i,f} = \frac{2\pi}{h^3}\,[m_{gi,f}(E_{i,f})]^{3/2} \cdot E_{i,f}^{1/2}\,dE_{i,f}.$$

$$(4.1.18)$$

Since $E_f \gg E_i$, the structure of the conduction band plays the dominant role. Introducing $m_{gf}(E_f)$ in place of m_f, assuming on the
basis of Eq. (4.1.13) that $E_f \approx E$ (when $m_i > m_f$), and using Eq.
(4.1.18), we obtain the following expression for the effective density of states participating in direct transitions [this expression re-

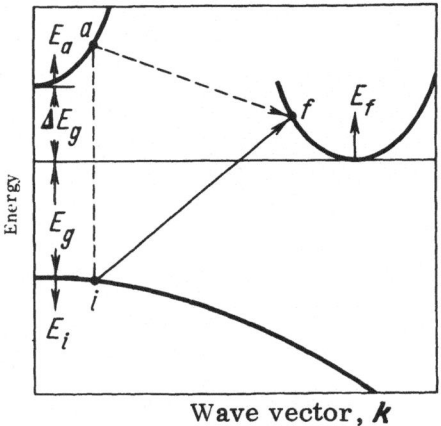

Fig. 4.4. Energy band scheme explaining an indirect $i - f$ transition through an intermediate state a.

places Eq. (4.1.16)]:

$$N_{\text{eff}}\,(E)\,dE = \frac{2\pi}{h^3}\left[\frac{m_i}{m_i + m_{gf}\,(E)}\right]^{3/2}[m_{gf}\,(E)]^{3/2}\,E^{1/2}dE\,. \qquad (4.1.19)$$

The absorption coefficient is now

$$\alpha\,(E) = \text{const} \cdot N_{\text{eff}}\,(E). \qquad (4.1.20)$$

Hence, it follows that a deviation of the experimentally observed $\alpha\,(E)$ curve from Eq. (4.1.17) for direct transitions near the absorption edge indicates an effective nonparabolicity of the conduction band, which represents a dependence of the "virtual" effective mass on the energy.

In heavily doped semiconductors, when one of the bands is degenerate in the Fermi sense, the states near an extremum of this band are filled with carriers. Therefore, the absorption edge shifts in the direction of higher energies. This is known as the Burshtein shift [16].

By way of example, we shall consider a p-type degenerate semiconductor which obeys Eq. (4.1.17). The minimum energy of electrons in the valence band of this semiconductor is equal to the Fermi level energy. On the basis of our discussion, we may conclude that the spectrum does not depart from the nondegenerate case at all energies $E_i > \mu$. Near $E_i = \mu$, there should be a rapid fall of the absorption to zero (Fig. 4.3).

Thus, the extrapolation of the absorption curve $\alpha^2 = f(h\nu)$ from $E_i \gg \mu$ to 0 always gives the value of the gap between the extremal points of the initial and final bands, and such extrapolation cannot be used to determine the Fermi level position. If we determine experimentally the whole $\alpha^2 = f(h\nu)$ curve, then its shift with respect to zero on the energy axis is generally still not equal to μ. It follows from Eq. (4.1.13) that this shift is $(1 + m_i/m_f)$ times larger than μ. For example, for germanium the shift may be an order of magnitude greater than the Fermi level energy.

The shift of the absorption edge of degenerate n-type germanium is almost equal to μ in the conduction band because the electron energy in the band is, as indicated by Eq. (4.1.13), almost equal to the energy given by Eq. (4.1.14).

Indirect Transitions. A scheme illustrating indirect transitions is given in Fig. 4.4. A transition from an initial state in a band i to a final state in a band f can be regarded as a two-stage process: an electron of momentum k_i close to zero undergoes, under the action of a quantum of light of energy $h\nu$, a direct transition to a virtual state in the conduction band and is then scattered by some scattering center so as to reach a final state f with a momentum k_f. In heavily doped semiconductors, the most likely scattering centers are impurity ions.

The residence time of electrons in the virtual state is negligible, and therefore the law of conservation of energy applies to the transition as a whole. The total probability of an indirect transition $|M_t|^2$ is governed by the probability of a direct transition $|M_d|^2$ and by the matrix element representing the probability of electron scattering $|M_s|^2$:

$$| M_t |^2 = \frac{| M_d |^2 | M_s |^2}{| \Delta E_g + E_a - E_f |^2}. \tag{4.1.21}$$

If we assume that M_s is independent of the energies of the initial and final electron states, we find that the energy dependence of the total probability of an indirect transition is governed only by the denominator in Eq. (4.1.21).

We shall consider silicon for which $E_{g0} - E_g \geq 1.5$ eV [17] (Fig. 1.5), i.e., the absorption edge satisfies the inequality

$$\Delta E_g = E_{g0} - E_g \gg E_a, E_f. \qquad (4.1.22)$$

In this case, the denominator in Eq. (4.1.21) may be assumed to be equal to $(E_{g0} - E_g)$ and, consequently,

$$|M_t| = \text{const} |M_d|.$$

As in the case of direct transitions, we obtain the following expression for the absorption coefficient:

$$\alpha_{if}(E) = \frac{n}{N_c} \frac{dN}{dt} = \text{const} |M_d|^2 N_{if}(E), \qquad (4.1.23)$$

where $N_{if}(E)$ is the density of state pairs in the initial and final bands. The law of conservation of energy is in this case:

$$E = E_i + E_f. \qquad (4.1.24)$$

We shall assume first that M_d is independent of the energy. For spherical bands with a quadratic dispersion law, the average number $\overline{N}_{if}(E)$ is:

$$\overline{N}_{if}(E) = \left(\frac{2\pi}{h^3}\right)^2 (2m_i)^{3/2} (2m_f)^{3/2} \int \int E_i^{1/2} E_f^{1/2} \delta(E - E_i - E_f) \, dE_i dE_f. \qquad (4.1.25)$$

Using the notation

$$\rho_i = \frac{2\pi}{h^3} (2m_i)^{3/2} \text{ and } \overline{\rho}_f = \frac{2\pi}{h^3} (2m_f)^{3/2},$$

we obtain, after first integration,

$$\overline{N}_{if}(E) = \overline{\rho}_i \overline{\rho}_f \int_0^{E_{i,\max}} E_i^{1/2} (E - E_i)^{1/2} \, dE_i. \qquad (4.1.26)$$

For a heavily doped degenerate n-type semiconductor, $E_{i,\max} = E - \mu$.

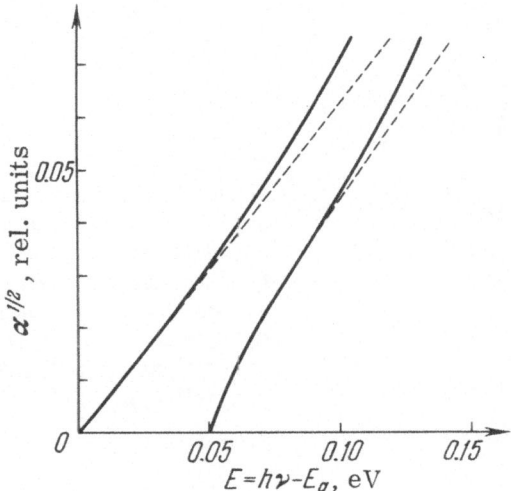

Fig. 4.5. Dependence $\alpha^{1/2} = f(E)$, calculated from
Eq. (4.1.23) for $\mu = 0.05$ eV and from Eq. (4.1.31)
for $\mu = 0$ with an effective mass given by Eq. (4.1.29)
and $a = 10$. The dashed curves represent Eqs. (4.1.27)
and (4.1.28) for a constant effective mass.

Integrating Eq. (4.1.26) with this upper limit, we obtain:

$$\bar{N}_{if}(E) = \frac{E^2}{8}\bar{\rho}_i\bar{\rho}_f\left[\frac{2}{E}\left(1 - \frac{2\mu}{E}\right)(E\mu - \mu^2)^{1/2} + \sin^{-1}\left(1 - \frac{2\mu}{E}\right) + \frac{\pi}{2}\right].$$

$$(4.1.27)$$

If there is no degeneracy ($\mu = 0$), the above expression becomes:

$$\bar{N}_{if}(E) = \frac{\pi}{8}\bar{\rho}_i\rho_f E^2.$$
$$(4.1.28)$$

In fact, the dispersion law for silicon is more complex and the
transition probability may depend on the energy. Consequently,
we should not expect the experimental data to agree with this simple
theory.

Dubrovskii [13, 14] has considered indirect transitions in silicon, introducing, by analogy with direct transitions, a "virtual"
effective mass m_{gd}, which represents an allowance for the usual
increase in the density of states because of the band nonsphericity

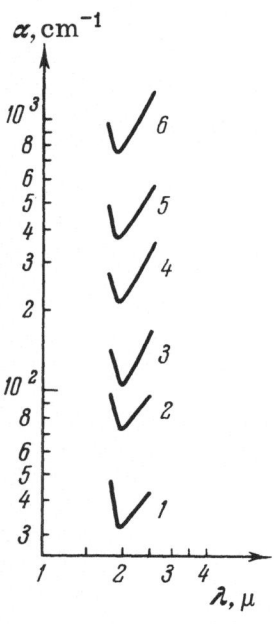

Fig. 4.6. Absorption coefficient of germanium heavily doped with arsenic (T = 300°K). The numbers alongside the curves indicate different electron densities (in cm^{-3}) at T = 300°K: 1) $5 \cdot 10^{18}$; 2) $8 \cdot 10^{18}$; 3) $1.2 \cdot 10^{19}$; 4) $2.3 \cdot 10^{19}$; 5) $4 \cdot 10^{19}$; 6) $6.2 \cdot 10^{19}$.

and for an additional increase in the density of states because of the energy dependence of the transition probability.

The dependence, obtained by Dubrovskii, of the "virtual" effective mass on the energy has the form

$$[m_{gd}(E_i)]^{3/2} = \bar{m}_{gd}^{3/2} + a\,(\bar{m}_{gd}E)^{3/2}; \qquad \bar{m}_{gd} = m_d,\qquad (4.1.29)$$

where

$$m_d^{3/2} = m_{i_1}^{3/2} + m_{i_2}^{3/2} + m_{i_3}^{3/2}.$$

In this case, the spin-orbital splitting of the third band is ignored.

The introduction of a variable mass indicates that the density of states in the band for an energy E_i measured from the top of the band is equal to the density of states for the same energy in a parabolic band with an effective mass given by Eq. (4.1.29). The total density of state pairs is obtained by integrating Eq. (4.1.26) after replacing $\bar{\rho}_i$ with $\bar{\rho}_i(1 + a E_i^{3/2})$:

$$N_{if}(E) = \bar{\rho}_i\bar{\rho}_f \left\{ \frac{E^2}{8} \left[\frac{2}{E}\left(1 - \frac{2\mu}{E}\right)(E\mu - \mu^2)^{1/2} + \sin^{-1}\left(1 - \frac{2\mu}{E_i}\right) + \right.\right.$$

$$+ \frac{\pi}{2} \Big] + \frac{4a}{105} \left[8E^{7/2} - \mu^{3/2}(35E^2 - 42E\mu + 15\mu^2) \right] \Big\}. \qquad (4.1.30)$$

In the absence of degeneracy ($\mu = 0$), this expression becomes

$$N_{if}(E) = \bar{\rho}_i \bar{\rho}_f \left(\frac{\pi}{8} E^2 + \frac{32}{105} aE^{7/2} \right). \qquad (4.1.31)$$

Figure 4.5 shows the dependences $\alpha^{1/2}(E)$ for silicon, calculated by Dubrovskii using Eqs. (4.1.23), (4.1.27), and (4.1.30) for $\mu = 0.05$ eV, and using Eqs. (4.1.28) and (4.1.31) for $\mu = 0$. The quantity a in Eq. (4.1.30) is an adjustment parameter, which should be selected by the experimenter. A suitable selection of the value of a, giving the best agreement with experiment, makes it possible to determine the nature of the dependence of the "virtual" effective mass on the energy. Analysis of indirect transitions in heavily doped silicon is made easier by the simplification of Eq. (4.1.21). In the case of germanium, the quantity $E_{g0} - E_g$, which occurs in the denominator, is only 0.15 eV [17] and, therefore, the inequality (4.1.22) is not satisfied by germanium. Consequently, in the case of germanium it is necessary to allow for the energy dependence of the probability of indirect transitions, governed by the denominator in Eq. (4.1.21). Such an allowance has been made by Hartman [18].

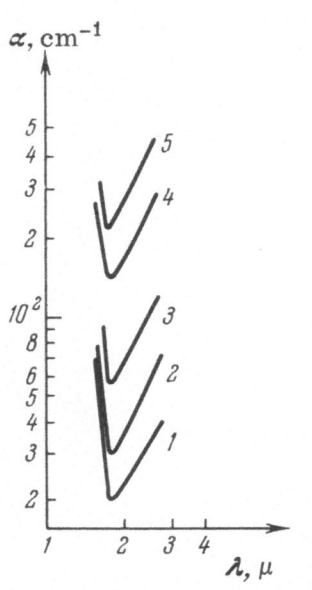

Fig. 4.7. Absorption coefficients of germanium heavily doped with arsenic (T = 4.2°K). The numbers alongside the curves have the same meaning as in Fig. 4.6.

Experimental Investigations of Absorption in Heavily Doped Germanium. The most detailed investigations of the absorption of light have been carried out on germanium samples heavily doped with As [19-21] and with Ga [22].

We shall consider first the results for n-type germanium. Figures 4.6 and 4.7 show the spectral

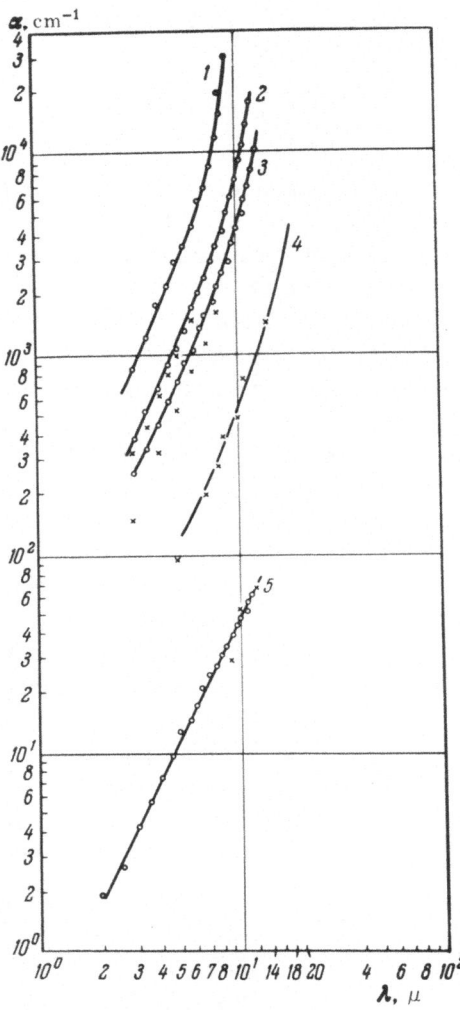

Fig 4.8. Spectral dependence of the absorption coefficient for the free-carrier absorption in n-type germanium   Electron density at 300°K (cm^{-3}): 1) $3.6 \cdot 10^{19}$; 2) $1.8 \cdot 10^{19}$; 3) $1.35 \cdot 10^{19}$; 4) $4.8 \cdot 10^{18}$; 5) $8 \cdot 10^{17}$.

TABLE 4.1

N, cm^{-3}	$5 \cdot 10^{18}$	$8 \cdot 10^{18}$	$1.2 \cdot 10^{19}$	$2.3 \cdot 10^{19}$	$4.0 \cdot 10^{19}$
$\Delta_{000} \cdot 10^3$, eV	28	32	38	52	60
$\Delta_{111} \cdot 10^3$, eV	59	57	66	79	93

dependences of the absorption coefficient, found by Pankove [21].
These results were deduced from measurements of the transmission of light through samples, allowing for the multiple reflection
of light from the surfaces. The nature of the spectral dependence
indicated two absorption processes: the absorption by free carriers
at long wavelengths and the absorption in the region of the fundamental absorption edge at short wavelengths. In order to isolate
the second absorption process, it was necessary to extrapolate the
data on the free-carrier absorption (the right-hand branches of the
curves in Figs. 4.6 and 4.7) in the direction of shorter wavelengths
and to subtract the extrapolation values from the values of α found
experimentally for these short wavelengths. The errors committed
in such an extrapolation could be very considerable. Therefore,
several investigations have been made of the law governing this
extrapolation, i.e., of the spectral dependence of the absorption co-

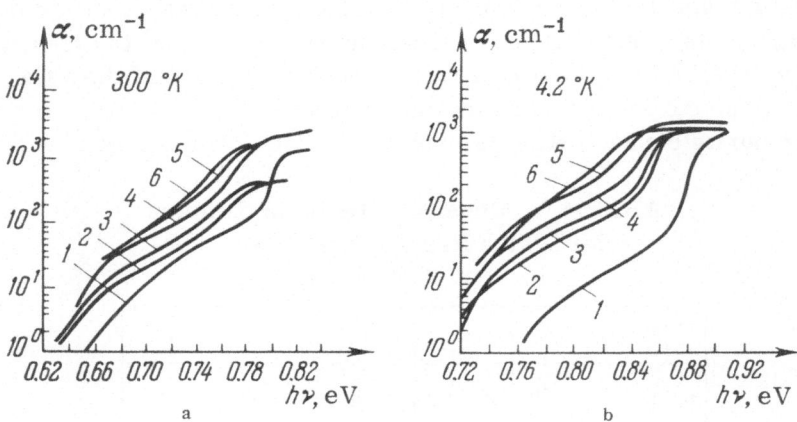

Fig. 4.9. Spectral dependence of the absorption coefficient for interband transitions in germanium doped with arsenic [21]. The numbers alongside the curves
have the same meaning as in Fig. 4.6

TABLE 4.2. Narrowing of Energy Gaps at $k = (0, 0, 0)$ and at $k = (1, 1, 1)$ in Heavily Doped Germanium [23]

Impurity	P	P	P	P	P	As
N, cm^{-3}	$2.4 \cdot 10^{18}$	$4.5 \cdot 10^{18}$	$9.6 \cdot 10^{18}$	$1.95 \cdot 10^{19}$	$4.3 \cdot 10^{19}$	$2.17 \cdot 10^{19}$
$\Delta_{000} \cdot 10^{3}$, eV (300° K)	16	25	25	37	66	40
$\Delta_{000} \cdot 10^{3}$, eV (200° K)	19	25	28	41	53	39
$\Delta_{000} \cdot 10^{3}$, eV (80° K)	11	19	25	38	51	39
$\Delta_{111} \cdot 10^{3}$, eV (300° K)	13	25	24	35	50	41
$\Delta_{111} \cdot 10^{3}$, eV (200° K)	21	33	34	45	65	54
$\Delta_{111} \cdot 10^{3}$, eV (80° K)	27	31	40	47	70	54

efficient for the free-carrier absorption. The data on this type of absorption in germanium samples heavily doped with various donor impurities are presented in Fig. 4.8. The same figure includes the spectral dependence of the absorption coefficient of a sample with a low concentration of arsenic atoms. These data, as well as those in Figs. 4.6 and 4.7, indicate that the nature of the dependence of the free-carrier absorption coefficient on the wavelength is not affected by doping and, right up to the highest impurity concentrations, we may assume that $\alpha \propto \lambda^2$.

Allowing for the free-carrier absorption, Pankove [21] obtained the spectral dependence of the absorption coefficient for interband transitions (Fig. 4.9). These curves show that the energy gap corresponding to a direct transition from the top of the valence band to the conduction-band minimum at $k = (0, 0, 0)$, decreases when the impurity concentration is increased. The forbidden band width

TABLE 4.3. Density of Holes in Samples for Which Results Are Given in Figs. 4.10 and 4.11

Sample No.	1	2	3	4	5
p, cm^{-3}	$2 \cdot 10^{15}$	$2 \cdot 10^{17}$	$6 \cdot 10^{17}$	$1.5 \cdot 10^{18}$	$6.3 \cdot 10^{18}$
Sample No.	6	7	8	9	10
p, cm^{-3}	$8.7 \cdot 10^{18}$	$1.6 \cdot 10^{19}$	$2.5 \cdot 10^{19}$	$3.6 \cdot 10^{19}$	$7.3 \cdot 10^{19}$

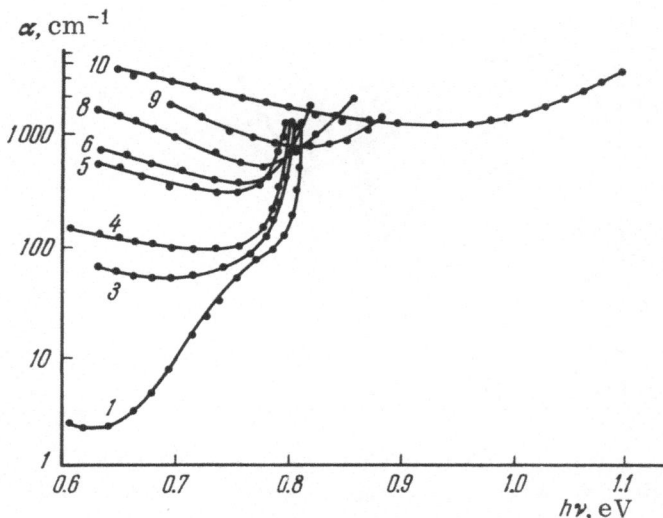

Fig. 4.10. Absorption coefficient of germanium heavily doped with gallium (293°K).

for indirect transitions from the top of the valence band to the conduction-band minimum at $k = (1, 1, 1)$ is reduced even much more by strong doping. The qualitative values of this reduction, Δ_{000} and Δ_{111}, found by Pankove [21] at 4.2°K, are listed in Table 4.1.

Similar data on the spectral dependence of the absorption coefficient of germanium heavily doped with phosphorus have been obtained by Haas [23]. Haas has also found that the energy gaps for $k = (0, 0, 0)$ and $k = (1, 1, 1)$ decrease when the doping level is increased (Table 4.2). However, absolute values of Δ_{000} and Δ_{111} found by Haas differ from those obtained by Pankove.

The most probable cause of the discrepancy between Pankove's and Haas's data is the indeterminacy of the extrapolation of the free-carrier absorption in the fundamental absorption region. However, the discrepancy may also be due to the different influence of phosphorus and arsenic impurities on the band structure of heavily doped germanium. Moreover, the results of these two workers are difficult to compare because they have been obtained at different temperatures.

We shall now consider the experimental results on the absorption in heavily doped p-type germanium. The experimental data, reported in [22], on gallium-doped germanium are presented

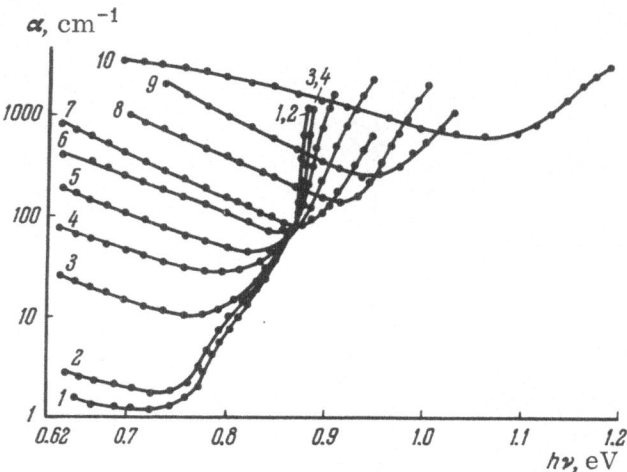

Fig. 4.11. Absorption coefficient of germanium heavily doped
with gallium (80°K).

in Figs. 4.10 and 4.11. The hole densities in the investigated samples are listed in Table 4.3.

The curves denoted by 1 in Figs. 4.10 and 4.11, which represent a lightly doped sample at 293 and 80°K, clearly exhibit indirect and direct interband transition regions, as well as the free-carrier absorption region. An increase in the hole density alters the parts of the curves corresponding to these three regions. The free-carrier absorption increases when the degree of doping is increased, and it reaches values comparable with the fundamental direct-transition absorption when gallium concentration becomes 10^{19} cm^{-3}.

It is evident from Figs. 4.10 and 4.11 that even when $p \approx 6 \cdot 10^{18}$ cm^{-3}, the direct and indirect transition regions are indistinguishable. Moreover, at these carrier densities, the absorption coefficient begins to depend less strongly on the frequency of the incident light. This can be seen more clearly from Fig. 4.12, which presents the same data as Fig. 4.11 but after the subtraction of the free-carrier absorption [22].

When the degree of doping is increased, the threshold energy corresponding to the absorption edge increases. This is explained in [22] by pointing out that in degenerate germanium electrons are

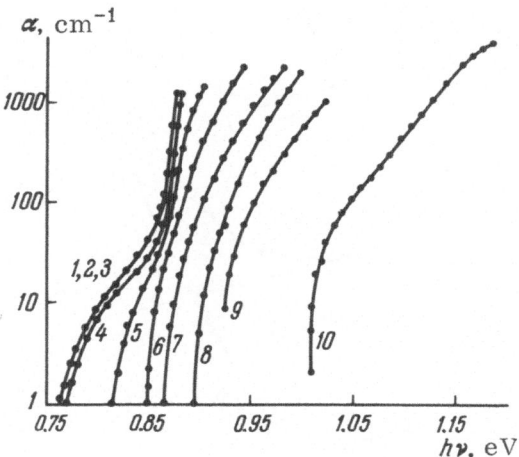

Fig. 4.12. Absorption coefficient for interband transi-tions in germanium doped with gallium [22] (80°K).

transferred by indirect transitions not from the top of the valence band but from the Fermi level (to within kT) in the conduction band, because all the energy levels $E > \mu$ are free of electrons. There-fore, in this case the energy corresponding to the absorption edge should increase by an amount equal to μ. In the direct transition case, the threshold energy should increase by a larger amount, which can be estimated from the law of conservation of energy [22]:

$$\frac{k^2}{2m_n^*} + \frac{k^2}{2m_p^*} + E_{g0} = h\nu.$$

If we bear in mind that the second term for degenerate p-type ger-manium is equal to μ, it follows that

$$h\nu = E_{g0} + \mu\left(1 + \frac{m_p^*}{m_n^*}\right).$$

Hence, the threshold energy should increase by a considerable amount because $m_p^* > m_n^*$. For germanium, this increase is al-most 10μ (and not 2μ as assumed in [22]). These qualitative con-siderations give a rough picture of the true change of the absorp-tion edge with the doping level. Since the inequality of Eq. (4.1.22) is not satisfied by germanium, the interpretations of the experi-

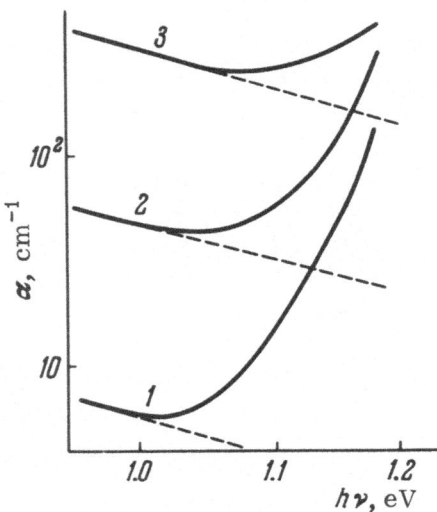

Fig. 4.13. Absorption spectrum of silicon doped with arsenic [13]. Electron densities (cm^{-3}): 1) $1.7 \cdot 10^{18}$; 2) $6 \cdot 10^{18}$; 3) $2.2 \cdot 10^{19}$.

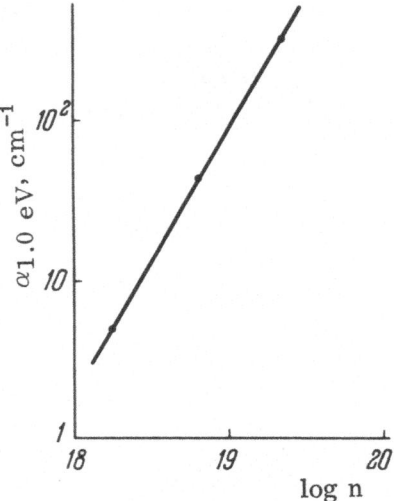

Fig. 4.14. Dependence of the absorption coefficient for quanta of 1.0 eV energy on the electron density in n-type Si, plotted from the results in Fig. 4.13.

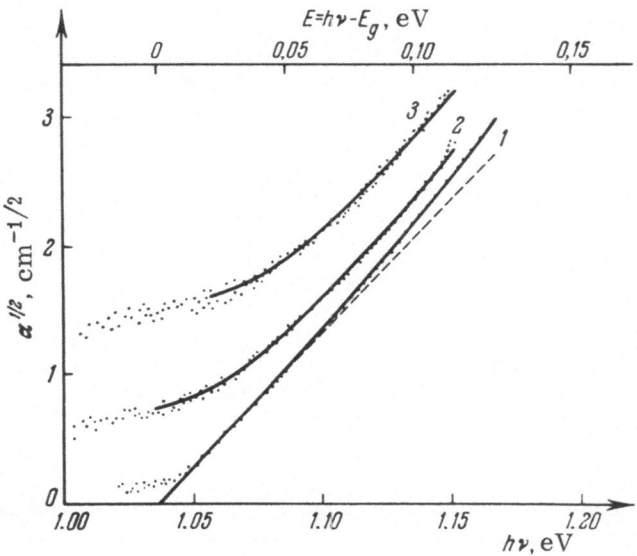

Fig. 4.15. Fundamental absorption edge of silicon heavily doped
with arsenic. 1) $1.7 \cdot 10^{18}$ cm^{-3}; 2) $6 \cdot 10^{18}$ cm^{-3}; 3) $2.2 \cdot 10^{19}$ cm^{-3}.

mental data for n-type [21, 23] and p-type germanium [22] may be
inaccurate.

Experimental Investigations of Absorption in
Heavily Doped Silicon. Figure 4.13 shows the dependence
$\alpha = f(h\nu)$, obtained at room temperature for n-type silicon samples
with high arsenic concentrations [13]. The frequency dependence
of the free-carrier absorption of heavily doped samples is described by the relationship $\alpha \propto \nu^{-3.5}$ It is evident from Fig. 4.14
that the free-carrier absorption increases with the electron density
as $n^{1.7}$, which is sufficiently close to the dependence [Eq. (4.1.7)]
expected in the case of the scattering of electrons by impurity ions.
However, the observed frequency dependence $\alpha \propto \nu^{-3.5}$ represents
impurity scattering without allowance for the screening of ions by
electrons [Eq. (4.1.8)]. Since an allowance for the screening is
essential in the case of heavily doped semiconductors, it is not clear
why the observed dependence should obey the law obtained without such an allowance. After extrapolation to the fundamental absorption edge and the subtraction of the extrapolated values from
the total values of the experimentally determined α, Dubrovskii [13]

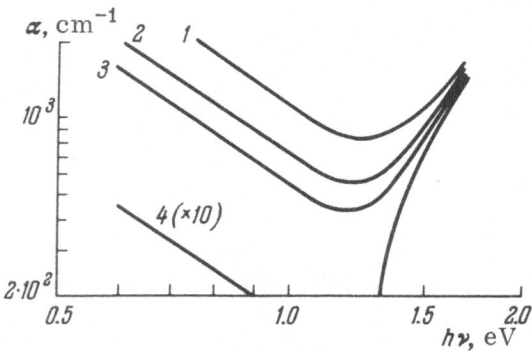

Fig. 4.16. Absorption spectrum of silicon heavily doped
with boron (300°K). 1) $2 \cdot 10^{20}$ cm^{-3}; 2) $9.7 \cdot 10^{19}$ cm^{-3};
3) $6 \cdot 10^{19}$ cm^{-3}; 4) $8 \cdot 10^{18}$ cm^{-3}.

obtained the results presented in Fig. 4.15. All the curves in this
figure have rectilinear parts at high values of hν. The forbidden
band width, found by the extrapolation of the rectilinear part of
curve 1 to zero absorption (E = hν − E$_g$ = 0), is 1.036 eV.

At first sight, the extrapolation of the two other curves indicates a decrease in the forbidden band. However, as pointed out
in [13], the extrapolation of these curves is unjustified. In the energy range hν < 1.036 eV, these curves exhibit absorption varying
slowly with the frequency. This additional absorption makes it impossible to estimate the forbidden band width by simple extrapolation of the rectilinear parts of curves 2 and 3 in Fig. 4.15.

Dubrovskii demonstrated by direct experiments [13] that this
additional absorption is due to the scattering of light in a sample.
Such is observed only in heavily doped n-type silicon samples with
electron densities of at least 10^{18} cm^{-3}. The intensity of the scattered light increases when the carrier density is increased.

The scattering may be due to second-phase precipitates which
can form in silicon at very high impurity concentrations [24]. The
continuous curves 2 and 3 in Fig. 4.15 represent the theoretical dependences $\alpha^{1/2} = f(h\nu)$. Curve 2 is obtained by adding the "absorption coefficient" representing scattering (this coefficient is 0.58
cm^{-1} for hν = 1.05 eV and varies with the frequency as $\alpha \propto \nu^{4.6}$) to
the coefficient calculated from Eq. 4.1.31 using a "virtual" effective
mass given by Eq. (4.1.29) and a parameter $a = 4.5$.

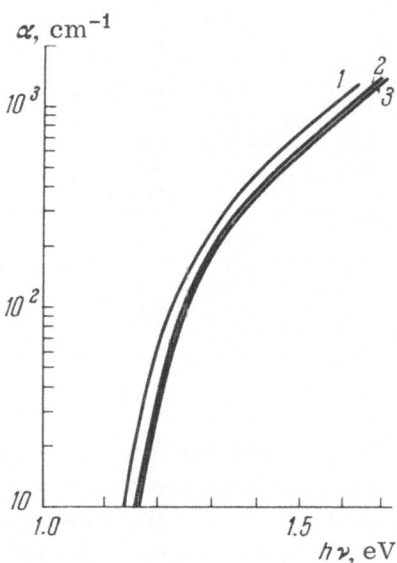

Fig. 4.17. Fundamental absorption edge of silicon heavily doped with boron. 1) 10^{18} cm^{-3}; 2) $6 \cdot 10^{19}$ cm^{-3}; 3) $2 \cdot 10^{20}$ cm^{-3}.

The continuous curve 3 is obtained similarly but in this case the "absorption coefficient" representing the scattering is now equal to 2.45 cm^{-1} for $h\nu = 1.05$ eV and the absorption coefficient for indirect transitions is calculated allowing for the degeneracy, i.e., using Eq. (4.1.30) with $\mu = 0.02$ eV. The values of $\bar{p}_i$, $\bar{p}_f$ and a, required in this calculation, are found from the rectilinear part of curve 1, which represents a relatively lightly doped sample. The continuous curves 2 and 3 obtained in this way are in good agreement with the experimental points.

Thus, the "virtual" effective mass method, proposed by Dubrovskii in the form of Eq. (4.1.29), is very successful. The agreement with experiment shows that the model of indirect transitions between a parabolic conduction band and a nonparabolic valence band, combined with the scattering by impurity ions, provides a correct description of the absorption in silicon.

The spectral dependence of the absorption coefficient for boron-doped p-type silicon samples is shown in Fig. 4.16. In contrast to n-type silicon, the parts of the curves corresponding to the

TABLE 4.4. Values of Absorption
Cross Section σ at $\lambda = 9 \mu$
and of the Power Exponent
in the Law $\alpha \propto \lambda^p$

Compound	n, cm^{-3}	$\sigma = \alpha/n$, cm^2	p	Source
InSb	$(1—3) \cdot 10^{17}$	2.3	2	[60]
InAs	$(0.3—8) \cdot 10^{17}$	3.6	3.0	[61]
InP	$(0.4—4) \cdot 10^{17}$	4.5	2.6	[55]
GaAs	$(1—10) \cdot 10^{17}$	4.7	3.1	[62]
AlSb	$(0.4—4) \cdot 10^{17}$	15	2	

free-carrier absorption obey the law $\alpha \propto \nu^{-2}$. Extrapolation to
larger values of $h\nu$ and the subtraction of the extrapolated quantities from the measured total absorption coefficient gives the spectral dependence $\alpha(h\nu)$ for the fundamental absorption (Fig. 4.17).
An increase of the hole density to 10^{20} cm^{-3} shifts the absorption
edge in the direction of higher energies by an amount approximately
equal to 0.03 eV.

Unfortunately, because of the large errors in the determination of small values of the absorption coefficient against a background of a strong free-carrier absorption, the absorption edge
cannot be determined directly at $\alpha < 10$ cm^{-1} for samples containing more than 10^{19} cm^{-3} holes. In this range of values of α, the
form of the $\alpha(h\nu)$ curve changes considerably in the degenerate case
and the changes are similar to those for n-type silicon.

Experimental Investigations of Absorption in
Heavily Doped $A^{III}B^V$ Semiconductors. An important
characteristic of the majority of $A^{III}B^V$ compounds is their small
effective electron mass in the conduction band. This tends to produce a strong degeneracy even at relatively low impurity concentrations. A consequence of this is a large Burshtein shift, i.e., optical transitions of electrons take place from the valence band to
the Fermi level in the conduction band, which rises with an increase
in the degree of doping. Experimental observations indicate an apparent broadening of the forbidden band. This effect has been observed for InSb [48], InAs [49-51], GaAs [52], and other $A^{III}B^V$
compounds.

A decrease in the forbidden band width due to the overlap of the impurity and intrinsic bands in $A^{III}B^V$ compounds can be observed only under conditions of small Burshtein shifts and this is possible if a heavily doped crystal is compensated. Such experiments have been carried out on InSb [53], InAs [54], InP [55], and GaAs [56, 57].

Another important property of $A^{III}B^V$ compounds is the conduction band nonparabolicity, because of which the absorption does not obey Eq. (4.1.17). A theory of the absorption with an allowance for the band nonparabolicity has been developed by Kane [58].

The free-carrier absorption in heavily doped $A^{III}B^V$ compounds is as important as in germanium or silicon. The law $\alpha \propto \lambda^p$ has been established experimentally for n-type $A^{III}B^V$ crystals. The values of p lie between 2 and 3.0 (cf. Table 4.4 [59]). These values indicate the considerable participation of the optical phonons in the scattering of electrons (p = 2.5 [64]) and a small role played by the scattering by ionized impurities (p = 3.5). The free-carrier absorption in p-type $A^{III}B^V$ crystals is usually independent of the wavelength. This is due to the existence of several subbands in the valence band of the majority of $A^{III}B^V$ compounds. The absorption cross section σ_p is more than an order of magnitude larger than σ_n.

An investigation of the absorption of light by free carriers in samples doped to various degrees has shown that the absorption coefficient is proportional to the carrier density. However, the absorption coefficient for the free-carrier absorption is also inversely proportional to the effective mass and the mass begins to depend on the carrier density when the electron gas is strongly degenerate. Therefore, above a certain carrier density, the linear dependence of the absorption coefficient on the carrier density (impurity concentration) is no longer observed.

A more detailed discussion of the optical properties of $A^{III}B^V$ compounds is given in [65].

§4.2. Reflection of Light

Reflection Coefficient. The reflection of light from solids is represented by a reflection coefficient R, equal to the ratio of the intensities of the reflected and incident light.

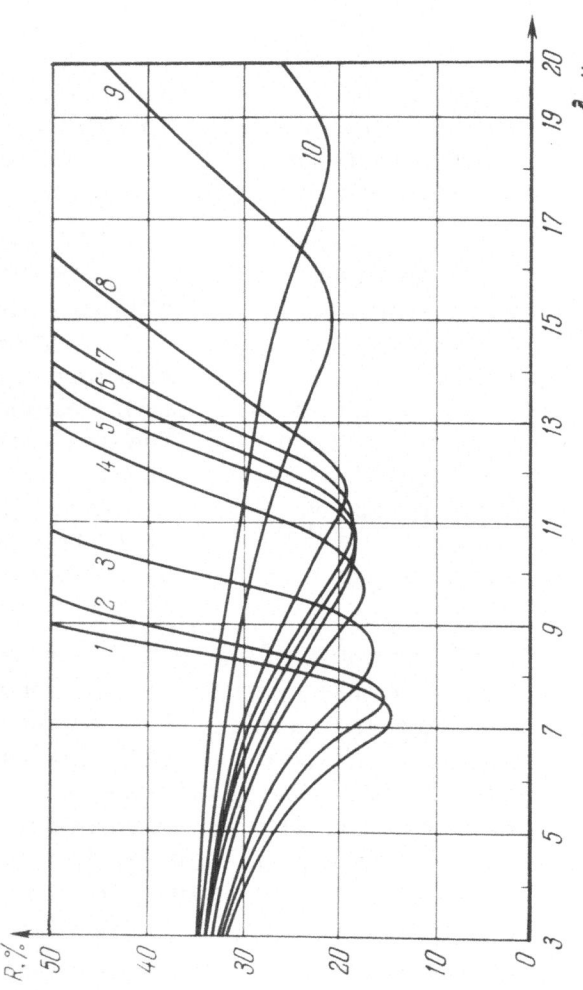

Fig. 4.18. Dependence of the reflection coefficient on the wavelength of incident light for germanium crystals with different arsenic concentrations (cm^{-3}): 1) $3.6 \cdot 10^{19}$; 2) $3.2 \cdot 10^{19}$; 3) $2.5 \cdot 10^{19}$; 4) $1.9 \cdot 10^{19}$; 5) $1.8 \cdot 10^{19}$; 6) $1.5 \cdot 10^{19}$; 7) $1.3 \cdot 10^{19}$; 8) $1.2 \cdot 10^{19}$; 9) $8.3 \cdot 10^{18}$; 10) $4.5 \cdot 10^{18}$.

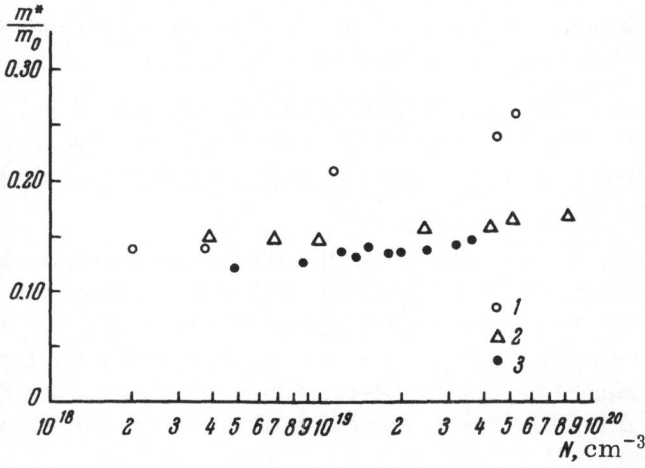

Fig. 4.19. Dependence of the effective electron mass on the concentration of arsenic in germanium crystals, plotted using data
taken from 1) [27]; 2) [28]; 3) [29].

TABLE 4.5.
Values for m_n^*/m_0
for Germanium
and Silicon Doped
with Various Donor
Impurities [28-30]

Semi-conductor Donor	Germanium	Silicon
As	0.160	0.267
P	0.166	0.300
Sb	0.140	0.258

The reflection coefficient is related to the refractive index and to the
absorption constant k by the well-known
Fresnel formula [17]

$$R = \frac{(n-1)^2 + k^2}{(n+1)^2 + k^2},\qquad (4.2.1)$$

where $k = \alpha/4\pi\lambda$ and λ is the wavelength
of the incident light.

Reflection in the Infrared
Region. Tolpygo [26] and later
Spitzer and Fan [25] suggested that the
effective carrier mass can be found
from measurements of the reflection
coefficient in the infrared region. If the absorption can be neglected,
the relationship between R and m* is

$$\left(\frac{1+\sqrt{R}}{1-\sqrt{R}}\right)^2 = \varkappa_0 - \frac{Ne^2\lambda^2}{m^*\pi c^2},\qquad (4.2.2)$$

where $\varkappa_0$ is the permittivity of an impurity-free semiconductor; N is the density of carriers in a sample.

The first application of this method to heavily doped germanium and silicon was reported in [27]: the effective electron mass is found to rise considerably when germanium is alloyed with arsenic.

In later investigations [28, 29], this conclusion was shown to be wrong. The reflection spectrum $R(\lambda)$ of germanium has the form shown in Fig. 4.18 and the values of m* for various carrier densities are shown in Fig. 4.19. We can see that, within the limits of the experimental error, the effective mass may be assumed to be constant. A similar result was obtained also in an investigation of heavily doped silicon [30].

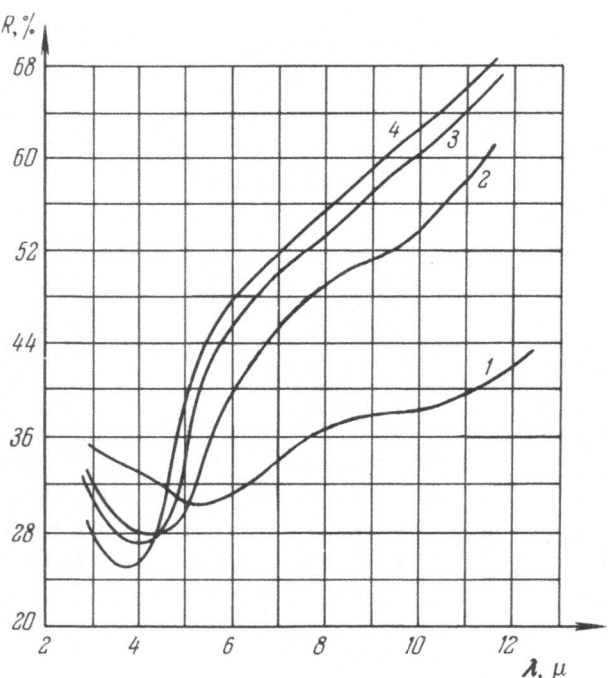

Fig. 4.20. Reflection coefficient of p-type germanium heavily doped with gallium (cm^{-3}): 1) $3.9 \cdot 10^{19}$; 2) $7.8 \cdot 10^{19}$; 3) $1 \cdot 10^{20}$; 4) $1.2 \cdot 10^{20}$.

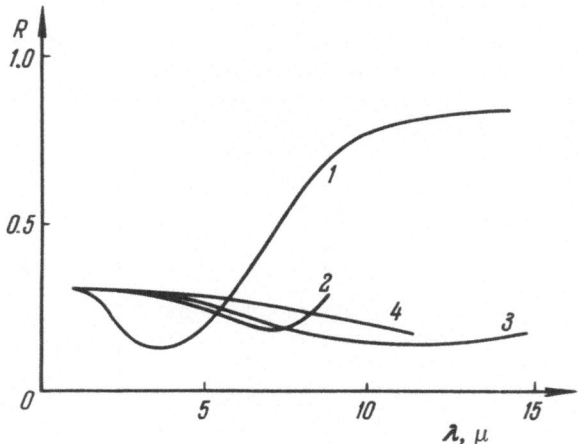

Fig. 4.21. Reflection of heavily doped silicon samples:
1) $2 \cdot 10^{20}$ cm^{-3} (B); 2) $5 \cdot 10^{19}$ cm^{-3} (B); 3) $2 \cdot 10^{19}$ cm^{-3}
(As); 4) $1 \cdot 10^{18}$ cm^{-3} (As).

An investigation of the reflection of light from germanium and
silicon samples, doped heavily with various impurities of the fifth
group in the periodic system [28-30], has indicated some differences between the values for the effective electron mass (Table 4.5).

The reflection spectra of heavily doped p-type germanium
samples, obtained by the author of the present monograph, are
shown in Fig. 4.20.

In contrast to those for n-type germanium, the $R(\lambda)$ curves
for silicon cannot be interpreted in a simple manner because of the
existence of several sub-bands in the valence band. The reflection spectra of silicon samples are presented in Fig. 4.21 (cf. [13]).

Reflection in Ultraviolet Region. Philipp and
Taft [31, 32] investigated the reflection spectra of germanium and
silicon and found maxima at energies of 2.5 and 4.4 eV for Ge and
at 3.5 eV for Si.

Phillips [33] suggested that the reflection maximum of Ge at
2.5 eV is associated with direct transitions from the valence to the
conduction band at the Brillouin zone boundary along the [111] di-
reaction between points L_3', and L_1 (cf. Fig. 1.4). This has made it possible to determine the forbidden band width at points other than $k = 0$.
Investigations of the ultraviolet reflection of a large number of

TABLE 4.6. Values of Effective
Electron Masses of $A^{III}B^V$ Crystals,
Found from Reflectivity Data

Compound	N, cm^{-3}	m_n^*/m_0
InSb	$3.5 \cdot 10^{17}$ $4.2 \cdot 10^{17}$ $1.2 \cdot 10^{18}$ $2.8 \cdot 10^{18}$ $4.0 \cdot 10^{18}$	0.023 0.029 0.032 0.040 0.041
InAs	$2.4 \cdot 10^{17}$ $1.3 \cdot 10^{18}$ $5.0 \cdot 10^{18}$	0.030 0.034 0.053
InP	$5 \cdot 10^{18}$	$0.2 - 0.26$
GaAs	$5 \cdot 10^{17}$ $5.4 \cdot 10^{18}$	0.079 0.086

materials [34-37] have yielded extensive information on the band structures.

The influence of strong doping on the structure of the ultra-violet reflection spectrum of germanium has been investigated by Cardona and Sommers [38]. They have found that a reflection maximum at 2.2 eV shifts in the direction of lower energies by an amount approximately equal to 0.03 eV at carrier densities of the order of 10^{19} cm^{-3}. They have observed such shifts in samples doped with donor and acceptor impurities.

On the other hand, in heavily doped silicon no shifts of the reflection maxima at 3.4 and 4.3 eV have been observed up to donor or acceptor concentrations of the order of 10^{20} cm^{-3} [13].

The transition $L_{3'} \rightarrow L_1$ in Ge takes place between the absolute minimum of the conduction band but does not involve the top of the valence band. It would seem that this situation would result in a different influence of the doping with donor and acceptor impurities on the minimum energy of this transition. In the case of n-type Ge, we should observe a considerable increase in the minimum transition energy, due to the Fermi degeneracy, i.e., the penetration of the Fermi level into the conduction band. The change in the minimum energy of the transition $L_{3'} \rightarrow L_1$ due to doping with acceptors should be solely due to the Coulomb interaction and a change in the lattice parameter, but these factors are small and there are no grounds for assuming that a change in the distribution of the extremal points of the allowed bands, due to the electron interaction and a reduction in the lattice parameter, will be different for n- and p-type materials. In fact, investigations of the fundamental absorption edge of Ge doped with donor [21, 23] and acceptor [22] impurities have shown that these effects are approximately equal. Thus, we should expect the energy position of the 2.2 eV maximum for n- and p-type Ge to shift by an amount equal to the Fermi level energy in n-type germanium. For a concentration of 10^{19} cm^{-3} at room temperature, this would represent ≈ 0.05 eV. Therefore, a decrease in the energy of this maximum in n- and p-type Ge by 0.03 eV, reported in [38], is unexpected. To explain this, it is suggested in [39] that in heavily doped germanium the curvatures of both allowed bands near the Brillouin zone boundary are the same along the [111] direction.

In view of this suggestion, an investigation of the reflection peak at 3.4 eV of heavily doped silicon would be of great interest. It is suggested in [39] that this peak corresponds to the transition $\Gamma_{25'} \rightarrow \Gamma_{15}$, but because the energies of the valence and conduction bands of silicon near k = 0 vary in opposite directions when the wave vector is increased, the absence of a shift of the reflection maximum at 3.4 eV cannot be due to the same band curvature as in the case of $L_{3'} \rightarrow L_1$ in Ge. On the other hand, analysis shows [13], that if the transition $\Gamma_{25'} \rightarrow \Gamma_{15}$ is responsible for the maximum at 3.4 eV, we should observe a shift by an amount

$$\Delta E = \frac{m_i + m_f}{m_f} \mu_i,$$

where m_i and m_f are the effective masses in the initial and final bands; μ_i is the Fermi level in the initial band.

An estimate for a sample of Si containing 10^{20} cm^{-3} of boron gives $\Delta E = 0.1$ eV. Such a shift can be easily measured. Therefore, the absence of a shift of the 3.4 eV maximum of heavily doped Si has forced Dubrovskii to regard this maximum as representing not the $\Gamma_{25'} \rightarrow \Gamma_{15}$ transition but rather the $L_{3'} \rightarrow L_1$ transition [13].

Reflection of Light from A^{III}B^V Crystals.

Studies of the reflection of light from n-type crystals of InSb [66,

67], InAs [67, 68], InP [55,69], and GaAs [62] have been carried out in order to determine the values of the effective electron mass. The main results of these investigations are given in Table 4.6.

§4.3.   Faraday Effect

The Faraday effect is the rotation of the plane of polarization of a light wave passing through a crystal parallel to an applied magnetic field. A detailed analysis of this phenomenon can be found, for example, in [40]. The rotation of the plane of polarization is represented by the angle of rotation θ, which is defined by [41]

$$\theta = \frac{\omega l}{2c}(n_R - n_L), \tag{4.3.1}$$

where ω is the angular frequency; l is the length of the sample; n_R and n_L are the refractive indices for the left-handed and right-handed circularly polarized components into which a plane-polarized wave can be resolved.

A positive value of θ represents the clockwise rotation of a wave propagated along the direction of the magnetic field away from the observer.

In the infrared range, the rotation of the plane of polarization may be due to free or bound carriers.

In heavily doped semiconductors, the Faraday effect is due to free carriers. A simple classical analysis of this case shows [42, 43] that θ has the form:

$$\theta = \frac{e^3 B N \lambda^2 l}{8\pi c^3 \varkappa_0 n m^{*2}}, \tag{4.3.2}$$

where N is the density of free carriers; B is the magnetic induction; $\varkappa_0$ is the permittivity of vacuum; λ is the wavelength, n is the refractive index of the light incident on a sample.

This expression is derived on the assumption that the following conditions are satisfied:

$$\omega^2 \tau \gg 1; \quad \omega^2 \gg \omega_c^2; \quad n^2 \gg k^2, \tag{4.3.3}$$

where k is the absorption coefficient of the sample; τ is the relaxation time; ω_c is the cyclotron frequency, defined by $\omega_c = eB/m^*$.

The relationships specified in Eq. (4.3.3) are satisfied by semiconductors in the range of wavelengths beyond the absorption edge.

A more rigorous treatment of the Faraday effect due to free carriers [44], carried out within the framework of the Boltzmann transport equation, gives an expression similar to Eq. (4.3.2), but the new expression allows for the band structure of the semiconductor. For semiconductors such as germanium, whose constant energy surface consists of four ellipsoids of revolution directed along [111], the expression for θ includes an anisotropy parameter K [45]:

$$\theta = \frac{Ne^3B\lambda^2 l}{8\pi^2c^3n\varkappa_0} \cdot \frac{K(K+2)}{3m_\parallel^2}. \tag{4.3.4}$$

For p-type germanium, the angle of rotation of the plane of polarization is

$$\theta = -\frac{e^3B\lambda^2 l}{8\pi^2c^3n\varkappa_0}\left(\frac{p_h}{m_h^2} + \frac{p_1}{m_1^2}\right), \tag{4.3.5}$$

where p_h and p_1 are the densities of heavy and light holes; m_h and m_1 are the masses of these holes. For bound electrons, the theory predicts (for $\omega > \omega_c$)

$$\theta = \frac{Ne^3\omega^2Bl}{2nc\varkappa_0(\omega_0^2 - \omega^2)^2 m^{*2}}, \tag{4.3.6}$$

which shows that the quadratic dependence of θ on the wavelength is no longer obeyed.

Characteristic features of the theoretical expressions for θ in the case of free-carrier absorption are the direct proportionality of θ to the carrier density and to the square of the wavelength of the incident light, as well as an inverse proportionality of θ to the value of $(m^*)^2$. In view of this, experimental investigations of the Faraday effect are of interest from two points of view: when the effective mass is known, the measured values of θ can be used to

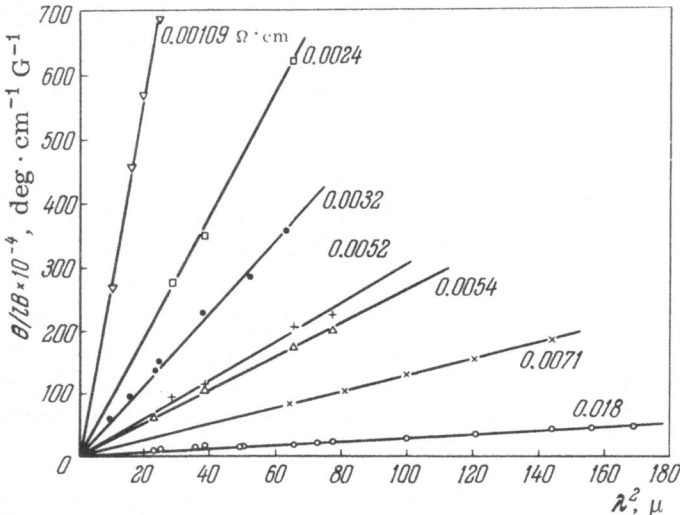

Fig. 4.22. Spectral dependence of the angle of rotation of the plane of polarization of light in germanium samples, according to [41]. The numbers by the curves indicate the resistivity at 300°K.

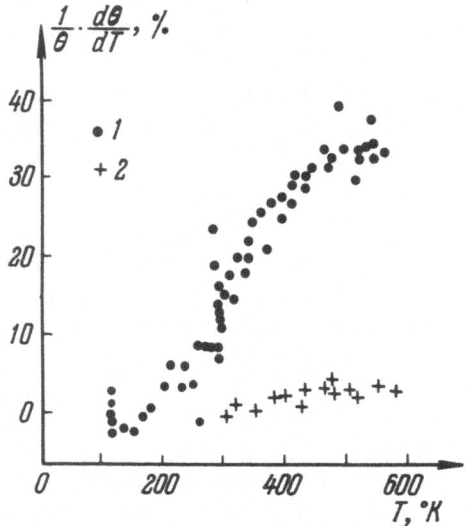

Fig. 4.23. Temperature dependence of the angle of rotation of the plane of polarization in germanium and silicon. 1) Germanium, $N = 5.8 \cdot 10^{17}$ cm^{-3}; 2) silicon, $N = 3.7 \cdot 10^{17}$ cm^{-3} ($T = 293°K$).

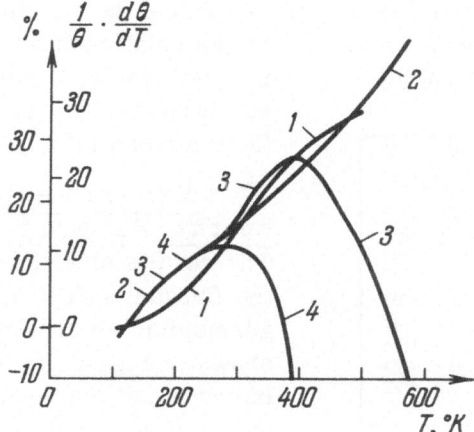

Fig. 4.24. Temperature dependence of the angle of rotation of the plane of polarization of $A^{III}B^V$ semiconductors (at $T = 293°K$). The scale on the right applies to curves 1 and 2; the scale on the left applies to curves 3 and 4. 1) InSb, $N = 0.75 \cdot 10^{18}$ cm^{-3}; 2) GaAs, $N = 1.2 \cdot 10^{17}$ cm^{-3}; 3) InAs, $N = 2.9 \cdot 10^{17}$ cm^{-3}; 4) InAs, $N = 3.3 \cdot 10^{16}$ cm^{-3}.

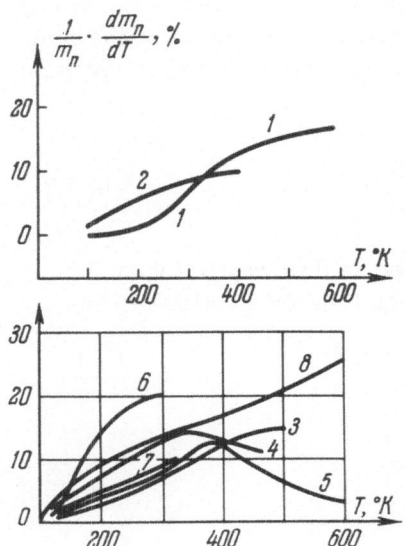

Fig. 4.25. Temperature dependence of the effective electron mass of Ge, Si, and $A^{III}B^V$ compounds, deduced from the Faraday effect. The numbers of the curves are explained in Table 4.7.

TABLE 4.7. Properties
of Samples for Which
Results Are Presented
in Fig. 4.25

Sample No.	Semi-conductor	N, cm^{-3} (293°K)
1	Ge	$5.8 \cdot 10^{17}$
2	Si	$3.7 \cdot 10^{17}$
3	InSb	$7.5 \cdot 10^{17}$
4	InAs	$3.7 \cdot 10^{16}$
5	InAs	$2.9 \cdot 10^{17}$
6	InP	$7.3 \cdot 10^{16}$
7	InP	$2.4 \cdot 10^{17}$
8	GaAs	$1.2 \cdot 10^{17}$

determine the carrier density N or the relationship between p_h and p_1; conversely, if the carrier density is known, we can find the effective mass m^*.

Experimental Results on the Faraday Effect.
The results of an investigation of the Faraday effect in doped n-type germanium single crystals are shown in Fig. 4.22. These results indicate that the quadratic dependence $\theta \propto \lambda^2$ is indeed obeyed. This dependence has been reported by many workers for a great variety of semiconducting materials. The temperature dependence of the Faraday effect, reported in [46, 47], is of considerable interest. This dependence indicates that the effective electron mass depends on temperature. The experimental results taken from [46, 47] are presented in Figs. 4.23 and 4.24.

Since the purpose of the investigations reported in [46, 47] was mainly to determine the dependence $m_n(T)$ rather than the value of m_n itself, only the temperature dependences of θ, N, and n were investigated carefully because the remaining quantities in Eq. (4.3.2) were independent of temperature (the thermal expansion of the sample was ignored). The temperature dependence of the effective mass was found from the following formula

$$\frac{1}{m_n}\frac{dm_n}{dT} = \frac{1}{2}\left\{\frac{1}{N}\frac{dN}{dT} - \left[\frac{1}{\theta}\frac{d\theta}{dT} + \frac{1}{n}\frac{dn}{dT}\right]\right\}. \qquad (4.3.7)$$

The values of $(1/m_n)\ dm_n/dT$ calculated in this way for various semiconductors are given in Fig. 4.25 and the properties of the samples used are listed in Table 4.7.

Chapter 5

Behavior of Impurities in Heavily Doped Semiconductors

§5.1. Preliminary Remarks

In the majority of cases with which semiconductor physics is concerned, the crystals contain relatively few impurities, which are separated by fairly large distances so that the interaction between them can be neglected, at least in the first approximation. In the heavy doping case, we must allow for the interaction between the impurities themselves, between the impurities and the host atoms, and between the impurities and the structure defects.

Heavily doped semiconductors can be regarded, even at the highest possible solubilities, as strongly diluted solutions. It follows from Figs. 5.1 and 5.2 that the maximum possible concentrations of donors are, respectively, $2 \cdot 10^{20}$ cm^{-3} in germanium and $2 \cdot 10^{21}$ cm^{-3} in silicon, representing 0.4 and 3.0 at.%, respectively. For this reason, we can apply some of the laws of the physical chemistry of ideal solutions to heavily doped semiconductors.

These laws are as follows.

1. The law of mass action, which is based on chemical thermodynamics; for reactions of the type

$$aA + bB + \ldots \rightleftharpoons mM + nN + \ldots \tag{5.1.1}$$

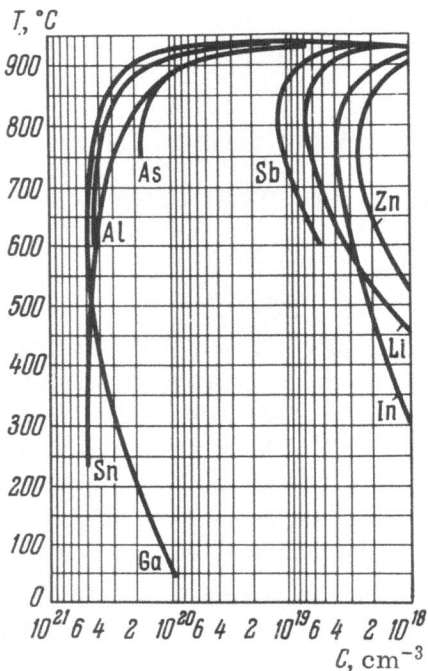

Fig. 5.1. Solubility of impurities in solid
germanium, according to [1].

this law is

$$\frac{C_M^m C_N^n}{C_A^a C_B^b \ldots} = K_C(T),\tag{5.1.2}$$

where C_A, C_B, ..., C_M, C_N, ..., are the concentrations of the components participating in the reaction; $K_C(T)$ is the equilibrium constant of the reaction, which depends only on temperature.

An important result follows from the law of mass action: under equilibrium conditions, the concentrations of all the substances taking part in a reaction are related so that a change in the concentration of one of them alters all the other concentrations. All the concentrations change until the equilibrium constant recovers its initial value.

The law of mass action given by Eq. (5.1.2) was derived originally for dilute solutions of weak electrolytes but it can be applied

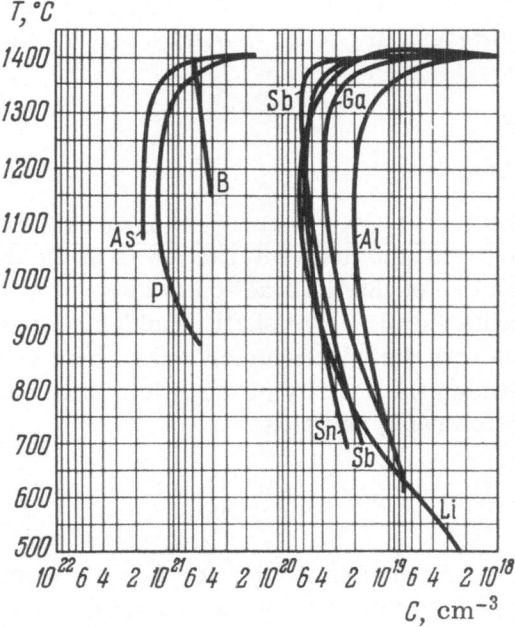

Fig. 5.2. Solubility of impurities in solid
silicon, according to [1].

also to more concentrated solutions if we introduce the concept of
activity. By the activity a_i, we shall understand the ratio of the
vapor pressure p_i of a given component in a given solution to the
vapor pressure p_i^0 of the pure component at the same temperature:

$$a_i = \frac{p_i}{p_i^0}. \tag{5.1.3}$$

2. Raoult's law, which is the ratio relating the relative
drop in the vapor pressure above the surface of a solution and the
relative concentration (molar fraction) of the solute in the solution:

$$\frac{p_0 - p}{p_0} = \frac{n}{N + n}, \tag{5.1.4}$$

where n and N are, respectively, the number of moles of the solute
and solvent, respectively.

3. Henry's law relates the solubility of a gas in a liquid x at a constant temperature to its pressure p:

$$x = Hp, \qquad (5.1.5)$$

where H is the Henry constant.

Main Interactions between Impurities. In lightly doped crystals, the main interaction is of the donor – acceptor type. The nature of this interaction is governed by the Coulomb attraction of unlike ions [2]. We shall consider a solute M, which is in equilibrium with a semiconductor and with an external (ambient) phase in which the activity of the solute M is constant. Then, we can have the following reactions:

$$M^L \overset{K_1}{\rightleftharpoons} M^S \overset{K_2}{\rightleftharpoons} M^+ + e, \qquad (5.1.6)$$

where M^L and M^S are un-ionized particles in the external and solid phases, and M^+ is the ionized solute in the solid phase.

From the law of mass action follows

$$K_1 = \frac{N_{M^S}}{N_{M^L}}, \qquad (5.1.7)$$

$$K_2 = \frac{n N_{M^+}}{N_{M^S}}, \qquad (5.1.8)$$

and hence we find

$$n N_{M^+} = N_{M^L} K_1 K_2 = K_3. \qquad (5.1.9)$$

It is evident from Eq. (5.1.8) that any increase in the electron density should, according to the law of mass action, reduce the solubility of M in the solid semiconductor phase. Conversely, a reduction in the electron density, which always takes place when acceptors are introduced, should increase the solubility of the substance M.

A more rigorous estimate of the influence of acceptors on the solubility of donors has been obtained in [3] using Eq. (5.1.9) and the relationship

$$np = n_i^2, \qquad (5.1.10)$$

which is valid in the absence of degeneracy of the electron and hole gases.

Another manifestation of this donor – acceptor interaction is the formation of ion pairs. This effect has been analyzed in [4]. A semiconductor is considered to be an analog of an aqueous solution, in which the formation of ion pairs is also observed. A solid semiconductor differs from an aqueous solution by its lower permittivity and the shorter distances between the ions.

The donor – acceptor interaction is practically absent in heavily doped crystals, because a high degree of compensation cannot be reached in such crystals even using special techniques. Another interaction is also absent in heavily doped crystals: this interaction is described by the usual ionization equilibrium

$$D^0 \rightleftharpoons D^+ + e. \qquad (5.1.11)$$

As demonstrated in the preceding chapter, in the heavy doping case all the dopant atoms are ionized and the above reaction cannot be shifted to the left by any change in temperature. This form of interaction in heavily doped semiconductors may be observed only for impurities forming deep levels in the forbidden band.

Little-Known Interactions. There have been suggestions of several interactions which can appear only in heavily doped crystals. One of them is the formation of compounds in semiconductors. Teltow [5] has suggested that in silicon a pair of two donor ions, e.g., phosphorus ions, can form a molecule similar to the hydrogen molecule H_2, i.e., a compound of the type $P^+:P^+$ may be formed. However, there has been no confirmation of such a molecule [6]. It has also been reported that Frisch has predicted theoretically the formation of polymeric aggregates BC by the reaction

$$AB + AC \rightleftharpoons AA + BC,$$

where the components B and C are dissolved in a substance A.

Assuming that A and B are mobile and there is no intramolecular interaction, Teltow has concluded that, if the pair concentration is 10^{19} cm^{-3} and the diffusion coefficient is 10^{-10} cm^2/sec, aggregates of radii of about 2000 Å may be formed. This reaction has not yet been confirmed experimentally.

Vacancy–vacancy, vacancy–atom, and vacancy–ion interactions have not been investigated much but they may be important in the heavy doping case, because the vacancy concentration is raised considerably by heavy doping. Thus, Longini and Greene [7] suggested that in the case of the total ionization of acceptor vacancies at temperatures such that $N_d \gg n_i$, the following relationship should be obeyed:

$$\frac{N_V}{N_{Vi}} \approx \frac{N_d}{n_i},$$

(5.1.12)

where N_{Vi} and N_V are the concentrations of acceptor-type vacancies in a pure semiconductor and in a donor-doped crystal, respectively.

Hence, it follows that if we introduce a sufficiently large number of donors, the number of equilibrium vacancies may increase by several orders of magnitude. These conclusions have been confirmed in [8] by an investigation of the self-diffusion of germanium in crystals heavily doped with arsenic ($N_{As} = 6 \cdot 10^{18}$ cm^{-3}) and, for the sake of comparison, in gallium-doped crystals ($N_{Ga} = 5 \cdot 10^{19}$ cm^{-3}). It is reported in [8] that the strongest diffusion is observed in n-type crystals whereas in p-type Ge the diffusion is slower even than in a pure semiconductor.

Investigations of Impurity Behavior in Heavily Doped Semiconductors. Investigations of the transport phenomena in heavily doped single crystals, carried out by Fistul' and his colleagues, have indicated the existence of certain anomalies in the properties of these samples. One such anomaly has been observed in the electrical conductivity of germanium heavily doped with phosphorus. It has already been suggested [9] that the dependences of the resistivity on the carrier density are identical for phosphorus-doped and arsenic-doped germanium. On the other hand, it is reported in [10] that these dependences are not identical when $n \geq 10^{19}$ cm^{-3}. This contradiction has stimulated Fistul' to carry

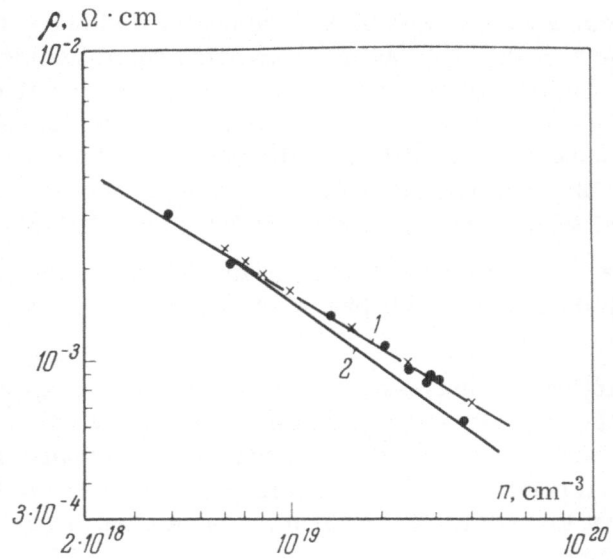

Fig. 5.3. Dependence of the electrical resistivity of heavily doped germanium on the electron density. 1) Arsenic-
doped germanium; 2) phosphorus-doped germanium [10];
black dots represent phosphorus-doped samples and crosses
represent arsenic-doped crystals, according to [11].

out a careful investigation of the electrical conductivity of germanium doped with phosphorus and to compare it with the electrical
conductivity of arsenic-doped germanium. The results of this investigation [11] are presented in Fig. 5.3. The straight line 1 in
this figure represents arsenic-doped germanium. The experimental points for such germanium samples are not shown in Fig. 5.3
in order to avoid overcrowding (the same line is shown with all its
experimental points in Fig. 3.15).

With the exception of one sample with a carrier density of
$3.8 \cdot 10^{19}$ cm^{-3} the points (black dots) for phosphorus-doped germanium all lie on the straight line 1 which represents arsenic-
doped samples. The electrical conductivity of the sample which
does not conform fits a curve given in [10]. Thus, with the exception of one sample, the results agree with [9] but not with [10]. It
is difficult to estimate the errors in these two investigations [9, 10].
A careful examination of a large number of phosphorus-doped samples, carried out later by Fistul', indicated some scatter of the

values of ρ for samples with the same carrier density. The majority of samples have values of ρ which coincide either with line 1 or with line 2 in Fig. 5.3, but there are also some samples whose resistivities lie between these two lines. Such samples also exhibit anomalies of other transport phenomena. Thus, for example, they have somewhat lower values of the thermoelectric power, as reported in [12]. Similar anomalies have also been observed for heavily doped silicon, as reported by Fistul' et al., in [13].

It is natural to assume that the observed anomalies are associated with the nature and behavior of impurities in heavily doped crystals.

Radioactivation, mass-spectroscopic, and metallographic analyses of the impurity composition of crystals, the one-probe method of investigating the inhomogeneity, infrared transmission, lattice thermal conductivity, and elastic limit studies have largely confirmed that the nature of the observed anomalies is associated with heavy doping.

§5.2. Experimental Investigations of Impurity Behavior in Heavily Doped Semiconductors

Polytropy of Doping Impurities. To account for the anomalous properties of heavily doped semiconductors, reported in §5.1, it is necessary to assume that in some heavily doped samples the carrier density (determined from the Hall effect) is less than the true impurity concentration. This disagreement, may, in principle, be due to several causes: a Hall factor whose value is greater than unity; incomplete ionization of impurity atoms; presence of a certain number of dopant atoms in a form not exhibiting donor properties.

The first of these possible causes must be rejected, since an analysis of the Hall factor $\mathscr{A}$, carried out allowing for the mixed scattering mechanism [14] (cf. Chap. 3), has shown that this factor practically never differs from unity for heavily doped samples.

The second possibility must also be rejected because (cf. Chap. 1) there are no shallow impurity levels in the forbidden band of a heavily doped crystal and therefore incomplete ionization of the dopant does not apply.

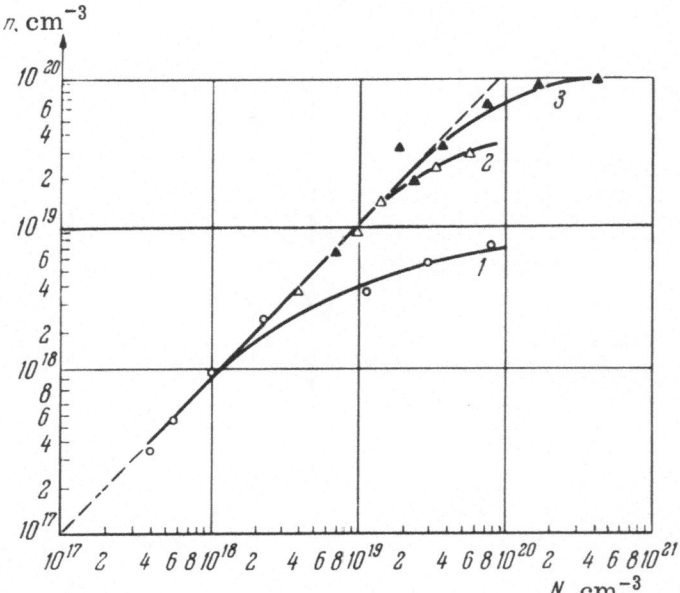

Fig. 5.4. Comparison of the carrier density with the concentration of As and Sb impurities in germanium and silicon. 1) Ge—Sb; 2) Ge—As; 3) Si—As.

It follows that we are left with the third possibility, i.e., we must check the validity of the ionization equation

$$N^0 = N^+ + e, \qquad (5.2.1)$$

where N^0 and N^+ are the concentrations of neutral and ionized dopant atoms, respectively.

This relationship has been checked experimentally on many occasions. Thus, it has been reported in [15] that it is satisfied well at impurity concentrations from 10^{15} to $5 \cdot 10^{17}$ cm^{-3}. In some earlier investigations, for example in [16], it has been found that the number of free carriers is not equal to the impurity concentration in heavily doped crystals, but this has been attributed to the inaccuracy of the results of chemical analysis with which the Hall data have been compared.

To check Eq. (5.2.1) for heavily doped germanium and silicon single crystals, Fistul' carried out parallel measurements of the

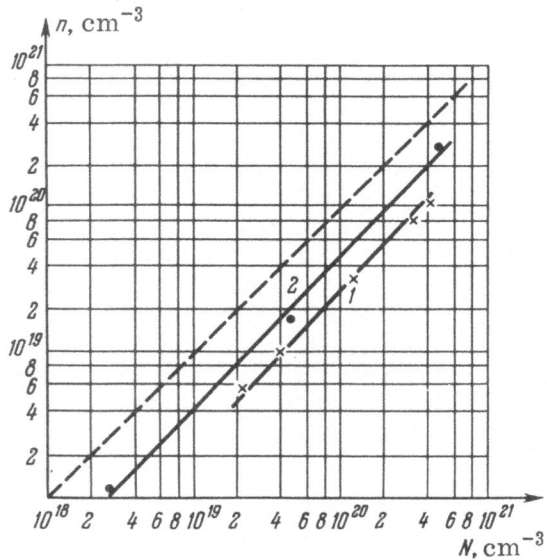

Fig. 5.5. Comparison of the carrier density and of the phosphorus impurity concentration in germanium: 1) data obtained by Fistul'; 2) results reported in [16].

Hall carrier density and of the impurity concentration, using the
radioactivation method in its gamma-spectroscopic form without
damaging the sample. The results of this analysis are compared
in Fig. 5.4. The ordinate gives the carrier density and the abscissa
shows the impurity concentration. It can be seen from this figure
that in all three investigated semiconductor-impurity systems, the
experimental data deviate from the straight line whose slope is 45°,
i.e., beginning from certain threshold values, the dopant concentration exceeds the carrier density. These threshold values and the
deviations from the straight line in Fig. 5.4 depend on the actual
growth conditions of a crystal. However, in general, there is a definite correlation with the solubility limit. The higher the solubility
limit of an impurity, the higher are the concentrations at which deviations from Eq. (5.2.1) are observed. It is important to note that
these deviations occur long before the solubility limit is reached.

The relationship between n and N is somewhat different for
the Si−P system. The results of the Hall measurements and of the
radioactive analysis for the phosphorus in this system are presented
in Fig. 5.5. We can see that the carrier density differs by a con-

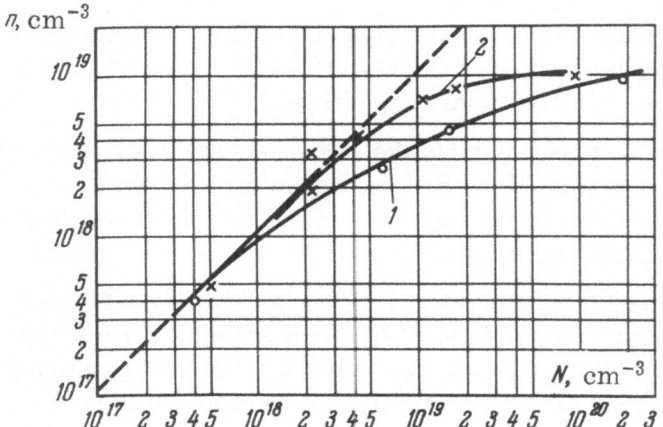

Fig. 5.6. Polytropy in gallium arsenide:
1) GaAs–Se; 2) GaAs–Te.

stant amount from the phosphorus atom concentration. It is interesting that earlier investigations [16], carried out using chemical
analysis, have also given the same result. The only difference between results from Fistul' and those given in [16] is the absolute
value of the discrepancy, which is undoubtedly due to the actual crystal growth conditions.

Unfortunately, the radioactivation analysis cannot be used to
determine phosphorus in germanium, and therefore it has not been
possible to determine the concentration of phosphorus atoms in
those samples for which the results are quoted in §5.1 (Fig. 5.3).
Other methods of analysis for phosphorus in Ge are inaccurate.
The results obtained show convincingly that at high impurity concentrations not all the impurities form a substitutional solid solution.
Some impurities are in states which do not exhibit donor properties. Thus, at high doping levels, the dopant may exist in a crystal
in several forms. This is known as the polytropy of the impurities.
A similar effect has been observed in germanium single crystals
doped with arsenic [10]. The polytropy in Si–As, Si–Sb, and in
Ge–Sb samples was first reported first in [17, 18].

Impurity Polytropy in Complex Semiconduc-
tors. Donor polytropy has been observed in gallium arsenide
doped with selenium [19] or tellurium [20]. The selenium content
has been determined by nephelometry, and the tellurium concentration–by a polarographic analysis using a mercury electrode.

Control experiments have established that the sensitivity of the polarographic method is about $4 \cdot 10^{17}$ cm^{-3} of tellurium, and that the measurement errors do not exceed 10%.

Figure 5.6 shows clearly the phenomenon of polytropy of tellurium and selenium. The effects of these two impurities differ in the slope of the curves and in values of the threshold concentration at which polytropy appears.

Ionization Equilibrium in Acceptor-Doped Semiconductors. We shall now consider the carrier density found from the Hall effect and the impurity concentration found by chemical analysis of p-type crystals. It is reported in [19, 21] that an investigation of the mobility and thermoelectric power indicated a disagreement between the hole density p and the acceptor concentration N, of the type p > N, i.e., opposite to that observed for donors. Similar disagreement is reported in [21, 22], where N was determined by a spectroscopic analysis of germanium heavily doped with gallium. Fistul' investigated similar samples using the radioactivation method [23]. He found that, even in the most heavily doped samples ($N = 5 \cdot 10^{20}$ cm^{-3}), the discrepancy between p and N does not exceed 20% and at lower concentrations it is less. However, even this small discrepancy can have different signs when measurements are repeated on the same sample. This indicates a random origin of the discrepancy.

The disagreement between the Fistul' results and those reported in [21, 22] is due to the different conditions which apply when the impurity being analyzed evaporates from the sample, in which it is present as a solution, and when it evaporates from a standard, which is a simple mechanical mixture of a host and an impurity (oxide). This type of error can be eliminated from spectroscopic analysis by using a graphite-base standard [24]. In such a standard, the oxide of the investigated impurity is reduced by carbon and its evaporation from the standard sample takes place under conditions approximately similar to those in the investigated sample.

Later [25] it was found that the values of p and N are equal for p-type GaAs doped with zinc.

The Concept of "Polytropy." It is worth mentioning why Fistul' introduced a new term to denote the existence of doping impurities in several forms. At first sight, the concepts of

polymorphism, allotropy, or the amphoteric state could be used
as well.

The first two concepts are used to denote structural differences between substances, such as monoclinic and triclinic sulfur,,
red and white phosphorus, etc. The different states of a dopant
may be associated with structure defects (dislocations or vacancies) but in practice they do not change the structure of the host
crystal. Therefore, the use of the term "polymorphism" and "allotropy" would have implied a "structural" meaning, which may be
unimportant.

The concept of amphoteric nature implies duality. It is used
to describe the existence of an element in the form of a cation in
one compound and in the form of an ion in another compound. For
example, in semiconductors, amphoteric properties are exhibited
by gold. Therefore, the application of the concept of amphoteric
nature to manifold forms would suggest oppositely charged states,
while in fact the second form of a dopant can be electrically inactive. Moreover, this second form may be a compound of the dopant with some third element, for example, with oxygen.

Possible Causes of Impurity Polytropy. Im-
purity polytropy in heavily doped semiconductor single crystals
may be due to various causes: the formation of short-range complexes of the M_xA_y type (here, M is a semiconductor and A is an
impurity) in disordered solid solutions; the precipitation of the dopant at various structure defects, for example, the formation of "impurity atmosphere" at dislocations; the presence of second-phase
occlusions; the formation of special impurity substructures of cellular type [26]; the presence of impurity atoms at interstices, etc.

We shall now consider the more important of these causes.

Formation of Complexes. We can use the experi-
mental data presented in Figs. 5.4-5.6 to obtain information about
the state of the impurity which does not exhibit donor properties.
We shall assume that the various states in which the doping impurity can exist are in equilibrium. We shall consider the most
likely reactions leading up to such equilibrium. We shall denote
the concentrations of impurity atoms exhibiting donor properties
by n, i.e., n is equal to the carrier density, and the total concentration of dopant atoms will be denoted by N, which is the quantity

found by chemical analysis. The simplest reaction is:

$$mA_1 \rightleftharpoons A_m, \tag{5.2.2}$$

where A is the chemical symbol of the dopant; m is the number of dopant atoms in a complex (quasi-molecule) which does not exhibit donor properties.

The equilibrium constant for this reaction, K_m is:

$$K_m = \frac{[A_m]}{[A_1]^m}. \tag{5.2.3}$$

The total concentration of the dopant N is

$$N = [A_1] + [A_m],$$

or, bearing in mind that $[A_1] \equiv n$ and using Eq. (5.2.3), we obtain:

$$N = n(1 + K_m n^{m-1}).$$

This expression can be conveniently reduced to the form

$$\log\left(\frac{N}{n} - 1\right) = \log K_m + (m - 1)\log n. \tag{5.2.4}$$

Thus, if the polytropy is due to the existence of impurity atoms in the form A_1 and in the form of complexes of the A_m-type, then N and n should be related by Eq. (5.2.4). This equation shows that by plotting the dependence of $(N/n - 1)$ on n using a logarithmic scale, we should obtain a straight line with a slope whose tangent is equal to $(m - 1)$.

It is also interesting to consider two reactions proceeding simultaneously:

$$\left.\begin{array}{r} mA_1 \rightleftharpoons A_m, \\ (m+1)A_1 \rightleftharpoons A_{m+1}. \end{array}\right\} \tag{5.2.5}$$

In this case, the total concentration N is:

$$N = n + mK_m n^m + (m+1)K_m K_{m+1} n^{m+1},$$

which can be transformed to:

$$\frac{\frac{N}{n} - 1}{n^{m-1}} = mK_m\left(1 + \frac{m+1}{m}K_{m+1}n\right).$$ (5.2.6)

We can find the value of m from the experimental values of N and n using an auxiliary curve

$$y = 1 + z.$$ (5.2.7)

Plotting this curve on a logarithmic scale and superimposing on it the experimental curves (5.2.6), calculated for various values of m, we can find m and mK_m from the shift of the curves along the axes. Other possible reactions, for example,

$$\left.\begin{array}{l} mA_1 \rightleftharpoons A_m, \\ (m+2)\,A_1 \rightleftharpoons A_{m+2}, \end{array}\right\}$$ (5.2.8)

and similar reactions can be dealt with the same way.

We must also consider the case when all the reactions of the Eq. (5.2.2) type take place simultaneously, i.e.,

$$\left.\begin{array}{l} A_1 \rightleftharpoons A_1^*, \\ 2A_1 \rightleftharpoons A_2, \\ 3A_1 \rightleftharpoons A_3, \\ \quad \cdot \quad \cdot \quad \cdot \quad \cdot \\ mA_1 \rightleftharpoons A_m, \\ \quad \cdot \quad \cdot \quad \cdot \quad \cdot \\ \infty A_1 \rightleftharpoons A_\infty. \end{array}\right\}$$ (5.2.9)

The total concentration of the dopant is:

$$N = \sum_{m=1}^{\infty} mK_m n^m.$$ (5.2.10)

If we assume that the probability of formation of A_m decreases when the value of m increases, we can then show that Eq. (5.2.10) becomes

$$N = \sum_{m=1}^{\infty} m\,(K'n)^m,$$ (5.2.11)

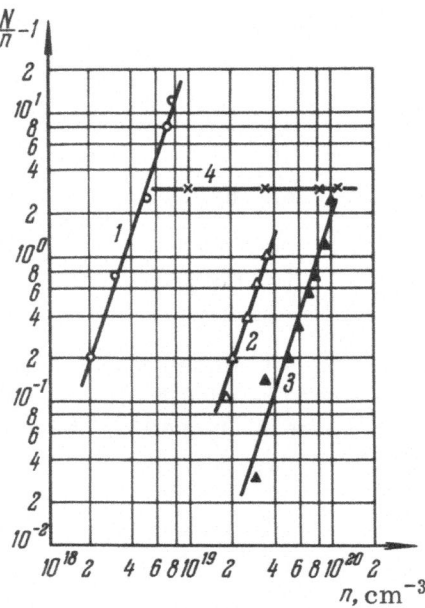

Fig. 5 7. Results of analysis of curves
in Figs. 5.4 and 5.5.

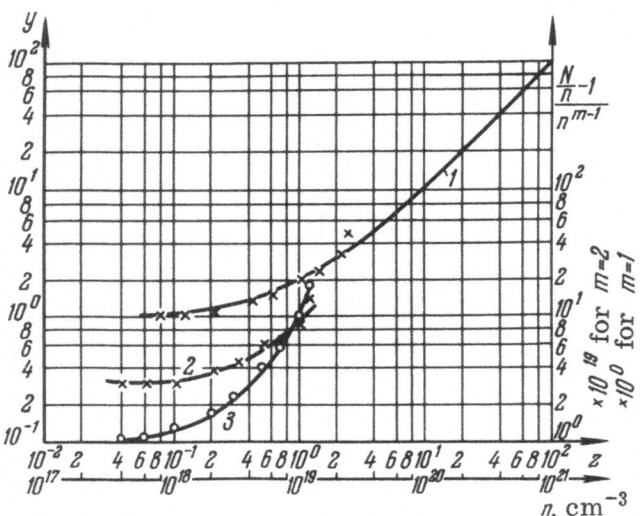

Fig. 5.8. Results of analysis of curves in Fig. 5.6.

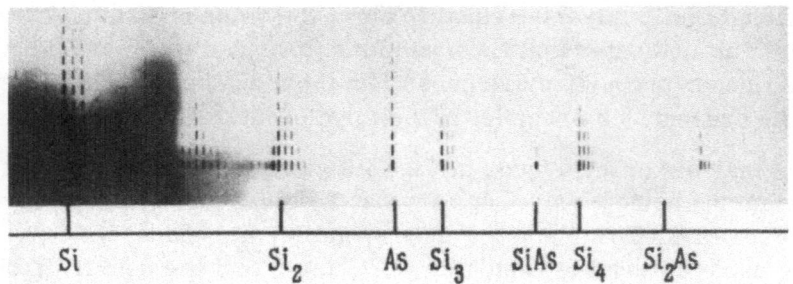

Fig. 5.9. Typical mass-spectrogram of a silicon single crystal doped with arsenic.

where

$$K' = \frac{K_{m+1}}{K_m}.$$

The expression (5.2.11) is simply a geometrical series, whose sum is

$$\log\left(1 - \sqrt{\frac{n}{N}}\right) = \text{const} + \log n. \qquad (5.2.12)$$

The above expression represents a straight line sloping at an angle of 45° when the dependence of $(1 - \sqrt{n/N})$ on n is plotted on a logarithmic scale.

Thus, the form of A_m complexes is found by analyzing the curves of Figs. 5.4-5.6 using Eqs. (5.2.4), (5.2.6), and (5.2.12). Such an analysis shows that the curves of Fig. 5.4 for germanium and silicon obey satisfactorily Eq. (5.2.4). The results are given in Fig. 5.7. For the Ge−Sb, Ge−As, and Si−As systems (lines 1, 2, 3), the tangent of the slope angle is equal to 3 and consequently m = 4. This means that the polytropy in germanium and silicon is due to the presence of complexes Sb_4 and As_4, in addition to Sb and As atoms in a substitutional solid solution. The absence of other complexes, for example, Sb_2, As_2, Sb_3, As_3, etc., can be easily demonstrated by showing that the analysis of the curves in Fig. 5.4 does not satisfy Eq. (5.2.12), which should be obeyed in this case.

The same figure (Fig. 5.7) shows the results of an analysis of the experimental data for the Si − P system (line 4), which has also been carried out using Eq. (5.2.4). Since the tangent of the

slope angle is in this case equal to zero, the value of m is equal to
unity. This indicates that those atoms of phosphorus which do not
exhibit donor properties are present in the monatomic form. They
may be present, for example, in the form of interstitial atoms.

Analysis of the curves in Fig. 5.6 shows that the best results
are obtained using Eq. (5.2.6). Figure 5.8 gives an auxiliary curve
1, representing Eq. (5.2.7). For Se in gallium arsenide, we can
show, using in turn the values m = 1, 2, 3, ..., that the best fit is obtained using m = 2. The dependence then obtained

$$\left(\frac{N}{n} - 1\right)/n^{m-1} = f(n) \qquad\qquad (5.2.13)$$

is represented by curve 2 in Fig. 5.8. Shifting this curve parallel
to the coordinate axes, we can make it fit satisfactorily curve 1.
The same figure includes curve 3, obtained on the assumption that
m = 1. Shifting this curve along the coordinate axes, we can easily
show that it cannot be made to coincide with curve 1. The results
are even poorer if it is assumed that m ≥ 3.

Similarly, we can show that the experimental and auxiliary
curves do not match if we use various values of m in a reaction of
the (5.2.8) type.

Thus, for the GaAs − Se system we must assume that selenium
is present in GaAs in three forms: Se_1, Se_2, and Se_3. Atomic selenium is present in an ordinary substitutional solid solution and
exhibits donor properties. Se_2 and Se_3 are most likely to be present in the form of compounds of the $GaSe_2$ and Ga_2Se_3 type. Vieland
and Kudman have explained the curve in Fig. 5.6 by the presence
of compounds of gallium and selenium in GaAs [19]. A similar attempt to deal with the GaAs − Te system has not been successful.
The reason for this is not yet clear.

This analysis of the causes of the polytropy should be regarded as tentative. It does not prove the existence of complexes
in crystals, but at least it indicates that the presence of such complexes may explain the discrepancy between the carrier density
found from the Hall coefficient and the impurity concentration found
by chemical analysis.

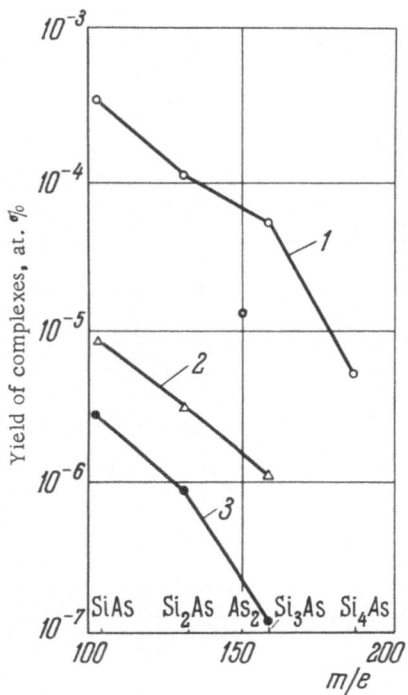

Fig. 5.10. Presence of complexes in the mass spectra of silicon single crystals heavily doped with arsenic. Carrier density n (cm^{-3}): 1) $9.6 \cdot 10^{19}$; 2) $7.7 \cdot 10^{19}$; 3) $2.7 \cdot 10^{19}$.

A direct confirmation of the formation of slightly different complexes has been obtained in an investigation of the mass spectra of silicon, which we shall now discuss in detail.

Mass-Spectroscopic Investigation of Heavily Doped Silicon and Germanium Single Crystals [27, 28]. In addition to the lines representing As$^+$ ions, the mass spectrograms of silicon include a line representing $(SiAs_2)^+$ ions but its intensity is extremely low, of the order of $10^{-8}\%$, i.e., it is close to the limit of the sensitivity of the mass spectrometer.

The mass spectra include clear lines corresponding to $(SiAs)^+$, $(Si_2As)^+$, and $(Si_3As)^+$ ions. The intensity of the lines decreases in the order in which they have just been listed. One such spectrogram is shown in Fig. 5.9. In addition to these complexes, the mass spectrum of silicon has lines representing complexes of the Si_m type, for m = 1, 2, 3, ..., 7, which shall be called clusters. Such clusters have also been discovered in the mass spectra of germanium. The mass spectrum of arsenic-doped germanium includes a line with m/e = 146. This line cannot be attributed with certainty to GeAs complex ions, since it is masked by the line of Ge^{72}Ge^{74} ions of the same mass.

Clusters of the Si_m type have been observed by other authors, who have analyzed semiconductors using a MS-7 mass spectrograph.

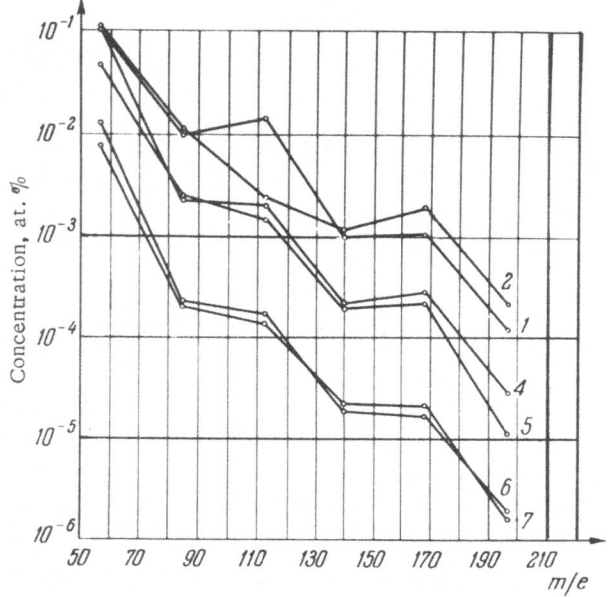

Fig. 5.11. Concentration of clusters, as indicated by the mass
spectra of silicon single crystals.

Sample No.	Electron density, cm^{-3}	Sample No.	Electron density, cm^{-3}
1 2	$1 \cdot 10^{15}$ $1.07 \cdot 10^{17}$	4 5 6.7	$2.7 \cdot 10^{19}$ $9.6 \cdot 10^{19}$ $\sim 1 \cdot 10^{20}$

Thus, it is reported in [29] that clusters are formed in the spark
of an ion source, but the tendency to form clusters is different for
different substances

Fistul' et al. [27] have established, by special experiments,
that complexes and clusters are present in real crystals. Thus,
for example, a change in the working conditions of an ion source
does not affect the yield of these clusters and complexes; however,
when samples are dissolved in a mixture of hydrofluoric and nitric
acids and the products are deposited on graphite electrodes, the
lines of complexes and clusters, present in the spectrum of the
original crystal, are no longer observed. Moreover, a preliminary

heat treatment alters, other conditions being equal, the concentrations of clusters and complexes recorded with a mass spectrometer.

It was interesting to investigate the yield of complexes and clusters as a function of the degree of doping. Such an investigation has been carried out on silicon samples [27, 28].

Figure 5.10 shows the yield of SiAs, Si_2As, and Si_3As complexes of various silicon crystals heavily doped with arsenic. For a carrier density $n < 10^{18}$ cm^{-3}, the total concentration of complexes is low and is more than 10^{-7} at.%. The yield of complexes is higher for higher concentrations of arsenic in silicon and it reaches $3 \cdot 10^{-4}$ at.% for $n \approx 1 \cdot 10^{20}$ cm^{-3}. Apart from the complexes already mentioned, a silicon sample containing the maximum possible amount of arsenic has exhibited a line at $m/e = 150$, representing As_2^+ ions ($\approx 10^{-5}$ at.%).

The yield of clusters also depends strongly on the concentration of the dopant in a crystal. However, this dependence is opposite to that observed for complexes. It is clearly evident from Fig. 5.11 that, when the impurity concentration is increased from 10^{15} cm^{-3} (curve 1) to 10^{17} cm^{-3} (curve 2), the concentration of clusters is practically unaffected. When the degree of doping is increased still further, the concentration of clusters begins to decrease considerably and at $n = 10^{20}$ cm^{-3} (curves 6 and 7) the fall in the concentration of clusters reaches almost two orders of magnitude. Further increase in the arsenic concentration again results in an increase in the cluster concentration. Similar dependence is also observed for germanium samples.

These mass-spectroscopic investigations of silicon indicate a correlation between the concentration dependences of the yield of clusters and complexes and the dislocation structure of single crystals: at low arsenic concentrations, silicon crystals contain dislocations and the yield of clusters is higher but there are no complexes. At an arsenic concentration of about $(4-5) \cdot 10^{18}$ cm^{-3}, the dislocations disappear, the yield of clusters decreases somewhat and the mass spectrum indicates the presence of complexes of the SiAs-type. This pattern becomes even clearer when the concentration of arsenic is increased still further up to 10^{20} cm^{-3}. Dislocations reappear in these crystals, the yield of clusters increases, the yield of complexes becomes less, but As_2 molecules are observed.

All these experimental data indicate that there is a close relationship between the formation of complexes and clusters and the
dislocation of single crystals.

Second-Phase Occlusions. The natural step in an
investigation of the polytropy of impurities is a direct observation
of complexes or chemical compounds, which may be present in the
form of second-phase occlusions. Such occlusions may be formed
during the growth of heavily doped single crystals, either during
the crystallization process or during the cooling of a grown crystal, because of the precipitation of a supersaturated solid solution.
The formation of supersaturated solid solutions is not very likely:
impurities have a retrograde-type solubility in germanium and silicon and, as a rule, the solubility of the impurities at low temperatures is higher than at the melting point of the host crystal [1]. In
fact, prolonged heat treatment at 400°C of silicon single crystals
doped with phosphorus and arsenic (concentrations up to $\approx 1 \cdot 10^{20}$
cm^{-3}) does not reduce the carrier density; such reduction would
have been observed if the solid solutions were supersaturated. The
formation of second-phase occlusions during the crystallization
process requires, as indicated by the phase diagrams of germanium
and silicon with group V dopants, that the concentration of the dopant in the system be sufficiently high in order to produce a melt of
composition close to the eutectic. For example, in Ge−As and
Si−As systems, in which compounds are formed, the eutectic melt
contains 41 and 40.5 at.% of arsenic, respectively; the concentrations of the dopant in the eutectic melts of Ge−Sb and Si−Sb are
even higher.

Concentrations of this order can, in practice be reached only
by very strongly nonuniform doping of germanium and silicon single
crystals.

The possibility of the presence of second-phase occlusions
has been investigated for silicon single crystals heavily doped with
arsenic, antimony, and aluminum. A metallographic analysis of
As-doped crystals, etched in a chromic etchant [30], has shown that
up to carrier densities of $(6-7) \cdot 10^{19}$ cm^{-3}, these samples have a
single-crystal structure with a periodic distribution of impurities
in the "growth bands," typical of ingots grown by the Czochralskii
method. At higher concentrations, a cellular substructure is observed and this structure is replaced, at $n > 1.5 \cdot 10^{20}$ cm^{-3}, with
a polycrystalline structure. In Sb-doped crystals, a cellular sub-

structure is observed even at $n \approx 6 \cdot 10^{18}$ cm^{-3}, but when the concentration is increased further the cellular structure does not spread and the crystals rapidly change to a polycrystalline structure A cellular substructure is not observed in Al-doped single crystals and the transition to the polycrystalline state takes place through multiple twinning, which begins in samples with a hole density of $\sim 4 \cdot 10^{17}$ cm^{-3}. The dislocation density in all the investigated crystals increases as the polycrystalline state is approached. The dislocation density increases particularly strongly in crystals doped with aluminum, which have the lowest value of the distribution coefficient. In addition to dislocation etch pits in Al- and Sb-doped single crystals, deep etch pits of unknown origin and of very large dimensions are also observed. The formation of such pits may be due to the presence of second-phase occlusions in the crystals. To observe such occlusions, some of the samples were treated in selective etchants, which affect silicon very little or not at all but which should dissolve easily the probable occlusions: a 5% solution of NaOH at room temperature for Al-doped samples; a mixture of HCl and HNO$_3$ (3:1) at room temperature for Sb-doped samples; a boiling solution of HCl (1:1) for As-doped crystals.

Second-phase occlusions can be seen easily in infrared radiation. This method has been used to detect such occlusions in silicon heavily doped with aluminum and antimony. In Al-doped crystals, these occlusions are larger and are observed even during optical polishing before etching. Subsequent etching makes it possible to observe numerous small occlusions. Second-phase occlusions are observed most clearly with an infrared microscope. The distribution of occlusions in the interior of a crystal obeys definite crystallographic rules: in the majority of cases the planes of preferential segregation of impurities are (111).

Measurements of the microhardness of silicon samples without occlusions have yielded values within the limits 950-1200 kg/mm^2 but at the points where occlusions have been observed in Al-doped crystals the microhardness decreases to 175 kg/mm^2, which is close to the published data on the microhardness of the Si−Al eutectic [31]. This indicates that the observed occlusions are aggregates of aluminum atoms. Considerable variations of the microhardness are observed also for Sb-doped crystals. Second-phase occlusions in silicon crystals heavily doped with Ga and Sb have been also observed by other workers [32, 33].

Second-phase occlusions have not been observed in arsenic-doped silicon single crystals, although these crystals exhibit a very nonuniform distribution of the microhardness

Impurity Atmospheres at Dislocations. The formation of impurity atmospheres at dislocations may be the result of the elastic interaction between the stress fields of impurity atoms and dislocations or it may be due to the electric (Coulomb) interaction Impurity atmospheres at dislocations have been discovered in infrared investigations of silicon single crystals heavily doped with antimony and aluminum [34]. The precipitation of doping impurities at dislocations results in a natural decoration of the dislocations and makes them visible when observed in the infrared microscope. Comparison of the infrared photomicrographs with the results of etching has shown that: 1) far from all dislocations are decorated with the dopant; 2) small precipitates of the second phase, which are not associated with dislocations, form pits of similar shape to the dislocation etch pits; 3) large second-phase occlusions are usually nucleated at dislocations and they spread from dislocations to other parts of the crystals.

Indirect proof of the existence of impurity atmospheres at dislocations has also been obtained in an investigation of the etch patterns of silicon single crystals heavily doped with phosphorus and arsenic. In dislocation-free single crystals, containing 10^{18}-10^{19} cm^{-3} of arsenic or phosphorus, very definite periodic inhomogeneities are observed in the impurity distribution. These inhomogeneities are much weaker in crystals with the same dopant concentration grown under the same conditions, but containing $\approx 5 \cdot 10^3$ cm^{-2} of dislocations. The reduction in the periodic inhomogeneity in the latter case is probably due to the precipitation of some of the dopant impurities at dislocations, which results in the equalization of the dopant concentrations in neighboring layers of a single crystal. A similar effect has been observed by Furuoga [35] in an investigation of a layered distribution of oxygen in lightly doped silicon single crystals with different dislocation densities.

Impurity Substructures. The search for second-phase occlusions in heavily doped silicon samples has led to the discovery of a cellular structure (Fig. 5.12) of the same type which has been described in [25] for low-melting-point metals.

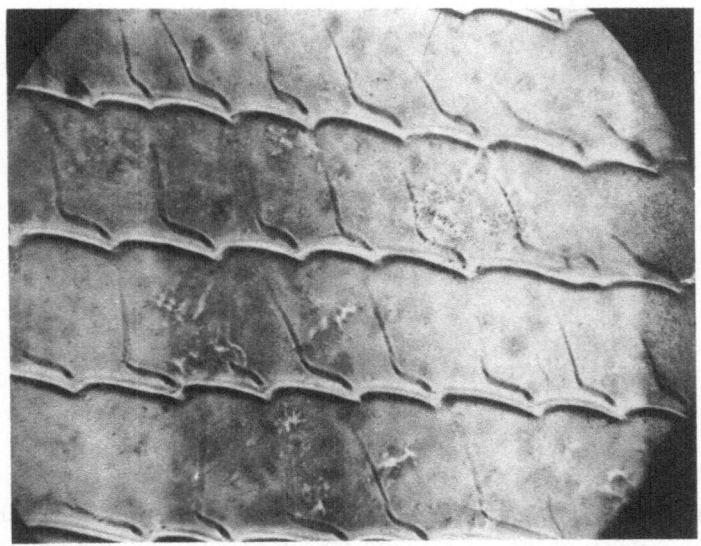

Fig. 5.12. Cellular structure in heavily doped silicon single crystal.

Before the appearance of a paper by Tyler et al. [18], the
cellular structure in semiconductor crystals had been reported
only for gallium–doped germanium [36]. Such a structure is the
consequence of the cellular form of the crystallization front.

The cellular shape of the front is due to the fact that, because
impurities are driven back away from the front, an equilibrium crystallization temperature gradient Γ appears near the front. At a certain value of this gradient (Γ_0) the front becomes unstable with respect to small perturbations of its shape. Usually [25, 36, 37], the
transformation of a plane front to a cellular one is attributed to
concentration supercooling of the melt, i.e., to the condition $\Gamma > G_1$,
where G_1 is the temperature gradient in the melt. A theoretical
analysis of the quantity Γ_0 has been carried out in [38] and will be
considered in Chap. 6.

Theoretical and experimental estimates show that the dopant
concentration at cell boundaries is considerably higher than the
average concentration in a single crystal [39, 40]. By way of example, Fig. 5.13 shows the distribution of the resistivity for a cross

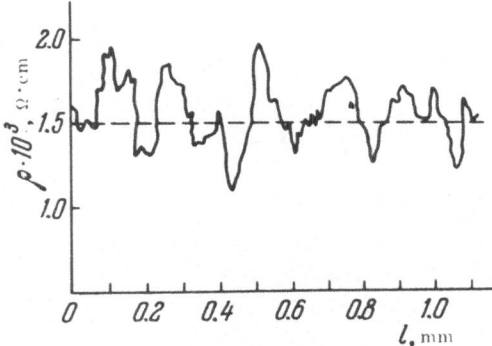

Fig. 5.13. Distribution of the resistivity along
a heavily doped silicon plate (measured with
a one-probe method).

section of a silicon single crystal doped with arsenic and exhibiting
a cellular substructure [41]. This single crystal had a relatively
low concentration of As and a relatively weakly developed cellular
structure; moreover, the carrier density in this crystal, determined from the Hall effect measurements, agreed with the impurity concentration found from the radioactivation analysis. Under
these conditions the probability of the presence of dopant impurities in an electrically inactive state was low and this made it possible to determine fairly accurately the nature of the distribution
of the dopant from the value of the electrical resistivity.

It can be seen from Fig. 5.13 that the distribution is periodic.
The resistivity has minima at the cell boundaries. The distance
between minima (200–250 μ) in the resistivity curve is practically
equal to the width of the cells (220–250 μ), found from a metallographic investigation of the sample. It follows from these results
that the dopant concentration at cell boundaries is almost twice as
high as in the interior. An even stronger inhomogeneity is observed for samples with a well-developed cellular structure, as
reported in [40].

The preferential segregation of the dopant along the cell
boundaries tends to favor the appearance of polytropy at these
boundaries. The tendency to form a cellular substructure and the
degree of inhomogeneity due to this substructure increase, other
conditions being equal, when the distribution coefficient of an im-

purity (for impurities with k < 1) and the impurity concentration in
a crystal are increased. From a theory given in [38], it follows
that the smaller the distribution coefficient of the dopant the lower
is the critical concentration at which a cellular substructure appears in a single crystal. This may account, to a considerable degree, for the observation that crystals doped with different impurities but grown under the same conditions exhibit polytropy at different impurity concentrations. It is also in agreement with the
empirical correlation between the onset of the deviation of the
curves in Fig. 5.4 from straight lines and the impurity distribution
coefficient.

Influence of Heat Treatment. As already men-
tioned, the polytropy phenomena appear in heavily doped germanium
and silicon well before the solubility limit. If some of the impurities present in a crystal do not form a substitutional solid solution (although such a solution is far from saturation), then such a
system is thermodynamically unstable. We can expect the carrier
density in such a crystal to increase with time because of the additional transitions of impurity atoms to the solid-solution state. Such
an effect has in fact been observed after the long storage of germanium samples heavily doped with arsenic; the samples were
stored at room temperature in a desiccator. Some of these samples exhibited an increase in the electron density up to 20-80%.
The greatest changes were observed in samples which initially had
shown the greatest deviation from the linear dependence in Fig. 5.4.
A similar, although somewhat smaller, increase in the carrier density was observed also in germanium samples heavily doped with
phosphorus [23].

The heat treatment of heavily doped single crystals may accelerate the formation of a solid solution by an impurity. This problem was investigated in [18]. A specially selected As-doped germanium single crystal, for which the results of the Hall-effect measurements and radioactivation analysis differed considerably, was
annealed in a hydrogen atmosphere at 870°C for 3 hours and then
was quenched in water. This heat treatment almost doubled the
carrier density in a crystal: from $2.5 \cdot 10^{19}$ to $4 \cdot 10^{19}$ cm^{-3} so that
the results of the Hall-effect measurements and of the activation
analysis became practically equal. A metallographic investigation

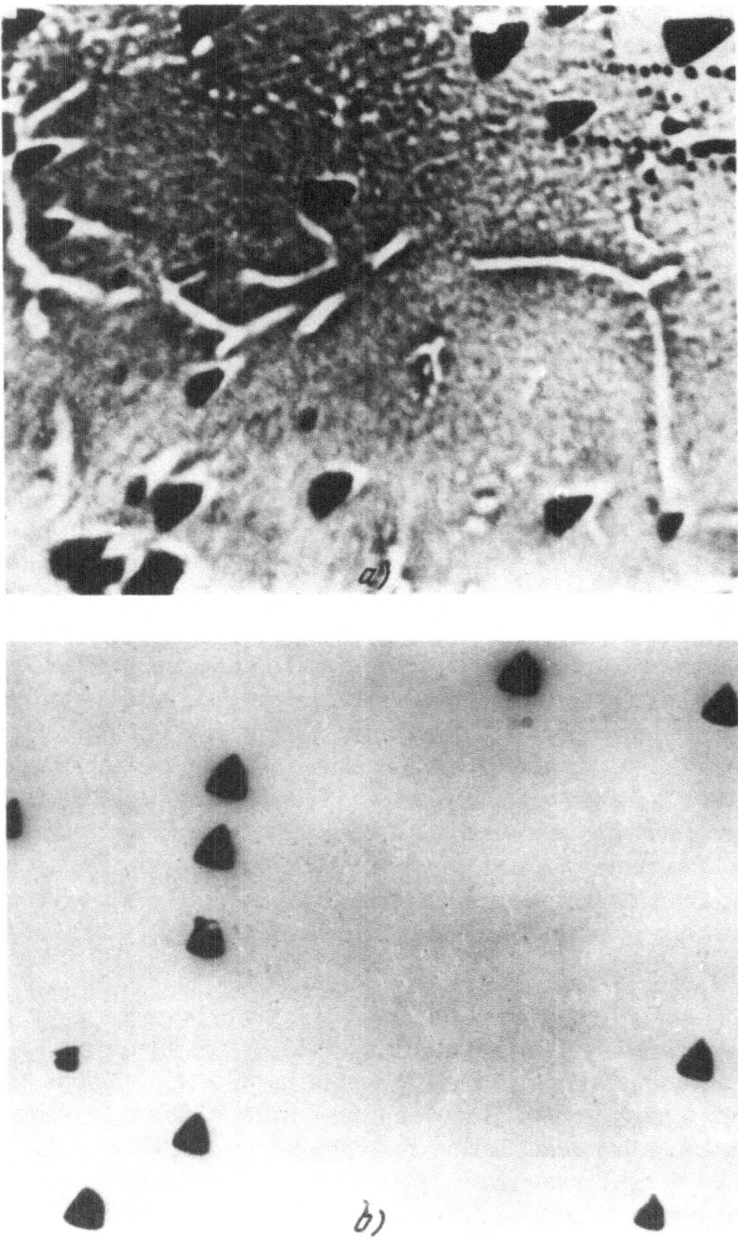

Fig. 5.14. Dislocation etch pits on the surface of a heavily doped germanium single crystal: a) before heat treatment; b) after heat treatment.

of this crystal before heat treatment showed that the dislocation density was 10^4 cm^{-2}. The dislocation etch pits had characteristic "tails" which joined the individual pits and formed a grid (Fig. 5.14a). After the heat treatment, this grid disappeared completely (Fig. 5.14b). The formation of "tails" was due to the precipitation of the dopant at dislocations and other defects. After the heat treatment, the impurities precipitated at defects formed a solid solution and this increased considerably the carrier density. The heat treatment also much reduced the concentration of clusters in the investigated samples: before heat treatment the concentration of Ge$_2$ and Ge$_3$ clusters was $3.4 \cdot 10^{-1}$ and $4.5 \cdot 10^{-4}$ at.%, but after heat treatment these values were $1 \cdot 10^{-3}$ and $3.8 \cdot 10^{-5}$ at.%, respectively.

These observations confirmed once more the close relationship between the polytropy and the dislocation structure of single crystals.

Silicon single crystals, doped with P and As and having a carrier density of $(8-9) \cdot 10^{19}$ cm^{-3}, were annealed in a hydrogen atmosphere at 1200°C for 24 hours and then quenched in water. In contrast to germanium, all these silicon samples had, before their heat treatment, a cellular substructure of the type shown in Fig. 5.12 but no dislocation etch pits. In this case, the increase in the carrier density after heat treatment did not exceed 10-12%. The carrier density in some of these silicon samples was not affected at all by the heat treatment. The cellular substructure remained in all the samples after the heat treatment, the concentration of clusters did not change, and the results of Hall-effect measurements and of the activation analysis still differed [41].

The polytropy of the impurities in these heavily doped silicon samples was associated mainly with the cellular substructure, which was not destroyed by the heat treatment.

Relationship between Polytropy and Some Physical Properties of Crystals. We shall now consider the transport phenomena in samples exhibiting impurity polytropy. Since the polytropy of arsenic-doped germanium samples is reduced considerably by heat treatment, it would be interesting to compare the transport phenomena before and after heat treatment. It is found that heat treatment reduces the electron mobility and the new value of the mobility (cf. Fig. 3.18) is in good agreement with the new value of the carrier density. This means that

the excess arsenic present in a form not exhibiting donor properties has a very small scattering cross section and has practically
no effect on the transport phenomena. This form of arsenic is probably electrically neutral.

The results are different for phosphorus-doped germanium.
Heat treatment of this material also increases the free-electron
density (and reduces the mobility), but the change in the mobility
is such that the results after heat treatment are represented by
curve 2 in Fig. 5.3, rather than by curve 1. This indicates that
phosphorus-doped germanium contains a second form of phosphorus which has a scattering cross section comparable with the
cross section of the ionized impurities. Unfortunately, the lack of
a strong effect of heat treatment on silicon makes it impossible to
determine fully the effect of polytropy on the scattering of electrons
in that semiconductor. However, the observed small changes in the
carrier density (10-15%) indicate changes in the mobility in full
agreement with the curves given in Fig. 3.24. This means that the
second form of arsenic, phosphorus, and antimony in silicon is probably electrically neutral.

The polytropy should give rise to levels in the forbidden band.
If the second form of a dopant is a complex M_xA_y or A_n, it follows
from general considerations that the energy positions of such quasi-
molecules can be roughly estimated as $E/\varkappa^2$, where E is the ionization energy of M_xA_y or A_n in the free state and $\varkappa$ is the permittivity of the crystal. Even if we assume that E is 20-30 eV, we
obtain values of $E/\varkappa^2$ equal to 0.078-0.11 eV. This is equivalent to
a temperature range 900-1300°K. Thus, few complexes should be
formed if a material is quenched from a sufficiently high temperature. This is in agreement with the experimental observations.

One of the methods of investigating deep levels associated
with complexes is a study of the excess current in a tunnel diode.
The value of the excess current is extremely sensitive to the presence of deep levels in the forbidden band of a semiconductor [44].

Experiments on quenched semiconductors have shown that
quenching greatly increases the excess current. This was first reported in [45] for tunnel diodes made of arsenic-doped germanium
and then in [46] for phosphorus-doped germanium diodes. These
results can be explained satisfactorily by the formation of complexes and the effect of heat treatment. Moreover, it follows from

the curves in Fig. 5.4 that, for the same impurity concentration
(other conditions being equal), the following relationship should be
satisfied by the values of the excess current: $i_{Ge-Sb} > i_{Ge-P}$
$> i_{Ge-As}$. This relationship has indeed been confirmed experimentally [47].

Since p-type crystals do not exhibit polytropy, we can easily
predict that heat treatment (quenching) should have no effect on the
excess current of tunnel diodes prepared from germanium doped
with acceptor impurities, such as boron, gallium, or aluminum.
Experimental results [47] have confirmed this conclusion.

Discussion of Experimental Investigation of
Impurity Polytropy. We shall now discuss qualitatively the
experimental results given in this chapter by considering the structure of a substance in all its three states: solid (crystal), melt
(liquid), and vapor (gas).

We shall consider first the structure of molten semiconductors. We shall mention only the latest ideas, which owe much to
Frenkel' [48], Danilov [49], and Shvidkovskii [50]. According to
these ideas, all the particles in a liquid can be divided – at any moment – into two groups.

The first group consists of particles vibrating about centers
which represent a statistically disordered quasi-crystalline lattice.
The short-range order is observed at distances of tens of angstroms.

The second group comprises particles moving in accordance
with the random walk laws (random structure).

Thus, the thermal motion in a liquid has the characteristics
of the thermal motion in a solid (vibrations) and gas (random motion). Any particle can go from the first to the second group and
conversely. Therefore, the structure of a liquid is the result of
dynamic equilibrium between particles of both groups.

The structure of a real liquid is governed, to a considerable
degree, by the kinetics of the establishment of this dynamic equilibrium; this problem has not yet been solved. We shall mention only
two important properties of liquids.

The first property is that structural units in a liquid need not
be atoms but complex aggregates of the molecular type. Thus, for
example, it has been demonstrated in [51] that multiatomic mole-

cules are present in simple nonmetallic liquids: N_2, O_2, O_3, P_4, S_8. The compositions of the vapors above molten melts Cu, In, Ga, Bi, Cd, Sn have been investigated [52], and it has been shown convincingly that the experimentally observed multiatomic aggregates are not products of some secondary reactions in the gaseous phase but reflect the structure of the molten metal (liquid).

The liquid state is intermediate between the vapor and solid phases and a liquid changes to a vapor or a solid in a continuous manner. Cooling a liquid intensifies the molecular forces tending to rearrange all the molecules to form a crystalline structure, which is typical of these forces at a given temperature. This is why the structure of a given substance is similar below and above the solidification point. This characteristic property of liquids allows us to assume that the preliminary stages of the crystallization process take place in the melt. In other words, the properties of a crystal are, to a great extent, determined already in the liquid phase.

At high temperatures, the energy of thermal motion increases and this motion becomes stronger than the intermolecular forces. The structure of a liquid tends to a random distribution of particles, i.e., to the structure of a gas.

There are many experiments [50] which demonstrate that the majority of liquids change continuously (without a marked change in the properties) to crystals or gases; this applies also to metals.

We shall use these two properties of liquids to discuss the experimental data presented in this chapter.

A molten semiconductor (e.g., silicon) has the structure of a liquid consisting of some quasi-crystals and randomly moving Si atoms. We shall assume that the quasi-crystals are clusters of the Si_m type, where $m \geq 1$. The binding energy of the particles in clusters with $m > 1$ is higher than that for a simple pair of vibrating particles $(m = 1)$. It follows from general considerations that the probability of formation of clusters should decrease at higher values of the number m.

If clusters do exist in the melt, the difference between the binding energies of different clusters will be manifested in the different crystallization temperatures of a crystal and subcrystals from the clusters. In other words, the crystallization temperature

of clusters T_{0c} should be higher than the crystallization temperature of silicon itself, T_0. Therefore, when a single crystal is pulled
by the Czochralskii method, we obtain a crystal in which Si_m clusters are "frozen in."

Such occlusions are structural defects in the silicon lattice.
In mass-spectroscopic analysis using a spark ion source, Si_m clusters present on the surface are likely to enter the spark discharge
channel in the form of complete molecules, because such molecules
have higher binding energies. Consequently, the mass spectrum indicates the presence of $Si_2 - Si_8$ clusters (Fig. 5.11). When doped
(for example with arsenic) silicon crystals are grown from the melt,
the melt may contain not only quasi-crystals of the Si_m type but also
$SiAs_m$ and Si_mAs complexes and – at very high concentrations –
even As_m clusters. When a semiconductor crystallizes, only a
small fraction of these clusters is retained in the solid state. Crystallization of a doped material unavoidably reduces the concentration of Si_m clusters because the dopant replaces not only silicon
atoms at normal lattice sites, but also atoms in Si_m clusters. When
the degree of doping is increased, the concentration of Si_nAs_m complexes should increase and the concentration of Si_m clusters should
decrease. This is in agreement with the mass-spectroscopic data
(Figs. 5.10 and 5.11). Complexes of the Si_nAs_m type may be regarded as nuclei of the second phase, although they do not have visible boundaries and consequently a silicon single crystal represents
a single phase. However, at very high concentrations these nuclei
may grow and may be precipitated in the form of occlusions, which
have been observed experimentally by metallographic analysis and
by infrared transmission.

Thus, complexes may be the forerunners of chemical
compounds.

Clearly, As_m and Sb_m (or $SiAs_m$, $SiSb_m$) clusters exhibit
donor properties only if the value of m is odd.

When the value of m is odd, all these clusters have an excess
electron, which is likely to be located in the conduction band because of its low ionization energy. In the case of even values of
m, the "excess" electrons bind atoms into helium-like molecules
and therefore these electrons do not fill the conduction band. This
has been found in an analysis of the curves presented in Fig. 5.4.
Unfortunately, the limitations of the MS-7 mass spectrometer have

prevented Fistul' from detecting ions with such high values of m/e as 296 for As_4 or 484 for Sb_4.

The continuous variation of the properties in the transition from one state to another should not be understood too literally. Experiments on Bi [50] have shown that the ordering of atoms in liquid bismuth differs basically from the ordering in the solid state. A somewhat similar situation occurs also in phosphorus-doped silicon; although the vapor phase contains phosphorus clusters P_m [53], there are no such clusters in solid silicon (otherwise it would be difficult to explain the difference between the curves in Figs. 5.5 and 5.4).

We have already mentioned that clusters in solids are also structure defects. Naturally, quasi-crystals in liquids do not have lattices with completely filled sites. Shvidkovskii [50] reports that "in the majority of substances the volume increases by about 10% at the melting point, which represents one hundred vacant sites in the crystal lattice of a cube whose edge is equal to ten average interatomic distances. The presence of vacant sites naturally produces disorder in the atomic distribution (compared with the ideal order in a crystal lattice) and consequently the height of the potential barrier for interparticle interactions may be lowered locally." It follows that there should be a close relationship between the behavior of impurity atoms and structure defects (dislocations and vacancies) in a host lattice. In fact, such a relationship has been established experimentally. This problem will be considered in detail in the next chapter.

The explanation of the phenomenon of polytropy given here is acceptable only if clusters do exist in the melt. However, this has been demonstrated only for nonmetallic liquids. There are many indirect experiments which indicate that the same is true of metallic liquids. However, the experiments carried out so far have not given us the final answer to the molecular structure of metals. Therefore, the explanations put forward here should be regarded as an attempt to present a unified picture of the experimental data on liquids, gases, and solids. Such an attempt is basically in the nature of a statement of the problem rather than its solution. The true mechanisms of the phenomena resulting in the impurity polytropy are very complex and involve fundamental aspects of solid-state and liquid state physics.

Unfortunately, considerable difficulties are encountered in investigations of the polytropy. The problem is to detect "quasi-occlusions" in a crystal when this crystal is still in the single-phase state, as judged from the conventional point of view. Such occlusions can hardly be detected by x-ray methods since even 1% of an impurity cannot be found by this method. The electron microscopy is also unsuitable because of the small dimensions of these occlusions. One may hope to obtain some results by local analysis of the secondary x-ray emission when a crystal is bombarded with electrons [54], provided a sufficiently high resolving power is available. At present, this method can be used to analyze volumes ≈ 1.0 μ^3, so that it can be used only as an auxiliary method in investigations of the impurity polytropy.

Useful information may be obtained from the scattering of x-rays, γ-rays, and possibly neutrons. Much is expected from the further developments of methods involving the transmission of ultrasound through crystals and liquids. Finally, it would be desirable to study the viscosity of melts near the crystallization temperature.

Careful investigations of the dependence of the vapor composition above a semiconductor-impurity system on the impurity concentration and temperature could also be a valuable aid.

It would also be necessary to interpret the results obtained in a quantitative manner. For this purpose, it would be necessary to consider the thermodynamics and kinetics of the crystallization process, bearing in mind that liquids are not monomeric but contain certain amounts of various dimeric, trimeric, and polymeric clusters.

A correct approach may possibly be to develop Frenkel's treatment of heterophase fluctuations [48], on which the conventional theory of spontaneous crystallization is based [49]. The rate of nucleation of a polymeric ("m-meric") complex in the liquid phase can be written in the form similar to the rate of nucleation of ordinary crystallization centers [49]:

$$g_m = N_i^1 \exp\left[-\frac{E_m}{k_0 T}\right] \exp\left[-\frac{c\sigma_m}{T\,(\Delta T_m)^2}\right]. \qquad (5.2.14)$$

Here, E_m is the activation energy for the transition of an atom to an "m-meric" complex; c is a constant; σ_m is a quantity analogous to the surface tension; ΔT_m is the concentration supercooling of the melt due to the crystallization of an "m-meric" complex: $-\Delta T = T - T_{0c}$.

Since, in the equilibrium state

$$N_m^1 = K_1 g_m, \qquad (5.2.15)$$

it follows that the concentration of an "m-meric" complex is

$$\frac{N_m^1}{N_1^1} = K_1 \exp\left[-\frac{E_m}{k_0 T}\right] \exp\left[-\frac{\omega \sigma_m}{T (\Delta T_m)^2}\right]. \qquad (5.2.16)$$

This expression has a maximum at a temperature higher than the true crystallization temperature because $T_{0c} > T_0$.

An experimental confirmation of this maximum would serve as a starting point for the development of a theory of crystallization of a liquid containing polymeric aggregates (clusters) and then a theory of crystallization of a doped (i.e., containing complexes) liquid.

Chapter VI

Preparation of Heavily Doped Semiconductors*

The properties of a growing crystal are governed primarily by the nature of the crystallizing substance and by the characteristics of the crystallization process; they are affected also by the growth conditions. The latter may depend strongly on the apparatus used. Moreover, the nature and concentration of the dopant may affect considerably the process of growth and the properties of an ingot. The latter two factors are particularly important in the case of heavily doped semiconductor single crystals.

In this chapter, we shall not consider the general problems of the theory and practice of the preparation of single crystals but we shall concentrate our attention on those characteristic features of the crystallization process which are associated with the presence of impurities in relatively high concentrations. Our discussion will be concerned with the growing of crystals from the melt, which is the most widely used method in the preparation of single crystals.

§6.1. Stability of a Smooth Crystallization Front

A characteristic feature of the growth of heavily doped semiconductor single crystals is that they frequently grow under conditions such that a smooth (planar) crystallization front becomes

*Chapter 6 was written by M. G. Mil'vidskii.

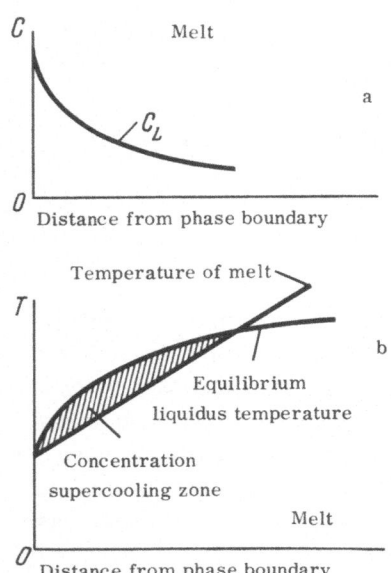

Fig. 6.1. Impurity (a) and equilibrium crystallization temperature (b) distributions at the boundary separating solid and liquid phases.

unstable. The main cause of such instability, resulting in the appearance of a cellular structure (Fig. 5.12), is the concentration supercooling due to a high impurity concentration in the melt. We shall consider this problem in detail.

The growth process is strongly affected by the distribution of a dopant in the melt at the boundary between the solid and liquid phases. When the distribution (segregation) coefficient k is less than unity (which is usually the case), the impurity is pushed out into the melt by the growing ingot. Under ideal mixing conditions, the impurity concentration in the melt should be uniform during the whole growth process.

In most of the practical techniques for growing crystals the mixing is far from complete and an impurity-rich layer forms at the surface of the growing ingot. In this case, the impurity concentration increases until a dynamic equilibrium is reached when the amount of the impurity expelled into the melt by the growing ingot becomes equal to the amount transferred away from the crystallization front by the diffusion and convection mixing. The impurity distribution at a liquid−solid phase boundary is shown schematically in Fig. 6.1a. Since the melt of any given composition has its own equilibrium crystallization temperature, we can use Fig. 6.1a to show schematically the distribution curve of the equilibrium crystallization temperatures in a transverse section through the melt (Fig. 6.1b). For a crystal to grow there should be a positive temperature gradient in the melt. Usually, the thermal conditions of growth are selected so that the real temperature of the melt is higher than the crystallization temperature (with the exception of a small supercooled region in the immediate vicinity of the phase boundary). When the distribution of the equilibrium crystalliza- tion temperatures is of the type shown in Fig. 6.1b, even when there is a positive temperature gradient in the melt, a region in which the melt temperature is lower than the liquidus temperature may appear at some distance from the crystallization front. In this region, the melt will be supercooled. This is known as "concen- tration supercooling," since it is the consequence of a change in the impurity concentration in the melt [1]. Concentration super- cooling is very unlikely in the growth of single crystals with low impurity concentrations. However, it may be observed in the prep- aration of heavily doped semiconductors.

The mathematical analysis of the concentration supercooling uses the impurity distribution in the melt. Impurity atoms ex- pelled into the melt by the growing ingot are transferred away from the crystallization front by the diffusion and convection taking place in the melt. The convection must be allowed for in the growth of crystals from stirred melts. Using the concept of a diffusion layer [2-5], we can apply the same mathematical analysis to melts stirred and not stirred during the crystal growth: the only difference is the choice of the boundary conditions.

We shall consider a one-dimensional steady-state case, mak- ing the following assumptions: 1) the diffusion of impurities in the solid state is negligible; 2) the liquidus and solidus curves in the range of concentrations of interest to us are linear and the distri- bution coefficient of impurities is constant and independent of the concentration; 3) the distribution of the true temperatures in the melt is linear.

The impurity distribution can be found by solving the diffu- sion equation [Eq. (6.1.1)], using the boundary conditions given by Eqs. (6.1.2) and (6.1.3):

$$D \frac{d^2C}{dx^2} + f \frac{dC}{dx} = 0. \tag{6.1.1}$$

Here, $C(x)$ is the impurity concentration in the melt at a distance x from the crystallization front; D is the diffusion coefficient of the impurity in the melt; f is the rate of growth of a crystal.

For the growth from the melt which is not stirred:

$$C(\infty) = C_L. \tag{6.1.2a}$$

For the growth from the melt which is stirred:

$$C(\delta) = C_L. \tag{6.1.2b}$$

In both cases, at the boundary separating the two phases, we have

$$D \frac{dc}{dx} + f(C_0 - C_S) = 0, \tag{6.1.3}$$

$$C_S = kC_0, \tag{6.1.4}$$

where C_0 is the impurity concentration in the melt at the phase separation boundary; C_S is the impurity concentration in the solid state.

The thickness of the diffusion layer δ can be calculated from the following expression [2]:

$$\delta = 1.6 \cdot D^{1/3} \cdot v^{1/6} \cdot \omega^{-1/2}, \tag{6.1.5}$$

where v is the kinematic viscosity of the melt; ω is the angular velocity of rotation of the crystal during growth.

The solutions of the equations just given have the following form:

a) for the growth from the melt which is not stirred [6]:

$$C(x) = C_L \left[1 + \frac{1-k}{k} \exp\left(-\frac{f}{D}x\right) \right], \tag{6.1.6}$$

$$C_0 = \frac{C_L}{k}; \tag{6.1.7}$$

b) for the growth from the melt which is stirred:

$$C(x) = \frac{kC_L}{k + (1-k) \exp\left(-\frac{f\delta}{D}\right)} \left[1 + \frac{1-k}{k} \exp\left(-\frac{fx}{D}\right) \right] \tag{6.1.8a}$$

and for $x > \delta$,

$$C_x = C_L. \tag{6.1.8b}$$

Then,

$$C_0 = \frac{C_L}{k + (1 - k) \exp\left(-f \frac{\delta}{D}\right)}. \tag{6.1.9}$$

Using Eqs. (6.1.4), (6.1.7), and (6.1.9), Eqs. (6.1.6) and (6.1.8a) can be reduced to:

$$C(x) = C_S + \frac{1 - k}{k} C_S \exp\left(-\frac{fx}{D}\right). \tag{6.1.10}$$

Here, $C(x)$ is expressed in terms of C_S instead of C_0 or C_L.

The liquidus temperature for any impurity concentration is found from the phase diagram

$$T_E(x) = T_0 + mC(x), \tag{6.1.11}$$

where m is the slope of the liquidus line (in our case $m = \text{const}$); T_0 is the melting point of the pure substance.

For a linear temperature distribution in the melt, the true temperature $T_A(x)$ is expressed in the form

$$T_A(x) = T_0 + mC_0 + G_L(x), \tag{6.1.12}$$

where G_L is the temperature gradient in the melt.

The magnitude of the supercooling $S(x)$ can be found from Eqs. (6.1.11) and (6.1.12):

$$S(x) = -m[C_0 - C(x)] - G_L(x). \tag{6.1.13}$$

A necessary condition for the concentration supercooling is $S(x) > 0$. Combined with Eq. (6.1.13), this condition is equivalent to the following inequality:

$$\left(\frac{dS}{dx}\right)_{x-0} > 0. \tag{6.1.14}$$

Using Eqs. (6.1.14), (6.1.10), and (6.1.13), we can obtain the conditions for the concentration supercooling in the growth from stirred and nonstirred melts:

$$C_S \geqslant C_{Sc}, \tag{6.1.15}$$

where

$$C_{Sc} = \frac{k}{1-k}\left[-\frac{D}{fm}\right]G_L. \tag{6.1.16}$$

In the presence of concentration supercooling, a smooth crystallization front becomes unstable [7, 8]. Any movement of the phase separation boundary in the direction of the melt, caused by random fluctuations, will extend to the supercooled region. A bulge formed in this way will expel the impurity into the melt and establish an impurity distribution similar to that at other points of the phase separation boundary (Fig. 6.2). This produces a concentration gradient between the tip of the bulge and its base. Consequently, impurities will flow from the tip of the bulge (where the impurity concentration is higher) and this flow will reduce the concentration near the point B (Fig. 6.2), which will make it possible for the crystal to grow at this point. At the same time, the impurity concentration near the base of the bulge will rise and the melting point will decrease. Due to a difference between the concentrations ahead of points A and B (Fig. 6.2), the crystal will grow at these points at the same rate, in spite of the fact that the temperature at point B is higher than that at point A. Thus, the bulge, formed at the phase separation boundary due to random fluctuations, becomes stable.

The height of the bulge is thus governed by the fact that it is moving in the direction of higher temperatures. However, the bulge cannot extend beyond the region which is defined by the equilibrium melting point of the pure substance. Another restriction on the height of the bulge is imposed by the rate of impurity diffusion from the vicinity of the tip of the growing bulge. The impurity diffusion

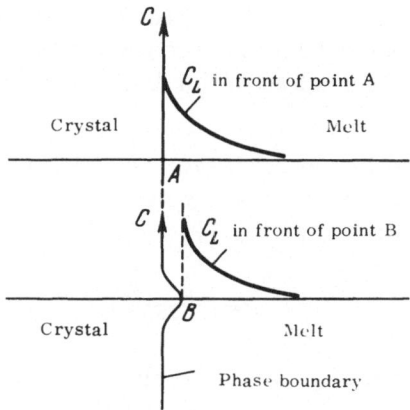

Fig. 6.2. Impurity distribution in the melt near a phase boundary with a bulge on its surface.

governs also the depression of the melting point at the base of the projection.

A bulge influences a very limited region of the crystallization front surface. Outside this region, other stable bulges may appear. Initially, these bulges are distributed at random along the phase separation boundary. However, as the number of such bulges increases, the "spheres of influence" of individual bulges begin to overlap and the crystallization front assumes the cellular structure (Fig. 5.12).

A more detailed theoretical analysis shows that the super-cooling of the melt is insufficient to disturb the stability of a smooth crystallization front. To determine the stability condition, Temkin [9] compared the heat transfer from a bulge with the heat transfer from a planar front. The problem has been solved by Voronkov [10] in a more general form, using the theory of small perturbations. A smooth front becomes unstable when

$$\Gamma > G = \frac{\varkappa_1 G_1 + \varkappa_2 G_2}{\varkappa_1 + \varkappa_2}, \qquad (6.1.17)$$

where Γ is the equilibrium crystallization temperature gradient; $\varkappa$ is the thermal conductivity; G is the temperature gradient at the phase separation boundary; the subscripts "1" and "2" refer, respectively, to the melt and to the solid crystal.

The instability criterion (the criterion of the formation of the cellular structure) does not include the temperature gradient in the melt G_1 but the average gradient G for both phases, which is larger than G_1 if $\varkappa_1 > \varkappa_2$.

Substituting into Eq. (6.1.17) the expression for Γ taken from [10], we can find the critical impurity concentration in the solid phase at which cells are formed [11]:

$$C_{Sc} = D \frac{Q \cdot k}{k_0 T_0^2 (1-k)^2 f} \frac{G}{f}. \qquad (6.1.18)$$

Here, Q is the heat of fusion per unit volume of the melt; k_0 is the Boltzmann constant.

Thus, the factors which govern the stability of a smooth crystallization front in the growth of heavily doped single crystals are: the thermal conditions at the phase separation boundary; the rate of growth; the nature and concentration of the dopant.

In the case of heavily doped single crystals of semiconducting compounds (for example, $A^{III}B^{V}$), the concentration supercooling and the cellular structure may appear due to a departure of the melt from the stoichiometric composition.

An experimental check of the theory of concentration supercooling in the crystallization of pure metals has been reported in [7, 12, 13]. The results obtained confirmed the theoretical conclusion that the critical impurity concentration in a crystal depends linearly on the ratio G_L/f, in agreement with Eq. (6.1.16).

There have been few investigations of the influence of concentration supercooling on the growth of semiconductor single crystals. Characteristic impurity substructures have been observed in germanium doped with bismuth [14, 15] and tin [16], with iron, manganese, and cobalt [17, 19], as well as with silicon [20] and nickel [21].

The first qualitative investigations of the influence of concentration supercooling on the growth of germanium single crystals heavily doped with gallium were carried out by Bolling et al. [22]. They grew crystals in horizontal graphite boats along the [321] direction and the crystallization front structure was investigated by the decantation method. The concentration of gallium in the melt varied from 0.014 to 0.75 at.%. Twins were observed in crystals containing more than 0.043 at.% of gallium. At higher gallium concentrations the density of twins increased and the phase separation boundary assumed a ridged structure (the greatest number of ridges was observed at twin boundaries). The separation boundary became cellular only when the concentration of gallium had reached 0.2 at.% and the cells were found to be well-developed pyramids. Theoretical estimates [cf. Eq. (6.1.16)] indicate that, under the growth conditions used in these investigations, a concentration supercooling zone should have formed in front of the phase separation boundary at gallium concentrations of the order of 0.01%. Since the cellular structure was not observed until gallium concentration had reached 0.2%, Bolling et al. concluded that the formation of bulges at the phase boundary in germanium is a difficult process and that the supercooling must be greater than in metals in order to observe the cellular structure. This may be due to the difference between the growth mechanisms of the majority of metals and of germanium, or due to a considerable increase in the separation boundary energy when bulges are formed in germanium.

The same germanium–gallium system has also been investigated using the Czochralskii method for growing the crystals from a stirred melt [23]. In contrast to Bolling et al., Bardsley et al., [23] have concluded that the cellular structure may appear in gallium-doped germanium single crystals when the critical supercooling gradient is small. The most likely cause of this sharp difference between the results of two teams may be a considerable indeterminacy in the selection of numerical values of the quantities which occur in the formula for the critical impurity concentration causing the appearance of the cellular structure; in particular, this applies to the values of the diffusion coefficients of impurities in the melt and of the thermal conductivity of semiconductors near the melting point, which have not been available until recently [11, 24-28]. A detailed investigation of the growth of heavily doped germanium and silicon single crystals under concentration supercooling conditions has been reported in [29]. Silicon single crystals have been grown by the Czochralskii method and germanium single crystals by the zone leveling method along the [111] direction. In both cases, measurements of the electrical conductivity and the Hall effect have been used to determine the critical impurity concentration in the ingot necessary for the appearance of a polycrystalline

TABLE 6.1. Doping Limits C_{Sl} of Germanium and Silicon with Various Impurities
($f = 1$ mm/min)

Crystal	Impurity	Doping limit C_{Sl}, cm^{-3}	Solubility limit C_S, cm^{-3}	$\|m\|$, deg/cm^3 according to [30]	k, according to [31]	D, cm^2/sec according to [3, 29]	C_{Sc}, cm^{-3}
Ge	Ga	$2 \cdot 10^{20}$	$5 \cdot 10^{20}$	$5.9 \cdot 10^{-21}$	0.037	$1.0 \cdot 10^{-1}$	$1.4 \cdot 10^{19}$
	In	$0.8 \cdot 10^{18}$	$4 \cdot 10^{18}$	$8.0 \cdot 10^{-21}$	0.001	$(1.0 \cdot 10^{-1})$	$1.2 \cdot 10^{17}$
	P	$3 \cdot 10^{19}$	$1.8 \cdot 10^{20}$	$(7.6 \cdot 10^{-21})$*	0.080	$(0.5 \cdot 10^{-1})$	$5.1 \cdot 10^{18}$
	As	$2 \cdot 10^{18}$	$1.3 \cdot 10^{19}$	$8.6 \cdot 10^{-21}$	0.020	$1.3 \cdot 10^{-1}$	$2.8 \cdot 10^{18}$
	Sb	$6 \cdot 10^{18}$	$4 \cdot 10^{20}$	$7.1 \cdot 10^{-21}$	0.003	$0.7 \cdot 10^{-1}$	$2.7 \cdot 10^{17}$
Si	B	$3 \cdot 10^{20}$	$6 \cdot 10^{20}$	$(9.3 \cdot 10^{-21})$	0.800	$2.4 \cdot 10^{-4}$	$4.3 \cdot 10^{20}$
	P	$8 \cdot 10^{19}$	$1.4 \cdot 10^{20}$	$(9.3 \cdot 10^{-21})$	0.350	$5 1 \cdot 10^{-4}$	$1.7 \cdot 10^{20}$
	Sb	$2.5 \cdot 10^{19}$	$6.7 \cdot 10^{19}$	$5.5 \cdot 10^{-21}$	0.023	$1.5 \cdot 10^{18}$	$6.4 \cdot 10^{18}$

*Calculated values are given in parentheses.

structure. This concentration has been regarded as the doping limit. The results obtained are collected in Table 6.1. The same table gives the published data on the maximum solubility of various impurities in germanium and silicon, as well as values of the impurity concentration necessary for the appearance of concentration supercooling, calculated using Eq. (6.1.16).

It follows from the results presented in Table 6.1 that the doping limits reached in practice are 2-10 times lower than the maximum solubilities of impurities. This indicates that the growth conditions depart considerably from equilibrium. The main factor which governs the doping limit is the concentration supercooling. In the majority of cases, the impurity concentration causing such supercooling is lower than the maximum doping limit. For a polycrystalline structure to appear, some concentration supercooling is necessary and the doping limit represents that impurity concentration at which the necessary value of the supercooling is reached. The opposite results obtained for phosphorus and arsenic in silicon are attributed in [29] to the insufficient impurity of the dopants.

We must mention at least two important deficiencies of the study reported in [29]. First, it is known [32] that at high donor concentration in a crystal not all the donors are electrically active and therefore the use of electrical measurements for the determination of the impurity concentration may result in considerable errors. Secondly, the value of C_{Sc} is identified with the doping limit,

Fig. 6.3. Initial stage of the development of the cellular structure in a longitudinal cross section of a heavily
doped silicon single crystal.

which is taken as equal to the impurity concentration in the direct
vicinity of a polycrystalline boundary. However, investigations of
other workers [11] have shown that a cellular structure may appear in an ingot much before the transition to the polycrystalline
state, i.e., at much lower concentrations of the dopant in a sample.
The difference between the critical concentration necessary for the
appearance of the cellular structure and the doping limit may be
very considerable and it may depend on the growth conditions. For
example, it has been reported in [11] that when arsenic-doped silicon single crystals were grown by the Czochralskii method at a pull
rate of 1.3 mm/min, the cellular structure appears at $C_S = 6.8 \cdot 10^{19}$ cm^{-3} and the doping limit is approximately $2.2 \cdot 10^{20}$ cm^{-3}.

An experimental check of the stability conditions for a smooth
crystallization front [Eq. (6.1.18)] is described in [11]. That investigation has been concerned with silicon single crystals, grown
by the Czochralskii method along the [111] direction and heavily

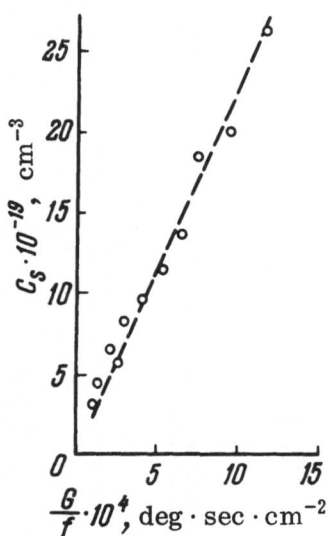

Fig. 6.4. Dependence of the critical impurity concentration on the growth conditions of silicon single crystals heavily doped with arsenic.

doped with aluminum, antimony, arsenic, or phosphorus. The gradients G_1 and G_2 have been measured by growing a thermocouple into a crystal. A series of measurements, using different growth rates and different shielding, has made it possible to determine the thermal conductivities $\varkappa_1$ and $\varkappa_2$ [26]. The impurity concentration in a crystal just ahead of the cells has been determined from the Hall effect, with a suitable correction for the phenomenon of polytropy, in accordance with [32]. Figure 6.3 shows a typical example of the initial stage of the development of the cellular structure along a longitudinal cross section (110) of a heavily doped silicon single crystal. The originally smooth "growth bands" have become wave-like, indicating a periodic deformation of the smooth crystallization front.

To check the proportionality of C_S and G/f, arsenic-doped crystals have been grown using various growth rates and gradients G. The results, presented in Fig. 6.4, show that the linear law is obeyed satisfactorily. Knowing the value of C_S for various impurities under given growth conditions, Eq. (6.1.18) can be used to find the impurity diffusion coefficients in the melt. These coefficients are $3 \cdot 10^{-4}$, $1 \cdot 10^{-4}$, $6 \cdot 10^{-5}$, and $5 \cdot 10^{-5}$ cm²/sec for Al, Sb, As, and P, respectively.

Analysis of the stability of a smooth front [10] shows that, initially, the front suffers periodic perturbation with a period d. The value of d represents the width of the cells in the initial stage of their development and is given by the expression (cf. [10])

$$d = 2\pi \sqrt[3]{\frac{2D}{k \cdot f \cdot G} \frac{\sigma T_0}{Q}}, \qquad (6.1.19)$$

where Q is the heat of fusion of the semiconductor, per 1 g of the melt.

The surface tension at the boundary between the solid and liquid phases, σ, which occurs in the above formula may depend on the impurity concentration. For not too low values of k, Eq.(6.1.19) is in good agreement with experiment. Thus, assuming by analogy with germanium [33] that $\sigma = 200$ erg/cm^2, we obtain—for $f = 1.3$ mm/min and $G = 35$ deg/cm—a value $d = 220$ μ for arsenic-doped silicon; under these conditions, the width of the cells is 200-250 μ.

There are no experimental data on the stability of a smooth crystallization front in the growth of heavily doped single crystals of other semiconductors, in particular semiconducting compounds. Nevertheless, the concept of the concentration supercooling may obviously be applied to other semiconductors. In fact, the cellular structure has been observed in the surface of the crystallization front of indium antimonide single crystals grown from indium- or gallium-rich melts [66].

The concentration supercooling plays a particularly important role in the growth of single crystals of easily decomposed semiconducting compounds, in which a deviation from stoichiometry may be due to the partial evaporation of the volatile component. The experimental data on gallium arsenide indicate that even slight deviations of the composition of the melt from stoichiometry reduce strongly the yield of single crystals [31].

§6.2. Some Features of Impurity Distribution

in Heavily Doped Semiconductors

A nonuniform distribution of impurities in a single crystal may be due to accidental and systematic factors. The accidental factors include uncontrolled changes during the process of crystallization due to imperfections of the apparatus used to grow single crystals. Specifically, they are: fluctuations of the power in the heater circuit in the absence of stabilization of the thermal conditions; asymmetry of the temperature field in the melt at the phase separation boundary; fluctuations of the growth rate because of the errors in the mechanical drive and lack of stabilization of the power supply, etc. Inhomogeneities due to accidental factors are not reproduced from one apparatus to another and are not even repeat-

able in successive runs in the same apparatus. To avoid accidental
inhomogeneities, it is necessary to ensure rigorously the stability
of thermal conditions and of the operation of the mechanical drives,
as well as a high quality of the manufacture and assembly of all
parts of the apparatus used to grow single crystals.

Systematic factors are due to characteristic features of the
single-crystal growth process itself. Here, we may include the
natural segregation of an impurity along the length of a growing ingot, the nonuniform distribution of the impurity in the interior due
to the periodic nature of the crystallization process, the appearance
of various crystallographic planes during growth, the shape of the
crystallization front, the nature and concentration of the impurity
itself, etc. Inhomogeneities of this type can be investigated and reduced to minimum by a suitable selection of the growth conditions.
It is well known that impurities with distribution coefficients differing from unity, are distributed nonuniformly along the length of an
ingot: impurities with $k > 1$ are collected at the front end of the
crystal and impurities with $k < 1$ are expelled to the near end of a
crystal. This natural segregation is particularly important in the
growth of heavily doped single crystals because a change in the impurity concentration in the melt or in the ingot alters the conditions
of stability of a smooth crystallization front and gives rise to departures from the single-crystal structure. For this reason, the
attention of investigators has been concentrated on the development
of those growth methods which can ensure a uniform distribution
of an impurity along the length of a single crystal. Many of the proposed methods are based on the maintenance of constant dopant concentration in the melt throughout the whole process. In the case of
the Czochralskii method, we must mention here the floating crucible
technique [34-37] and the supply of the melt with an ingot of the same
composition as the growing ingot [38-40]. The zone leveling technique is also widely used [41]. However, the majority of these techniques counteract changes in the impurity concentration in the melt
due to impurity segregation and not those due to evaporation. Therefore, such techniques are suitable only for doping with nonvolatile
impurities or when the evaporation of the impurity from the melt
is suppressed.

When doping with volatile impurities, the best method to use
is evaporation to compensate for the increasing concentration of
the impurity in the melt (when $k < 1$) [42-44]. This method is based

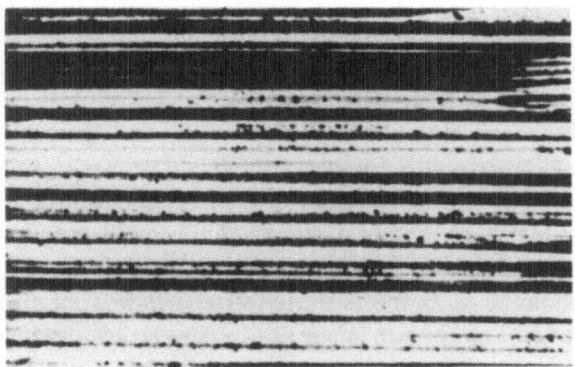

Fig. 6.5. Periodic inhomogeneity of the impurity distribution in "growth bands" in a longitudinal cross section of a
heavily doped silicon single crystal.

on a selection of those crystal growth conditions (growth rate, atmosphere in the furnace, stirring conditions, etc.) under which the
amount of the impurity expelled into the melt by the growing ingot
is, at each moment, equal to the amount of the impurity evaporat-
ing from the free surface of the melt.

Among other methods for growing single crystals with a uniform impurity distribution along the length, we must mention the
method in which the growth rate [3] or the rate of rotation of a
single crystal [46] follows a definite program. The method is based
on the dependence of the effective distribution coefficient of the doping impurity on the growth conditions. The programing is such that
a change in the impurity concentration in the melt due to the segregation is compensated by a corresponding change in the distribution coefficient.

More detailed investigations have shown that the impurity
distribution along the length of a growing crystal under real conditions is far from a smooth curve and is in fact periodic. This
periodic impurity distribution ("growth bands") is a serious source
of volume inhomogeneity; it has been investigated by many workers
[47–50]. Periodic fluctuations in the impurity content have been
observed in crystals grown by the Czochralskii method as well as
in crystals prepared by the horizontal zone or directional crystallization method. In pulled single crystals, several systems of
"growth bands" have been observed in addition to periodic inhomogeneities which are associated with the rotation of the crystal and

the crucible (these inhomogeneities have been observed at steps
equal to the quotient of the value of the growth rate and the corresponding rate of rotation); fine "growth bands" with a period of
3-20 μ have also been observed (Fig. 6.5). These have also been
found in crystals grown without rotation. The periodic inhomogeneity associated with the rotation of a crystal and a crucible is
evidently due to periodic fluctuations of the growth rate because
of the presence of an asymmetric temperature field near the phase
separation boundary. A more complex effect is responsible for the
fine growth bands observed in crystals prepared by various methods. These bands are obviously due to the periodic nature of the
crystallization process itself, one of whose external manifestations
may be oscillations of the phase boundary, as reported in [53]. In
this connection, it is worth mentioning the work of Voronkov [54],
who considers that intrinsic fluctuations of the crystallization front
are the result of deviations from thermodynamic equilibrium at the
phase boundary.

Some investigators [51, 55] have attributed periodic inhomogeneities to the concentration supercooling in the melt. We have
already shown that concentration supercooling is observed only at
high dopant concentrations and that it disturbs the stability of a
smooth crystallization front. However, growth bands are also observed in very pure single crystals. Nevertheless, the dependence
of periodic inhomogeneities in heavily doped single crystals on the
nature of the dopant and particularly on the value of the distribution
coefficient [49, 56] indicates that the accumulation of impurities in
the diffusion layer of the melt near the crystallization front may
play some role in the formation of growth bands In analyzing the
causes of periodic inhomogeneities, it is necessary also to bear in
mind a possible influence of the convection currents in the
melt, which result in fluctuations of the crystallization front
temperature [57].

One of the causes of a nonuniform distribution of impurities
in the interior of a single crystal may be the dependence of the effective distribution coefficient on the orientation of the crystallization
front surface. An investigation of the distribution of antimony in
germanium has indicated [13, 58] that a gradual change of the direction of growth from [100] to [110] and [111] alters the distribution coefficient on the average by 5-10% for crystal growth rates
up to 20 cm/hour. Investigations of the characteristic features of

the distribution of dopants in InSb single crystals have established that growth along the [111] direction sometimes produces a planar (111) facet of macroscopic dimensions when the phase separation boundary is convex in the direction of the melt. The effective distribution coefficients of Te, Se, S, and of other impurities in the (111) facet differ by a factor of several times from the distribution coefficients in neighboring regions of the crystallization front. This produces "channels" [59-61] and a considerable nonuniformity of the distribution of these impurities in the interior of single crystals. The appearance of channels is known as the "facet effect." A similar effect has been discovered in the growth of single crystals of germanium [50], silicon [62], and gallium arsenide [63]. It has been found that, under favorable conditions, any of the close-packed planes of the (111) system may give rise to the "facet effect." The possibility of the growth of facets is affected strongly by the shape of the crystallization front, the orientation of a crystal, the thermal conditions at the phase separation boundary, as well as the nature and concentration of the doping impurity. The magnitude of the "facet effect" for a given material depends strongly on the nature of the impurity. The same is true of identical impurities in different semiconductors.

A general theoretical treatment shows that a close-packed (111) facet appears at the phase separation boundary and the growth is maintained within this facet when the supercooling of the melt is locally stronger than in neighboring regions. Assuming a strong supercooling of the melt within the limits of the (111) facet, an attempt has been made [64] to explain the increase in the concentration of the dopant within a channel by a change in the composition of the solidifying phase due to a decrease in the crystallization temperature, following the solidus curve. Such an analysis shows that the magnitude of the "facet effect" should increase as the equilibrium distribution coefficient of impurities increases (for $k < 1$) which is rarely true in practice. It is more likely that the difference between the step-like (laminar) growth rates within a facet and at neighboring regions of the crystallization front and the influence of the adsorption of impurities on the surface layer of a growing crystal are the important factors [65]. Unfortunately, it is difficult to carry out an accurate experimental check of this theory because of the considerable indeterminacy of the values of the quantities which occur in the expression for the effective impurity distribution coefficient within a facet [65]. Nevertheless, this theory explains qualitatively many experimental observations. For example, the difference between the behavior of the donor and acceptor impurities in silicon and germanium single crystals (the former tend to exhibit more strongly the "facet effect") may be attributed to the difference between their adsorption energies due to the Coulomb interaction in the adsorbed layer. The relatively small change in the effective distribution coefficients of the dopants due to the "facet effect" in silicon, compared with the corresponding change in germanium and particularly in indium antimonide, is due to the weaker adsorption of impurity atoms on crystals with higher melting points. We must bear in mind that the possibility of specific adsorption of impurities on the surface layer of the crystallization front may govern not only the absolute value of the effect but may also affect considerably the kinetics of the growth of the facet itself.

These features of the dopant distribution in the interior of an ingot apply not only to doped single crystals. However, the specific features of the growth under concentration supercooling conditions produce inhomogeneities which are typical only of heavily doped semiconductors. These inhomogeneities include the segregation of the impurities along cell boundaries and the appearance of a second phase. We shall consider these two effects in more detail.

As already mentioned, the formation of bulges in the crystallization front produces a diffusion flow of impurities from the tip of the bulge to its base (cf. Fig. 6.2) and this increases the impurity concentration along the cell boundaries (for impurities with $k < 1$).

A theoretical analysis of the impurity distribution in a crystal with a cellular substructure grown from a stirred melt is given in [67, 68]. Assuming a steady-state impurity distribution at the phase separation boundary, the absence of diffusion in the solid state, and a constant impurity concentration within a cell, Tiller [67] obtained the following expression for the average value of the excess impurity concentration at a cell boundary (this applies to impurities with the distribution coefficient $k < 1$):

$$\overline{\Delta C} = \frac{k^0 D}{m(1 - k^0)} \left\{ \frac{+ [m C_\infty (1 - k^0)/k^0 D] + \dfrac{G_s(0)}{f}}{W + k^0/(1 - k^0)} \right\}, \qquad (6.2.1)$$

where k^0 is the equilibrium impurity distribution coefficient; D is the impurity diffusion coefficient in the melt; m is the slope of the liquidus line; f is the crystallization rate; C_∞ is the impurity concentration in the interior of the melt; $G_S(0)$ is the temperature gradient in the solid phase at the phase boundary; W is the ratio of the area of the cell boundaries to the total area of cells in the crystallization front.

Taking D from the first factor and placing it in front of the square brackets [bearing in mind that the quantity $C_\infty/k^0 = C_L(0)$ for unstirred melts under steady-state conditions is equal to the impurity concentration in the melt at the crystallization front and that $D/f = \delta$ is equal to the thickness of the diffusion layer [68]] we obtain:

$$\overline{\Delta C} = \frac{k^0}{m\,(1-k^0)} \left\{ \frac{+\,[mC_L(0)\,(1-k^0)] + G_S(0)\,\delta}{W + k^0/(1-k^0)} \right\}. \qquad (6.2.2)$$

We shall transform this formula to fit the case of growth from stirred melts. The stirring of the melt should reduce considerably the values of $C_L(0)$ and δ. The solution of the problem of the impurity distribution in the melt stirred by a rotating crystal has been solved for the steady-state case in [3] in the following form:

$$\frac{C_L(0) - C_S}{C_\infty - C_S} = e^\Delta, \qquad (6.2.3)$$

where $\Delta = f\delta/D$; C_S is the impurity concentration in the crystal, defined as $C_S = k^0 C_L(0)$.

From Eq. (6.2.3) we can find the value of $C_L(0)$ for stirred melts:

$$C_L(0) = \frac{C_\infty e^\Delta}{1 - k^0\,(1 - e^\Delta)}. \qquad (6.2.4)$$

The thickness of the diffusion layer δ for stirred melts is found from Eq. (6.1.5). After substituting the values $C_L(0)$ and δ into Eq. (6.2.2), the latter equation can be used to obtain a rough estimate of the degree of segregation of impurities along cell boundaries in crystals grown from stirred melts:

$$\overline{\Delta C} = \frac{k^0}{m(1-k^0)} \left\{ \frac{+\left[m(1-k^0) \frac{C_\infty e^\Delta}{1-k^0(1-e^\Delta)} \right] + G_S(0) \, 1 \, 6D^{1/3} \nu^{1\,6} \omega^{-1/2}}{W + \frac{k^0}{1-k^0}} \right\} . (6.2.5)$$

Mil'vidskii and Grishina [69] investigated the segregation of impurities along cell boundaries in silicon single crystals heavily doped with arsenic. A theoretical estimate of the value of $\overline{\Delta C}$ for samples with a regular cellular structure has been made using Eq. (6.2.5). For samples with a strongly developed irregular cellular structure, which is usually observed close to a dendritic structure, the value of $\overline{\Delta C}$ can be found, without committing a large error, using a simplified formula deduced in [67]. This formula has been derived on the assumption that, at a fixed temperature gradient in the solid and a constant crystallization rate, the concentration of the dopant in the melt is considerably higher than the critical value necessary for the concentration supercooling:

$$\overline{\Delta C} \approx \frac{C_\infty}{W + k^0/(1+k^0)}. \tag{6.2.6}$$

The results of such calculations are presented in Table 6.2.

The results of a theoretical estimate of the degree of segregation can be compared with the results obtained from the experimental data. An investigation of the structure of single crystals heavily doped with arsenic has shown that the cellular structure is of the block type. The tilt angles between blocks and fragments depend on the degree of development of the cellular structure and vary from 30" to several angular minutes [70].

The most likely cause of the block nature of the cellular substructure is a preferential segregation of impurities along cell boundaries. The higher concentration of the dopant in a cell boundary (for impurities with k < 1) may be a source of considerable local internal stresses, which may produce dislocation arrays. The appearance of dislocations due to a nonuniform distribution of impurities in a crystal has been discussed in [67, 71, 72]. Assuming a certain change in the impurity concentration in the region of a cell boundary, these workers have estimated the density of dislocations resulting from such a change and they have obtained good agreement with the experimental data.

TABLE 6.2.　Degree of Segregation of Impurities
along Cell Boundaries in Silicon Single Crystals
Heavily Doped with Arsenic

Sample No.			1 *	2 †	3 †	4 †
Average arsenic concentration in crystal C_S, cm^{-3}			$2.5 \cdot 10^{20}$ ‡	$7.9 \cdot 10^{19}$	$7.9 \cdot 10^{19}$	$6.0 \cdot 10^{19}$
Angle of tilt between blocks, ϑ			5'20″	2'00″	1'50″	1'10″
Linear dislocation density in boundary N_D, cm^{-1} after Eq. (6.2.8)			$4.0 \cdot 10^4$	$1.5 \cdot 10^4$	$1.4 \cdot 10^4$	$9 \cdot 10^3$
Degree of arsenic segregation along cell boundaries, $\overline{\Delta C}$	Theoretical estimate	Atomic fractions	0.019	0.0063	0.0063	0.0054
		cm^{-3}	$9.5 \cdot 10^{20}$	$3.1 \cdot 10^{20}$	$3.1 \cdot 10^{20}$	$2.7 \cdot 10^{20}$
	From value of ϑ	Atomic fractions	0.017	0.0065	0.0059	0.0039
		cm^{-3}	$8.5 \cdot 10^{20}$	$3.2 \cdot 10^{20}$	$3.0 \cdot 10^{20}$	$2.0 \cdot 10^{20}$

*Strongly developed irregular cellular structure.
†Regular cellular structure.
‡C_S =2.5 · 10^{20} cm^{-3}, according to activation analysis.

Knowing the dislocation density in a boundary, we can solve
the converse problem: we can estimate the degree of segregation
of an impurity along cell boundaries. We shall assume the cell
boundaries to be inclined, with small angles of inclination, and we
shall use a dislocation model of the boundary. Then, the angle of
tilt of the blocks can be used to determine easily the linear density
of dislocations in a boundary.

The distance D_d between dislocations in the boundaries is
defined as

$$\frac{b}{D_d} = 2 \sin \frac{\vartheta}{2}, \qquad (6.2.7)$$

and for small tilt angles $b/D_d \approx \vartheta$ [73]. Here, b is the Burgers vector of a unit dislocation (b = 3.84 · 10^{-8} cm); ϑ is the tilt angle. Hence,

the linear dislocation density in a boundary is

$$N_D = \frac{1}{D_d} \approx \frac{\vartheta}{b}. \qquad (6.2.8)$$

The values of the tilt angles between blocks in the investigated samples and the results of a calculation of the dislocation density in the boundaries, obtained using Eq. (6.2.8), are presented in Table 6.2.

The linear density of dislocations in a boundary is related to a local change in the impurity concentration $\overline{\Delta C}$ [67]:

$$N_D = \frac{1}{b}\left(\frac{\overline{\Delta C}\,\Delta\lambda}{\lambda} - \varepsilon_l\right), \qquad (6.2.9)$$

where ε_l is the elastic limit; λ is the atomic radius of solvent atoms (for example, Si); $\Delta\lambda$ is the absolute difference between the atomic radii of the solvent (Si) and the solute (As).

Assuming that all the macroscopic elastic stresses are relieved by the formation of dislocations (this condition is quite well satisfied in the case of cellular substructures), we can neglect the quantity ε_l in Eq. (6.2.9) and then

$$N_D \approx \overline{\Delta C}\,\frac{\Delta\lambda}{\lambda}. \qquad (6.2.10)$$

The results of a calculation of the quantity $\overline{\Delta C}$, using Eq. (6.2.10) with the value of N_D found from the tilt angles, are also given in Table 6.2. In this calculation, the following numerical values of the quantities occurring in the formula have been used: $b = 3.84 \cdot 10^{-8}$ cm; $\lambda_{Si} = 1.34 \cdot 10^{-8}$ cm; $\lambda_{As} = 1.46 \cdot 10^{-8}$ cm [74].

To obtain a direct estimate of the degree of segregation of an impurity along the cell boundaries, a single-probe method of measuring the electrical resistivity has been used. It is evident from Fig. 5.13 that the distribution of the electrical resistivity along a sample is periodic. The resistivity has minima at the cell boundaries. The distance between the minima in the resistivity distribution curve is practically equal to the cell widths, as determined in a metallographic investigation. The quantity ΔC in the region of

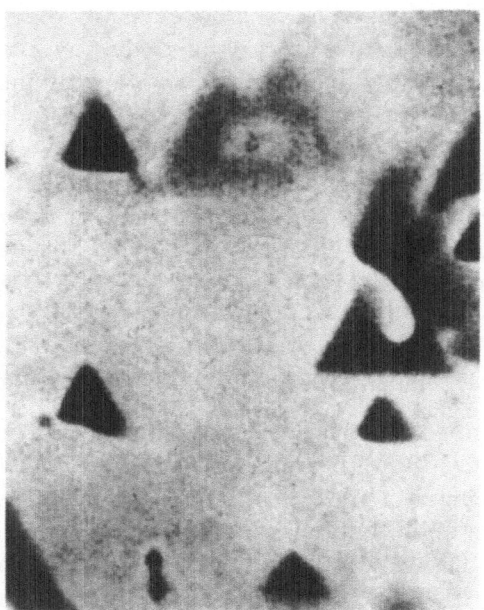

Fig. 6.6. Second-phase occlusions in heavily
doped silicon single crystals, observed in an infrared microscope (magnification: 20).

a cell boundary, calculated from the values of the electrical resistivity, is equal to $5.5 \cdot 10^{19}$ cm^{-3} and a theoretical estimate for the
same sample, obtained using Eq. (6.1.5), gives $\overline{\Delta C} = 1.9 \cdot 10^{20}$ cm^{-3}.

These results show that, for the majority of the investigated
samples, the results of a theoretical estimate of the degree of segregation are in good agreement with the values obtained from tilt
angles between blocks on the assumption of the dislocation nature
of the cell boundaries. For samples with a weakly developed cellu-
lar structure, the theory gives higher values of $\overline{\Delta C}$. Both theoretical and experimental estimates show that the concentration of the
dopant along the cell boundaries is considerably greater than the
average concentration in a crystal.

Second-phase occlusions are the extreme case of an inhomogeneity in a single crystal and are typical of the most heavily doped
ingots. Second-phase occlusions have been observed in silicon
single crystals heavily doped with aluminum, antimony [77, 75],
phosphorus [76], and gallium [78]; in germanium single crystals
doped with arsenic and gallium [79, 50]; and in several other ma-

terials. These occlusions have been observed by electrochemical methods, by chemical etching in various media, by x-ray diffraction microanalysis, by infrared microscopy, and by measuring the microhardness. For impurities which form a phase diagram of the eutectic type, second-phase occlusions are of the eutectic kind; for example, in silicon single crystals, aluminum and antimony are present in the elemental form [75, 77]. For systems in which the formation of semiconductor-impurity compounds is possible, the occlusions may be in the form of compounds. For example, germanium arsenide occlusions have been observed in silicon single crystals heavily doped with arsenic. A typical distribution of occlusions in a heavily doped silicon single crystal is shown in Fig. 6.6.

Second-phase occlusions may form during the growth of heavily doped single crystals either during the crystallization process or during the cooling of a grown crystal because of decomposition (precipitation) of a solid solution. The formation of supersaturated solid solutions is unlikely in the majority of the systems considered because impurities have a retrograde solubility in semiconductors and the solubility at low temperatures is usually little different from the solubility near the melting point of the host lattice [31]. The most likely cause of the formation of occlusions is a nonequilibrium capture by a growing crystal of molten droplets strongly enriched with impurities having very low distribution coefficients in the case of crystallization under conditions of a considerable concentration supercooling.

In theoretical discussions of the impurity distribution problems, the value of the distribution coefficient of the impurity is assumed to be constant. This approximation is justified for the growth of crystals with low impurity concentrations. A considerable amount of the dopant in the melt may change the melting point. The value of the distribution coefficient may then change in accordance with the phase equilibrium lines in the phase diagram. For systems forming ideal solutions, the temperature dependence of the distribution coefficient is described by the following relationship [80]:

$$k = \exp \frac{Q}{R}\left(\frac{1}{T_0} - \frac{1}{T_F}\right), \qquad (6.2.11)$$

where R is the universal gas constant.

This relationship is well satisfied by some systems: for example, this applies to the germanium−gallium system [80]. However, in other cases there are very considerable deviations from the ideal behavior. Anomalously large changes in the distribution coefficient in a certain range of dopant concentrations have been observed in the growth of zinc-doped single crystals of indium antimonide [81], gallium antimonide [82], and gallium arsenide [83]. In view of this, the values of the effective distribution coefficients in the case of heavily doped single crystals should be treated with caution and it is best to use the experimental values.

§6.3. Structure of Heavily Doped Semiconductors

When the concentration of a dopant is well below the solubility limit and a crystal is growing at a stable smooth crystallization front, an increase in the impurity concentration does not produce additional dislocations. In fact, in some cases an increase in the dopant concentration helps to obtain dislocation-free single crystals [84, 85].

Mil'vidskii, Stolyarov, and Berkova [85] investigated silicon single crystals heavily doped with boron, aluminum, phosphorus, arsenic, or antimony. These crystals were grown along the [111] direction. No special measures were taken to prevent the generation of dislocations: the dislocation density in seed crystals was $5 \cdot 10^3$-$1 \cdot 10^4$ cm^{-2}; crystals were grown without "constrictions" in a considerable axial temperature gradient and the crystallization front was bent appreciably. The dislocation density was determined by chemical etching, decoration with copper, and by x-ray diffraction methods. All three methods gave results which were in good agreement.

In p-type single crystals doped with boron and aluminum, dislocations were found along the whole ingot length. The situation was different in n-type single crystals: the dislocations disappeared at a certain dopant concentration. Phosphorus- and arsenic-doped crystals had no dislocations beginning from resistivities of $0.03 \, \Omega \cdot$ cm ($N = 5 \cdot 10^{17}$ cm^{-3}) and antimony-doped crystals were free of dislocations from $0.08 \, \Omega \cdot$ cm ($N = 1 \cdot 10^{17}$ cm^{-3}). An investigation of such single crystals showed that many of the dislocations which penetrated an ingot from the seed gradually moved to the surface; this reduced the dislocation density along the ingot and eventually led to a total disappearance of the dislocations. An increase in the initial concentration of the donor in a crystal accelerated, other conditions being equal, the process of the disappearance of dislocations. Crystals of the n-type remained dislocation-free over a wide range of dopant concentrations.

One of the factors which govern the climb of dislocations to the surface is an increase in the vacancy concentration. When dislocation-free silicon single crystals are required, it is recommended that, initially, a crystal be grown at a high rate in order to supersaturate it with vacancies [86]. Bearing in mind this vacancy mechanism of the climb of dislocations to the surface of a crystal, the results obtained in [85] can be explained as follows.

Vacancies in germanium and silicon act as acceptor centers. Consequently, the introduction of a donor or acceptor impurity may alter the vacancy concentration due to the displacement of the electron-hole equilibrium [87]. If the concentration of the dopant N is much lower than the intrinsic carrier density n_i in the range of temperatures at which plastic deformation is possible, the concentration of vacancies in an ingot is governed, other conditions being equal, by the thermal conditions during growth. When the values of N and n_i become comparable, the influence of the dopant should begin to be felt: the presence of donors should raise the concentration of the vacancies and the presence of acceptors should reduce it.

According to Longini and Greene [88] (cf. Chap. 5), in the case of complete ionization of the acceptor vacancies at temperatures such that $N_D > n_i$, the concentration of vacancies generated in the presence of a donor impurity is given by Eq. (5.1.12).

Thus, if a sufficiently large amount of a donor impurity is introduced into a crystal, the equilibrium concentration of the vacancies may increase considerably.

Vacancies help most effectively in the climb of dislocations in single crystals at fairly high temperatures, when the dislocations have an appreciable mobility. In the case of silicon this happens above 700°C. Calculations show that at 700°C the intrinsic carrier density in silicon n_i is $5 \cdot 10^{17}$ cm^{-3} and that it increases to $3 \cdot 10^{19}$ cm^{-3} at 1400°C. Therefore, an appreciable influence of donor impurities on the vacancy concentration, and consequently on the climb of dislocations during the growth of single crystals, is felt from

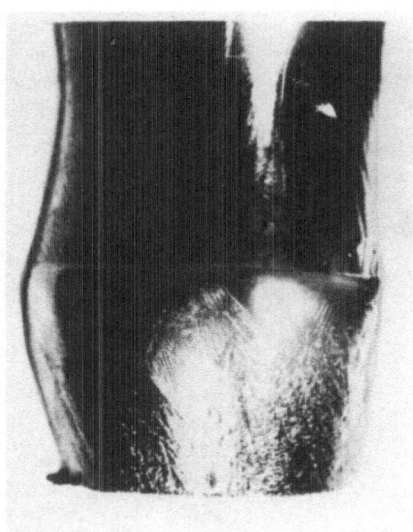

Fig. 6.7. Structure of the outer surface of a heavily doped silicon single crystal.

$n_i \approx 5 \cdot 10^{17}$ cm^{-3}, which is in good agreement with the experimental data. The hypothesis of a high concentration of vacancies in silicon single crystals heavily doped with donor impurities is supported also by the mechanical properties of the dislocation-free samples; the introduction of donor impurities softens n-type single crystals and lowers the activation energy of the dislocation motion, whereas acceptor impurities have the opposite effect [89, 90].

Heavily doped dislocation-free germanium single crystals have been obtained by the introduction of gallium [84]. The results obtained for these crystals show that the dopant concentration affects considerably the equilibrium vacancy concentration. When the penetration of dislocations from a seed to a growing crystal is restricted, the introduction of an acceptor impurity reduces the equilibrium vacancy concentration and the dimensions of the prismatic loops formed by the condensation of vacancies; this impedes the generation of new dislocations by thermal stresses. This observation is confirmed also by an investigation of dislocations in heavily doped n- and p-type germanium single crystals during their deformation.

The structure of heavily doped single crystals of germanium and silicon is considered in [24, 49, 70]. A characteristic feature of heavily doped single crystals is the considerable width of the "natural" faces on the surface of an ingot compared with undoped crystals, grown under the same conditions. At a certain concentration of the dopant in a crystal, which depends on the growth conditions, characteristic "striations" appear on the surface (Fig. 6.7). Such striations begin from the faces and spread to both sides of them forming a "tree." Initially, the striations are in the form of thin regular filaments. When the concentration of the dopant in a crystal is increased, the depth of the striations increases, they be-

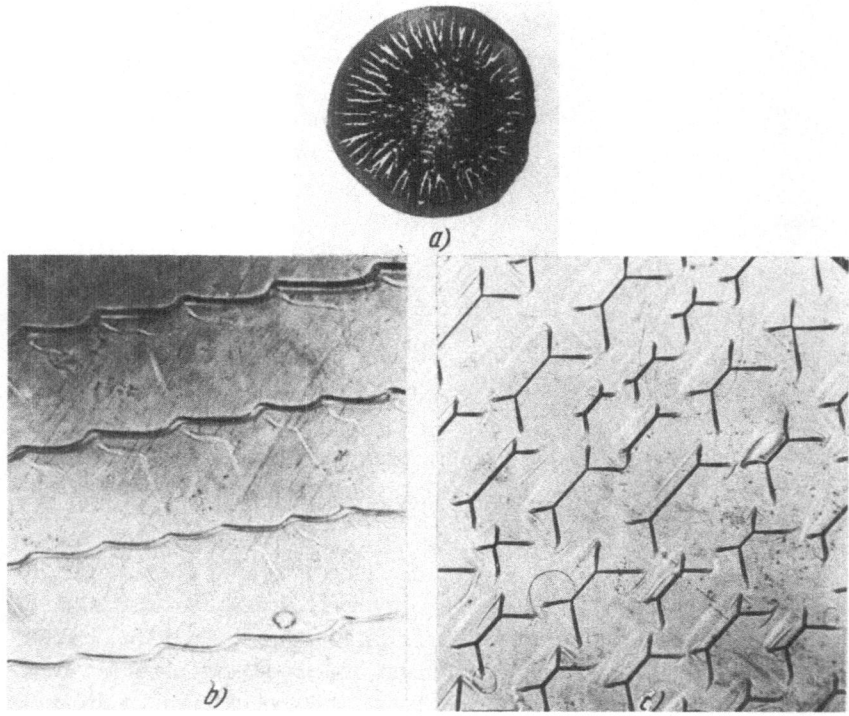

Fig. 6.8. Structure of heavily doped silicon single crystals (transverse cross sections): a) ridged structure; b) cellular structure for growth along [111]; c) cellular structure for growth along [511].

come less regular, change their direction, and are finally replaced by a dendritic structure. The surface at which such crystals are separated from the melt has a well-developed cellular structure.

The etching of heavily doped single crystals shows that the impurity distribution is periodic. The "growth bands" along longitudinal and transverse cross sections of such crystals are more pronounced than in lightly doped crystals. Heavily doped crystals also exhibit channels due to the facet effect. When the conditions approach a situation favorable for the development of the cellular structure, characteristic ridges appear against the background of the growth bands; these ridges are frequently regular, forming patterns with central symmetry along longitudinal and transverse cross sections (Fig. 6.8a). When the dopant concentration is increased, the number of such patterns in an ingot increases. Finally, they are replaced by the cellular structure. For a crystal-

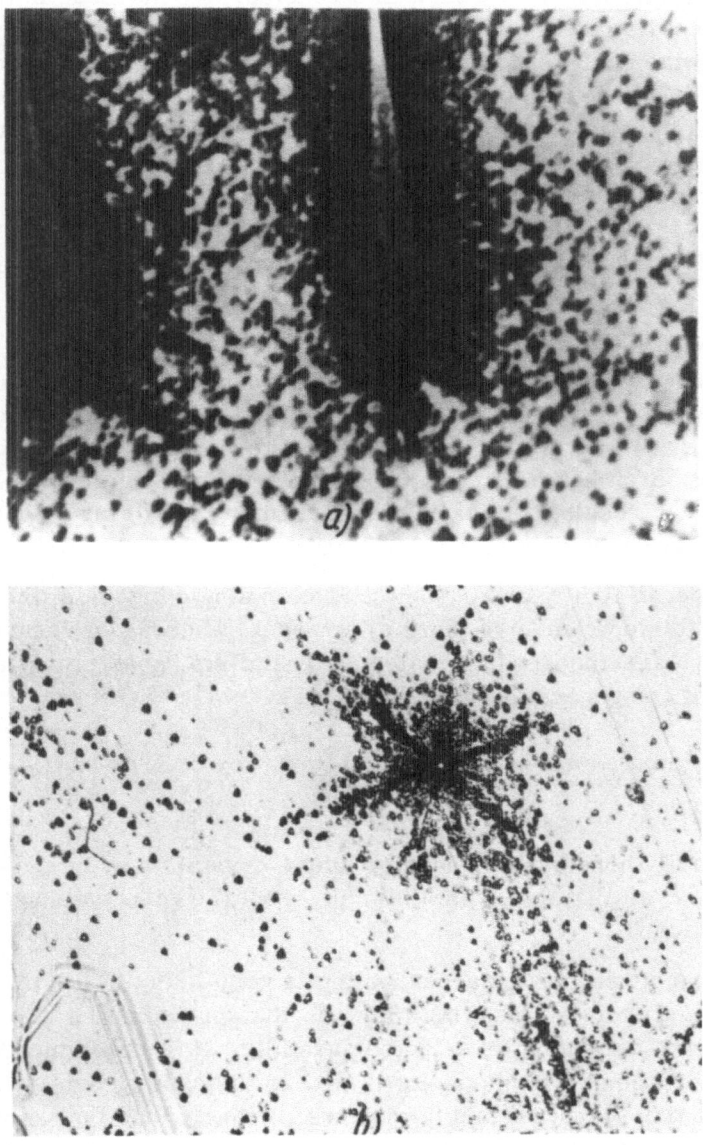

Fig. 6.9. Increase of the dislocation density near second-phase occlusions in heavily doped silicon single crystals. a) Si—Al; b) Si—Sb.
(× 150).

Second-phase occlusions observed in most heavily doped single crystals may also result in a considerable increase in the dislocation density in a sample. In this case, dislocations are generated by the considerable internal stresses near such occlusions. A strong increase in the dislocation density near such occlusions has been observed in silicon single crystals doped with aluminum and antimony [75], as well as in germanium single crystals doped with arsenic [92], gallium [50], and bismuth [93]. A characteristic feature of these dislocations is their concentration in the immediate vicinity of second-phase occlusions and their low density elsewhere in a crystal (Fig. 6.9).

Strong internal stresses due to the presence of a steep impurity concentration gradient may also produce a large number of dislocations by diffusion. A considerable increase in the dislocation density in silicon single crystals due to the diffusion of phosphorus or boron has been reported in several investigations [94-97]. Similar results have been obtained during the diffusion of zinc into gallium arsenide [98].

Thus, in lightly doped semiconductor single crystals, the main source of dislocations are thermal stresses, whereas in heavily doped crystals, the impurities themselves affect considerably the process of the generation of dislocations.

§6.4. Preparation of Heavily Doped Single Crystals

Heavily doped semiconductor single crystals may be prepared by crystallization from the melt, from the gaseous phase, and by diffusion.

Since concentration supercooling is very likely to occur in crystallizing from a melt with a high dopant concentration, the growth conditions must be selected carefully. It is recommended that crystals be grown more slowly than in the case of lightly doped semiconductors, that the melt be stirred strongly, and that an appreciable positive temperature gradient be established at the phase separation boundary [99, 100].

The most widely used method of growing heavily doped single crystals from the melt is the Czochralskii method. The correct selection of the dopant is important. The dopant should be easily dissolved in the semiconductor and should have a distribution coefficient close to unity; this makes it easier to introduce it into a crystal. From this point of view, gallium is the most suitable impurity for germanium; it is much more difficult to prepare germanium single crystals heavily doped with antimony, arsenic, or phosphorus. Boron, phosphorus, and arsenic are recommended for the preparation of heavily doped silicon single crystals. Zinc and tellurium should be used in the heavy doping of gallium arsenide and various antimonides.

The dopant may be introduced into the melt either in the elemental state or as an alloy (an alloy of the host semiconductor and the dopant) [99-100]. If the impurity is likely to evaporate easily from the melt, single crystals should be grown in an inert-gas atmosphere. Heavily doped single crystals can also be prepared by doping from the gaseous phase. In this case, a mixture of an inert gas or hydrogen and some volatile compound dopant (for example, a halide [101, 102] or a hydride [103]) is passed above the melt. The melt interacts with the gaseous phase and chemically reduces the compound to the elemental dopant. The reduced dopant is partly dissolved in the melt. By controlling the concentration of the dopant in the gaseous phase and the velocity of the gas flow, we can vary the concentration and the impurity distribution over wide limits, and we can prepare single crystals doped uniformly along the length. Silicon single crystals heavily doped with arsenic have been prepared by doping from a gaseous mixture of gallium and arsenic trichloride using the Czochralskii method [101]. Group V halides have been used in the preparation of heavily doped silicon single crystals by the floating zone method [102]; phosphine has also been used for the same purpose [103].

Single crystals of decomposing semiconducting compounds, such as arsenides and phosphides of group III elements, are prepared in hermetically sealed ampoules kept at a constant high temperature; in this case, the doping from the gaseous phase can be done using dopants in the elemental state. The high temperature of the ampoule walls ensures a sufficiently high vapor pressure in the working space and therefore in the crystal.

An interesting modification of the crystallization from the melt is the "solvent evaporation" method [104, 105]. In this method, the melt is enriched with the dopant to a considerable degree

(10-20%). The melt is kept at a constant temperature and the volatile dopant is gradually evaporated. Thus, the melt becomes supersaturated with respect to the semiconductor and the semiconductor crystallizes around a specially introduced seed. The seed may be rotated and gradually pulled upward. The advantage of this method is that it takes place under isothermal conditions, which help to make single crystals more uniform. This method has been applied to germanium and silicon single crystals, heavily doped with arsenic and antimony.

Heavily doped single crystals can also be grown from the melt using a temperature gradient [106] and directional crystallization in a boat. The latter method is widely used in the preparation of heavily doped single crystals of decomposing semiconducting compounds [107].

The method of crystallization from the melt has been applied successfully to germanium single crystals doped with gallium up to concentrations of $(1-2) \cdot 10^{20}$ cm^{-3}, with arsenic up to $(6-8) \cdot 10^{19}$ cm^{-3} [99, 100]; silicon single crystals doped with boron up to $(2-4) \cdot 10^{20}$ cm^{-3}, with arsenic and phosphorus up to $(1-1.5) \cdot 10^{20}$ cm^{-3} [29, 75]; gallium arsenide single crystals doped with zinc up to $(1-2) \cdot 10^{20}$ cm^{-3} and with tellurium up to $1 \cdot 10^{19}$ cm^{-3} [107, 108]; as well as heavily doped single crystals of other semiconductors.

Crystallization from the gaseous phase is widely used in the preparation of single-crystal epitaxial film. This method makes it possible to prepare heavily doped single crystals in a relatively simple manner. The dopant is introduced into the working space of a furnace either in the form of a volatile compound (a halide or a hydride) or in the elemental state. The addition of an elemental impurity is usually employed in the crystallization of semiconducting compounds with one volatile component. The impurity concentration in the gaseous phase is controlled by the gaseous mixture composition; in the case of elemental admixtures, it is controlled by the temperature of the source. Crystallization from the gaseous phase takes place at lower temperatures than in the melt method. Since the majority of the dopants have a retrograde solubility in the main semiconducting materials, the temperature of the substrate during the crystallization from the gaseous phase must be close to the temperature of maximum solubility of the impurity in the semiconductor: this ensures the maximum concentration of the impurity

in the grown crystal. The very slow rate of growth in crystallization from the gaseous phase, helps to increase the concentration of
the impurity in a sample. Crystallization from the gaseous phase
has been applied successfully to the preparation of heavily doped
single-crystal films of silicon [108], gallium arsenide [109, 110],
and other semiconductors.

Heavily doped single crystals can also be prepared by diffusion. In this case, we start from a single crystal grown by some
other method. The single crystal is doped by the diffusion of the
impurity at a high temperature. The diffusion usually takes place
in a special high-temperature furnace where a stable atmosphere
of the impurity vapor above the crystal is established. The impurity concentration in the sample is governed by its content in the
gaseous phase and by the diffusion temperature. Frequently, the
impurity is deposited directly on the sample by a chemical, elec-
trolytic, or evaporation method and then the crystal is annealed in
an inert-gas atmosphere. One of the main disadvantages of the diffusion doping method is the long duration of the process, since the
main electrically active impurities with shallow levels form substitutional solid solutions and have very low diffusion coefficients.
Nevertheless, doping by diffusion is widely used [111-114].

Chapter VII

Some Applications of Heavily
Doped Semiconductors

§7.1. Tunnel Diodes

<u>Principle of Operation of Tunnel Diodes</u>. Before the appearance of the tunnel diode, the semiconducting materials used in devices contained impurities in concentrations not exceeding 10^{16}-10^{17} cm^{-3}. The free-carrier density in such materials has been of the same order or even less. The relative positions of the Fermi level and of the band edges in such semiconductors are shown in Fig. 7.1. The energy structure of p−n junctions (under thermodynamic equilibrium conditions) formed in such crystals is shown in Fig. 7.2. The thickness of a p−n junction is given by the expression [1]

$$L = 1.05 \cdot 10^6 \left[\frac{\varkappa (U_C - U)}{2\pi e} \cdot \frac{N_d + N_a}{N_d \cdot N_a} \right]^{1/2} \text{cm.} \qquad (7.1.1)$$

where $\varkappa$ is the permittivity; e is the electron charge; N_d and N_a are the concentrations of ionized donors and acceptors on both sides of the p−n junction; U_C is the contact potential difference across the junction and U is the external voltage applied to the junction.

At impurity concentrations of 10^{16}-10^{17} cm^{-3}, the thickness of a p−n junction in germanium is approximately 10^{-4}-10^{-5} cm. In the absence of bias, the electric field $\mathscr{E}$ in such a junction is weak. For example, for germanium the electric field is $\mathscr{E} \approx U_C / L \approx 10^3$-$10^4$ V/cm.

317

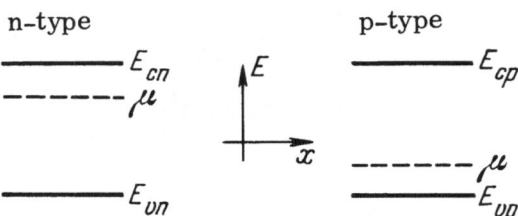

Fig 7.1. Position of the Fermi level
in a lightly doped semiconductor.

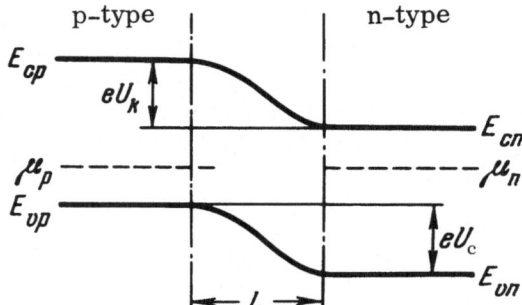

Fig. 7.2. Energy band scheme of a conventional
p−n junction.

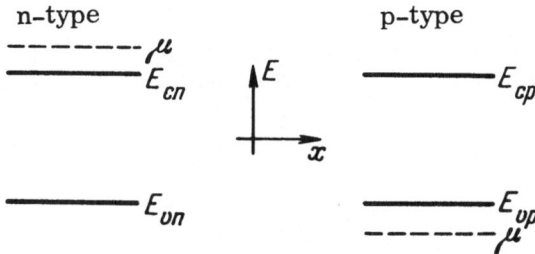

Fig. 7.3. Position of the Fermi level
in a heavily doped semiconductor.

p–type   n–type

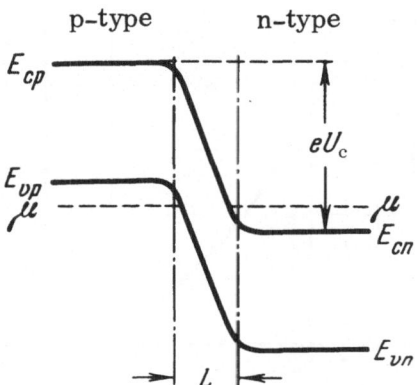

Fig. 7.4. Energy band scheme (in the absence
of bias) of tunnel p—n junction.

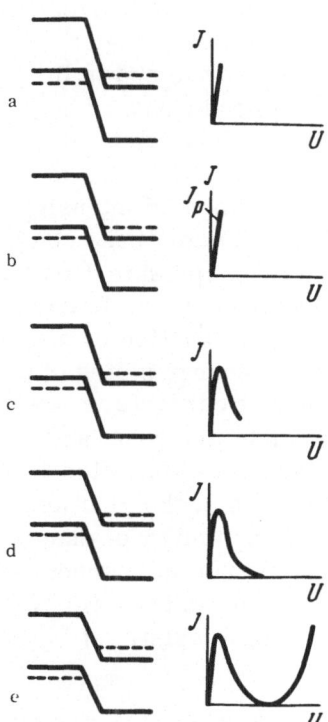

Fig. 7.5. Energy band scheme and
current—voltage characteristic of a
tunnel diode for various values of
positive bias.

Thus, a p—n junction in a
conventional diode is fairly thick
and the field intensity in it is low.
Under these conditions, electrons
may be transferred to a free band,
by overcoming an energy barrier,
only if they are supplied with an
additional energy. Strong fields
are created in a conventional
p—n junction only under high external voltages. Such strong
fields $(\mathscr{E} \approx 10^5$ V/cm) produce additional carriers whose number
increases considerably when the
field intensity is increased. Many
investigations (cf., for example,
[2, 3]) have shown that the most
likely mechanism of carrier generation in strong fields is impact
ionization. In addition to impact
ionization, strong fields may
transfer carriers directly from
a filled band to a free band. This
process is similar to the cold
emission of electrons from a metal. This mechanism is most apparent in tunnel diodes, which were
first proposed in 1958 by Esaki [4].

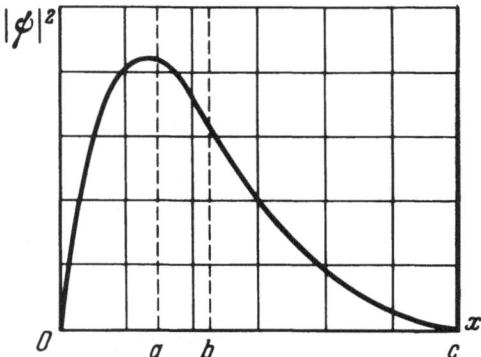

Fig. 7.6. Density of the probability of finding an electron in space. For the sake of simplicity, the one -
dimensional case is considered.

In contrast to conventional devices, tunnel diodes are made
of semiconductors containing 10^{18}-10^{20} cm^{-3} impurities. The energy spectrum of such a heavily doped semiconductor is shown in
Fig. 7.3. The band structure (under equilibrium conditions) of a
p−n junction prepared from a heavily doped semiconductor is presented in Fig. 7.4. In this case the thickness of a p−n junction is
much less: 100-200 Å [cf. Eq. (7.1.1)]. The electric field intensity
in a tunnel-diode junction is high: 10^{5}-10^{6} V/cm. Under these conditions, an electron in the conduction band on the n-type side does
not need additional energy to overcome the energy barrier. It may
reach the valence band on the other side of the p−n junction by a
tunnel transition [5], in which the initial and final energy states of
the electron are the same. We may assume that the currents across
the junction from right to left and from left to right are equal and
that the total current is zero. If a small forward bias is applied to
such a p−n junction, the Fermi levels on both sides of the junction
are no longer at the same energy (Fig. 7.5a). The number of electrons leaking from the right to the left is greater than the number
of electrons leaking in the opposite direction. A current now flows
through the p−n junction, the current-voltage characteristic of
which is shown in Fig. 7.5a on the right.

If the forward bias is increased to a value at which the Fermi
level μ_p becomes level with the band edge E_{cn}, the current increases and reaches a value J_p (Fig. 7.5b). Any further increase
in the applied voltage reduces the current (Fig. 7.5c) since the elec-

trons penetrating the barrier are facing forbidden levels. Next, when the band edges E_{vp} and E_{cn} have a common level, the current decreases to zero (Fig. 7.5d). A further increase in the bias produces the usual diode current through the junction (Fig. 7.5e). We can easily show that the application of a reverse bias produces a strong increase in the current.

Thus, the current through a tunnel diode is the sum of the tunnel current and the usual diffusion current. In real devices, the tunnel current never decreases to zero. In fact, its lowest value may be one quarter of the value of J_p as, for example, in silicon tunnel diodes [6]. This point will be considered later.

Tunnel Effect in Semiconductors. The tunnel effect is a quantum-mechanical phenomenon. It can be understood by considering Fig. 7.6. This figure shows the thickness of a barrier in the usual diode $(a-c)$ and in a tunnel diode $(a-b)$. We can see that due to the narrowness of the tunnel barrier, the probability of finding a particle to the right of this barrier is not equal to zero.

The leakage coefficient P for a particle tunneling from left to right through the barrier is given by

$$P = \frac{|\psi_{Qa}|^2}{|\psi_{bc}|^2}, \tag{7.1.2}$$

where ψ_{0a} and ψ_{bc} are the amplitudes of the wave functions of the particle on each side of the barrier, whose values are found from Schrödinger's equation.

The tunnel effect mechanism in a semiconductor (but not in a p—n junction) was first considered qualitatively by Zener [7]. An expression for the leakage coefficient is given correctly in [8]:

$$P = \exp\left(-\frac{\pi}{2e\hbar\mathscr{E}} \sqrt{2m^*}\, E_g^{3/2}\right), \tag{7.1.3}$$

where m* is the effective electron mass and E_g is the forbidden band width of a given semiconductor.

This expression assumes a triangular barrier. The actual shape of the barrier is not important because the de Broglie wave-

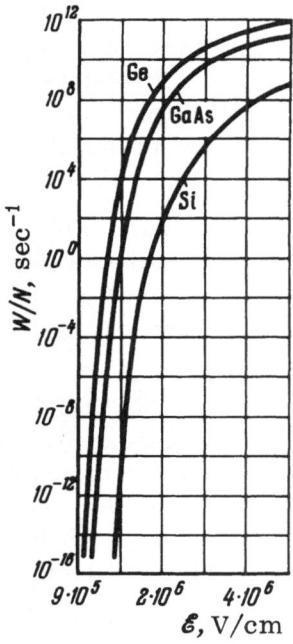

Fig. 7.7. Probability of a tunnel transition of an electron as a function of the field intensity in a p−n function.

length of the tunneling particles should be much less than the width of the barrier, i.e., the electric field should not vary over a distance equal to the de Broglie wavelength λ. We can easily show that λ is approximately 30 Å, while the thickness of the p−n junction is 100 Å, i.e., the quasi-classical conditions are not satisfied. Nevertheless, a quasi-classical approach allows us to understand more clearly the main features of the effect.

In order to obtain an expression for the number of electrons, W, leaking through a barrier per unit volume and per unit time, we must multiply Eq. (7.1.2) by the number of valence electrons N per unit volume and by the frequency of collisions of particles with the barrier. Thus, we finally obtain [6]:

$$W = N \frac{ae\mathcal{E}}{2\pi\hbar} \exp\left[-\frac{\pi}{2e\hbar\mathcal{E}} \sqrt{2m^*} \, E_g^{3/2}\right], \qquad (7.1.4)$$

where a is the lattice period.

It follows from this expression that a strong tunnel effect is observed in semiconductors with small values of E_g and m^*. By way of example, Fig. 7.7 shows the results of calculations using Eq. (7.1.4) for three semiconductors [9].

The theory of the tunnel effect in semiconductors has been developed further by Franz and Tewordt [10], and particularly by Keldysh [11, 12]. Keldysh has considered a three-dimensional problem and obtained an expression for W in the following form [11]:

$$W = \frac{1}{4} N \frac{a^3 e^2 \mathcal{E}^2 \sqrt{m_\parallel}}{\pi\hbar^2 \sqrt{E_0}} \exp\left[-\frac{\pi}{2e\hbar\mathcal{E}} \sqrt{m_\parallel} \, E_0^{3/2} + f(\gamma)\right], \qquad (7.1.5)$$

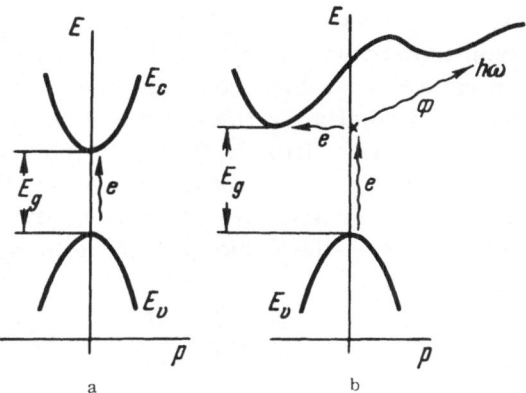

Fig. 7.8. Tunnel effect in different semiconductors:
a) direct tunnel transitions; b) indirect tunnel
transition.

where E_0 is the characteristic energy of the interband electron
transition; $f(\gamma)$ is a factor which governs the angular dependence;
$m_{\parallel}$ is a formal quantity, defined as

$$m_{\parallel} = \frac{1}{\sum\limits_i (\cos \gamma_i)^2 / m_i} , \qquad (7.1.6)$$

where m_i^{-1} are the principal values of the tensor $(m_{ik}^*)^{-1}$, and γ_i are
the angles between the direction of the field and the principal axes
of this tensor.

The expression (7.1.5) differs from (7.1.4) by a different field
dependence of the pre-exponential factor and by an angular dependence in the exponential function. However, the most important difference is in the treatment of the quantities E_0 and $m_{\parallel}$. These
quantities assume the values E_g and m^* only if the highest state
in the valence band and the lowest state in the conduction band have
the same value of the crystal momentum (Fig. 7.8a). This is known
as a direct transition. If these states do not have the same value
of the crystal momentum, then the quantity E_0, which represents
the red edge of the absorption of light in the given semiconductor,
will always be greater than E_g. Consequently, the values of the
critical fields corresponding to perceptible values of W should be
greater than those predicted by Eq. (7.1.4).

Keldysh has also shown that if there is any interaction which alters the electron momentum, then the transition of an electron to the conduction band is possible even when the maximum of the valence band is shifted relative to the minimum of the conduction band (Fig. 7.8b). Such a transition is known as indirect. Interactions which alter the electron momentum may be in the form of collisions of electrons with one another, or with impurity atoms or phonons. In the latter case, the value of W is defined as follows [12]:

$$W \propto \left(\frac{e\hbar\mathscr{E}}{\sqrt{2m^*_{\|}}\, E_g^{3/2}} \right)^{5/2} \exp \left[f(\gamma) - \frac{4}{3} \frac{\sqrt{2m^*_{\|}}}{e\hbar\mathscr{E}} (E_g - \hbar\omega)^{3/2} \right] \times$$

$$\times \left\{ \bar{N}(T) + [1 + \bar{N}(T)] \exp \left(-\frac{4}{3} \frac{\sqrt{2m^*_{\|}E_g}}{e\hbar\mathscr{E}} \hbar\omega \right) \right\}, \qquad (7.1.7)$$

where ω is the phonon frequency; $\overline{N}(T) = (e^{(\hbar\omega/kT)} - 1)^{-1}$. In contrast to Eq. (7.1.5), $m^*_{\|}$ is here the reduced effective mass of an electron and a hole.

Quantitative Treatment of the Tunnel Effect in a p – n Junction. From general considerations (Fig. 7.5), Esaki [4] has concluded that the tunnel current from the conduction band in the n-type region to the valence band in the p-type region, $J_{c \to v}$ is proportional to: 1) the probability of the tunnel effect; 2) the probability $f_C(E)$ of finding an electron in the conduction band to the right of the junction (Fig. 7.5) at an energy level E; 3) the probability $[1 - f_V(E)]$ that this level is vacant in the valence band to the left of the junction; and 4) the energy state densities in the bands, $\rho_C(E)$ and $\rho_V(E)$:

$$J_{c \to v} \propto \int_{E_c}^{E_v} W_{c \to v} f_c(E) [1 - f_v(E)] \rho_c(E) \rho_v(E) \, dE. \qquad (7.1.8)$$

Similarly, for a tunnel transition from the valence band on the left to the conduction band on the right, we have:

$$J_{v \to c} \propto \int_{E_c}^{E_v} W_{v \to c} f_v(E) [1 - f_c(E)] \rho_c(E) \rho_v(E) \, dE. \qquad (7.1.9)$$

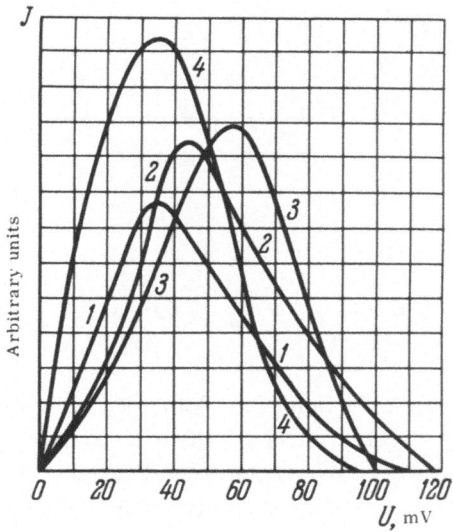

Fig. 7.9. Calculated dependences of the current on the voltage across a p—n junction in a tunnel diode: 1) according to Esaki; 2) according to Price and Radcliffe; 3) according to Ivanchik; 4) according to Kane.

The total Esaki current is the difference between these two currents. Moreover, Esaki assumes that the bands are parabolic and that $W_{c \to v} = W_{v \to c}$, then

$$J \propto \int_{E_c}^{E_v} W \left[f_c(E) - f_v(E) \right] \sqrt{(E - E_c)(E_v - E)} \, dE. \qquad (7.1.10)$$

This expression, based only on general physical considerations, has been given by Esaki without proof and therefore it cannot be regarded as a quantitative formula for the current–voltage characteristic of a tunnel diode. Nevertheless, its simplicity and clarity make it possible to explain the physical nature of the tunnel effect.

If the p- and n-type regions are degenerate to the same degree and if the Fermi level within the bands is separated by $2k_0T$ from E_V and E_C, respectively, Eq. (7.1.10) transforms to (7.1.11) [14]:

$$J = - A \frac{(E_v - E_c)^2 \left[1 - \exp\left(\frac{eU}{k_0 T}\right) \right]}{[m + n] e^{\frac{a}{2}} + \left[1 + \exp\left(\frac{eU}{k_0 T}\right) \right]}, \qquad (7.1.11)$$

where

$$m = \exp\left(-\frac{\mu_c - E_c}{k_0 T} \right), \quad n = \exp\left(\frac{E_v - \mu_v}{k_0 T} \right), \quad a = \frac{E_v - E_c}{k_0 T}.$$

Esaki's ideas have been developed by Ivanchik [15], who has used the Thomas–Fermi method [16] to calculate the potential in a p–n junction between two degenerate regions. Knowing this potential, Ivanchik has calculated the tunnel current but his expression is fairly cumbersome and will not be given here. It is sufficient to mention that Ivanchik's formula, which is given in the form of an integral, can be represented by a polynomial of the eighth power, in order to show how difficult it would be to compare it with the experimental results. In a qualitative sense, this formula represents satisfactorily the dependence of the tunnel current on the bias (Fig. 7.9).

Ivanchik's treatment deals only with the direct tunnel transitions.

The quantum-mechanical problem of the tunnel current through a p–n junction has been considered by Price and Radcliffe [16]. Their solution for the direct transitions is in the form of an integral

$$J = \frac{2\pi e m_r^* P \mathscr{E}}{a h^3} \int [f_c(E) - f_v(E)] g(E_1) g(E_2) \, dE, \qquad (7.1.12)$$

where m_r^* is the reduced effective mass, given by

$$m_r^* = \frac{m_p^* \, m_n^*}{m_p^* + m_n^*}. \qquad (7.1.13)$$

The factors g(E) are equal to zero outside the region of the band overlap and inside this region they are equal to unity.

In the band overlap region, the integral of Eq. (7.1.12) can be integrated between the limits of zero and $(E_v - eU)$:

$$J = \frac{2\pi e m_r^* k_0 T E_g P}{a h^3 L} \{ k_0 T [\ln (1 + e^{\frac{E_v - \mu - eU}{k_0 T}}) +$$

$$+ \ln (1 + e^{\frac{-\mu - eU}{k_0 T}}) - \ln (1 + e^{\frac{E_v - \mu - 2eU}{k_0 T}}) - \ln (1 + e^{-\frac{\mu}{k_0 T}})]\}. \quad (7.1.14)$$

If we substitute into Eq. (7.1.14) the expressions (7.1.1) and (7.1.3), which define L and P, the formula describing the current – voltage characteristic becomes difficult to compare with experiment.

Comparing Esaki's, Ivanchik's, and Price – Radcliffe's curves, shown in Fig. (7.9), we can see that all of them give qualitative agreement with the experimentally observed current – voltage characteristic of tunnel diodes. Unfortunately, the absence of numerical values for the quantities in these formulas makes it difficult to compare quantitatively these formulas and the experimental data.

The approach of Kane [17] and of Bonch-Bruevich [18] seems more fruitful.

Kane has assumed that the electric field in the p – n junction of a tunnel diode is uniform. Then, at absolute zero, the expression for the tunnel current density becomes :

$$j = \frac{e m^*}{18 h^3} \exp \left\{ - \frac{\pi (m^*)^{1/2} E_g^{3/2}}{2 \sqrt{2} \hbar \mathscr{E}} \right\} \int \exp \left(- \frac{2 E_\perp}{\bar{E}_\perp} \right) dE \, dE_\perp, \quad (7.1.15)$$

where

$$E_\perp = \frac{\hbar^2 (k_y^2 + k_z^2)}{2 m^*}, \quad (7.1.16)$$

$$\bar{E}_\perp = \frac{\sqrt{2} \hbar \mathscr{E}}{\pi (m^*)^{1/2} E_g^{3/2}}. \quad (7.1.17)$$

The factor in front of the integral represents the tunnel leakage coefficient for electrons. In Kane's theory, this factor is assumed to be constant.

It follows that the form of the current – voltage characteristic depends on the function D(U), which is used by Kane to denote the integral in Eq. (7.1.15). Kane represents the function D(U)
analytically in the form of a large number of formulas describing
the various parts of the whole current – voltage characteristic of a
tunnel diode. Presented graphically all these parts join to form a
continuous smooth curve D(U). By way of example, Fig. 7.9 shows
the dependence D(U) for

$$E_{cn} - \mu_n = E_{vp} - \mu_p$$

of direct tunnel transitions.

Kane has also considered indirect transitions. The relevant
treatment is given in [17].

In his earlier paper [18], Bonch-Bruevich considered not the
whole current – voltage characteristic but only its extremal points.
Thus, the peak voltage U_p, corresponding to the maximum of the
tunnel current, is represented by Bonch-Bruevich by the following
formula

$$U_p = \frac{1}{6e} \{ 2(E_{vp} - \mu_p) - (\mu_n - E_{cn}) +$$

$$+ \sqrt{[2(E_{vp} - \mu_p) + (\mu_n - E_{cn})]^2 + 4(E_{vp} - \mu_p)(\mu_n - E_{cn})} \}. \quad (7.1.18)$$

From the point of view of the experimenter, this approach is
most convenient. It gives simple and convenient expressions and
the quantities occurring in these expressions have clear physical
meaning and definite numerical values.

A complete theory of the current – voltage characteristic of
a tunnel diode for direct and indirect transitions, for T = 0 and T ≠ 0,
with allowance for an inhomogeneous field distribution in the p – n
junction, is given in the later papers of Bonch-Bruevich and Sere-
bryannikov [71, 72].

Semiconductors Used in Tunnel Diodes. It fol-
lows from our qualitative description of the principle of operation
of tunnel diodes that degenerate semiconductors must be used in
such diodes. At room temperatures, all the impurities in such semiconductors are ionized [19, 20] and therefore we can easily find the
degree of doping necessary for the tunnel effect. This can be done

TABLE 7.1

Semi-conductor	$\frac{m^*}{m_0}$	E_g, eV	neq, cm^{-3}	$(N_d)min$, cm^{-3}
Ge	0.15	0.65	$4 \cdot 10^{19}$	$3.5 \cdot 10^{18}$
Si	0.27	1.10	$1.4 \cdot 10^{20}$	$8.5 \cdot 10^{18}$
GaAs	0.06	1.35	$7 \cdot 10^{19}$	$9.0 \cdot 10^{17}$
InSb	0.04	0.18	$6 \cdot 10^{17}$	$4.8 \cdot 10^{17}$

using Eq. (2.3.4), which relates the carrier density to the Fermi level. In an approximate estimate, we may assume that the degeneracy begins when $\mu_n = E_c$, i.e., $\mu^* = 0$. The last column of Table 7.1 lists the minimum donor concentrations necessary to obtain a tunnel p⁻n junction in various semiconductors. In practice, higher degrees of doping are required.

The preparation of such heavily doped semiconductors (cf. Chap. 6) is not an easy problem. Heavy doping can be obtained with relatively few impurities which have high solubility limits. This problem has been investigated quite thoroughly in the case of germanium and silicon (cf. Figs. 5.1 and 5.2) but for $A^{III}B^V$ compounds (GaAs, InSb, GaSb, etc.) the impurity solubility limits are not yet known accurately. All that is known [27-29] is that the best results are obtained using GaAs + Sn and InSb + Te n-type systems, and GaAs + Zn, InSb + Zn, InSb + Cd p-type systems.

The tunnel effect has also been observed in gallium arsenide doped with sulfur and selenium [21], but in these cases there is a danger of forming chemical compounds of these elements with GaAs.

The properties of a crystal govern the density of the tunnel current not only through the value of the carrier density (impurity concentration) but also through such parameters as the forbidden band width and the effective mass, since these quantities occur in the expression for the probability of the tunnel effect [Eq. (7.1.3)]. Material with low effective masses and small values of E_g require fewer impurities to reach the same current density. Comparative data on m* and E_g for various materials are given in Table 7.1 [22].* The fourth column of Table 7.1 gives the values of

---

* These considerations are only qualitative because it is not clear which of the effective masses is important in the tunnel effect. Moreover, the effective masses of $A^{III}B^V$ compounds depend strongly on the impurity concentration.

$n_{eq} = N_d N_a / (N_d + N_a)$, which are needed to reach a tunnel current density of 10^4 A/cm^2 [23]. These data indicate that gallium arsenide and indium antimonide are the materials to be preferred.

Formation of a p – n Junction. To form a tunnel p – n junction, it is not sufficient to use a heavily doped semiconductor. The impurity concentration on the other side of a p – n junction should also make the electron gas degenerate. A high impurity concentration should be combined with a sudden (step-like) fall of the concentration in the region of the junction. Only in this case can the junction be sufficiently thin and the strong field necessary for the tunnel effect be produced in it.

Among the known methods for the preparation of p – n junctions (diffusion, alloying, etc.), the most suitable is the alloying method. In this method [24] a metal drop, containing an acceptor impurity, in the case of an n-type crystal or a donor impurity in the case of a p-type crystal, is alloyed to a crystal.

A high impurity concentration in the alloyed region is reached by suitable selection of the metal used in the alloying. It follows from Figs. 5.1 and 5.2 that the most suitable metals for alloying to n-type germanium are Ga and Al; As is best for p-type germanium and Al, B, and As are best for silicon. In practice, one uses not pure metals but alloys such as InGa [24], SnGa [25], or, in the case of silicon, AlB [6].

Naturally, the concentration of the element causing inversion of the sign of conduction in the alloyed region is less than the concentration at the solidus curve (Figs. 5.1 and 5.2). A rough estimate of the impurity concentration in the alloyed region can be obtained for this case by means of graphs (Figs. 7.10 and 7.11), which were obtained in [26, 27] on the assumption that the solid solutions formed by alloying binary alloys to germanium are ideal. The quantity A, plotted along the abscissa in these figures, represents the depth of alloying l, given by

$$l = A \frac{G}{S}, \tag{7.1.19}$$

where G is the weight of the metal drop used in the alloying; S is the area occupied by this drop after alloying.

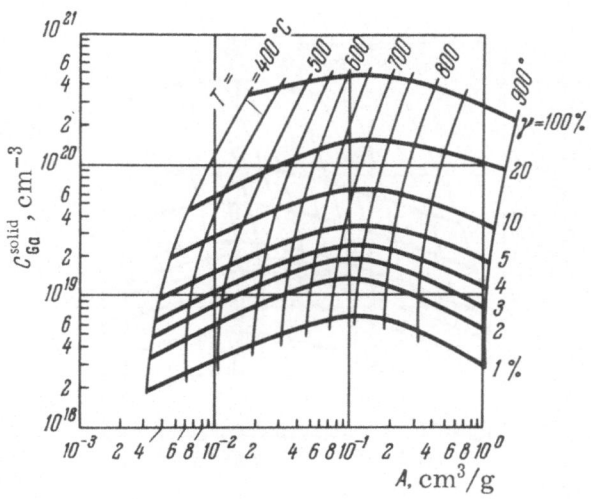

Fig. 7.10. Distribution of gallium in the alloyed region of a germanium p—n junction (alloying with In—Ga).

The quantity γ in Figs. 7.10 and 7.11 represents the concentration of Ga (in %) in the InGa and SnGa alloys; T is the alloying temperature.

The curves in Figs. 5.1 and 5.2 and the data in Figs. 7.10 and 7.11 apply only to slow equilibrium cooling processes of the molten zone in a crystal. In practice this never happens. In fact, it has been demonstrated in [27, 28] that a strong tunnel effect is obtained only if rapid cooling is employed. This observation can be explained quite simply: strong cooling increases considerably the impurity concentration in the recrystallized region. It is practically impossible to estimate this impurity concentration but the knowledge of it is very important, since it occurs in the expression for n_{eq}.

The duration of alloying much affects the quality of a tunnel p—n junction because if the alloying is prolonged, impurities diffuse to distances sufficient for considerable broadening of the p—n junction [9].

For a junction 100-150 Å thick, the diffusion to a distance ΔL, equal to 10 Å, can still be regarded as permissible. Then, if the

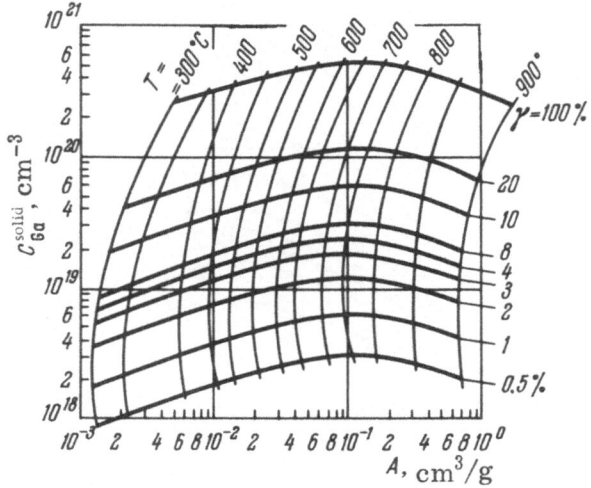

Fig. 7.11. Distribution of gallium in the alloyed region of a
germanium p−n junction (alloying with Sn−Ga).

diffusion obeys Fick's law [29]:

$$D\frac{\partial^2 N}{\partial x^2} = \frac{\partial N}{\partial t},\qquad(7.1.20)$$

where D is the diffusion coefficient, the maximum permissible alloying duration t_{max} can be estimated from the relationship

$$\frac{Dt_{max}}{(\Delta L)^2} \leqslant 1.\qquad(7.1.21)$$

Figure 7.12 gives the values of t_{max} as a function of the alloying temperature for several semiconductor – impurity systems.
If the alloying is carried out correctly, the alloyed region is uniform (Fig. 7.13a). Poor alloying produces two types of defect in
the junction structure (Figs. 7.13b and 7.13c).

When these defects are present the current through a crystal
has a component of non-tunnel origin, due to the existence of contacts between the metal and the n-type semiconductor, which shunt
the p−n junction. The contribution of this current to the total current may be altered by etching the junction. Consequently, etching may affect the quality of the junction, which can be conveniently

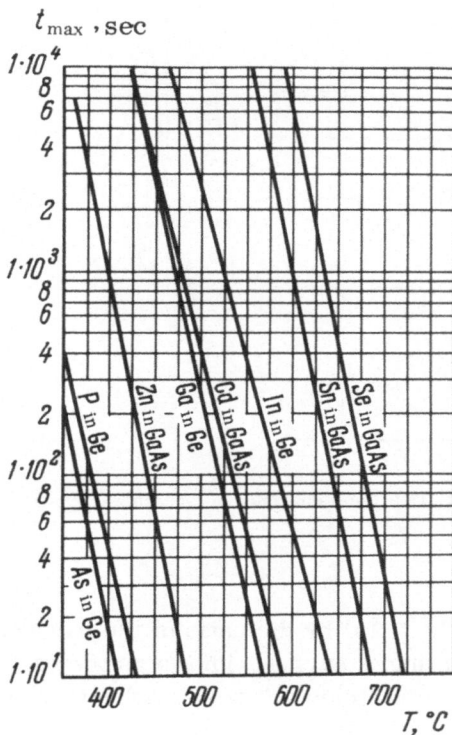

Fig. 7.12. Permissible duration of alloying
of impurities at various temperatures.

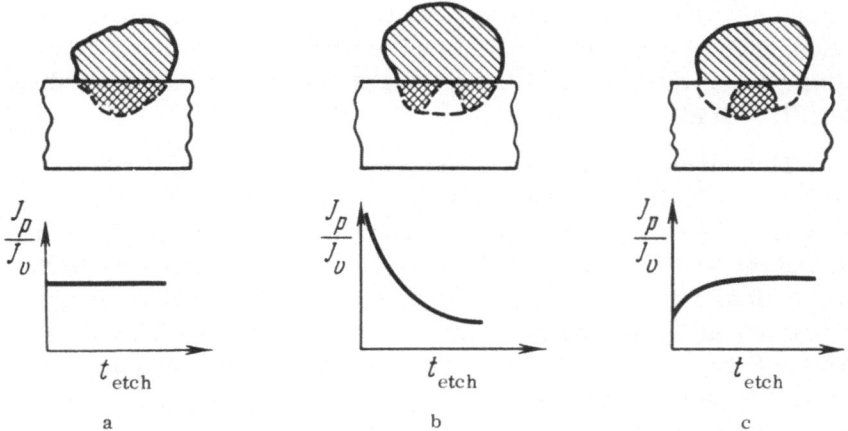

Fig. 7.13. Structures of the alloyed region in alloyed p−n junctions.

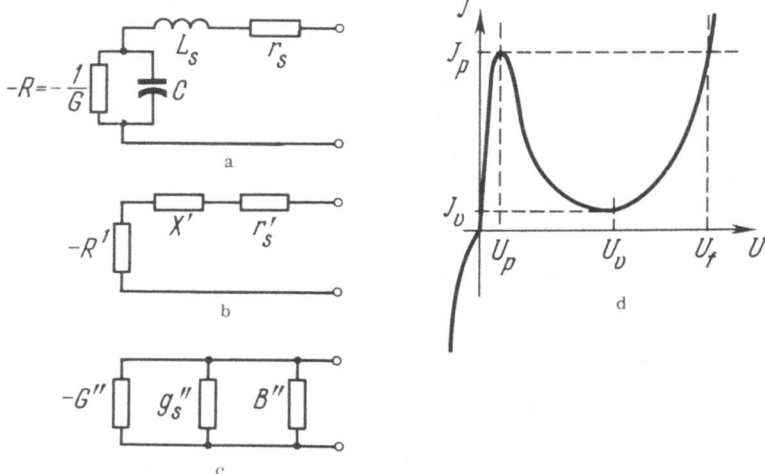

Fig. 7.14. Equivalent circuit and current—voltage characteristic of a tunnel diode.

represented (cf. Fig. 7.13) by the ratio of the currents J_p/J_v, where J_v is the minimum tunnel current, known as the valley current (Fig. 7.14d). Such effects of etching have been observed experimentally [9, 30].

Equivalent Circuit of a Tunnel Diode. To determine the parameters which can be used to represent a tunnel diode, we shall consider the equivalent circuit suggested by Sommers [23] (Fig. 7.14a). In this circuit, C is the p—n junction capacitance; r_s is the loss resistance; L_s is the diode inductance; R is the negative resistance, given by the angle of the slope of the rising part of the current–voltage characteristic of the tunnel diode (Fig. 7.14d).

The current—voltage characteristic of the tunnel diode is of the N-type. An important feature of this characteristic is the one-to-one correspondence between the current passing through the diode and the voltage applied to it; another important property is the point with an infinite resistance lying between the negative and positive branches (in this respect, the tunnel diode differs from devices with the S-type characteristic, where the voltage is a unique function of the current and the resistance at the corresponding point passes through zero).

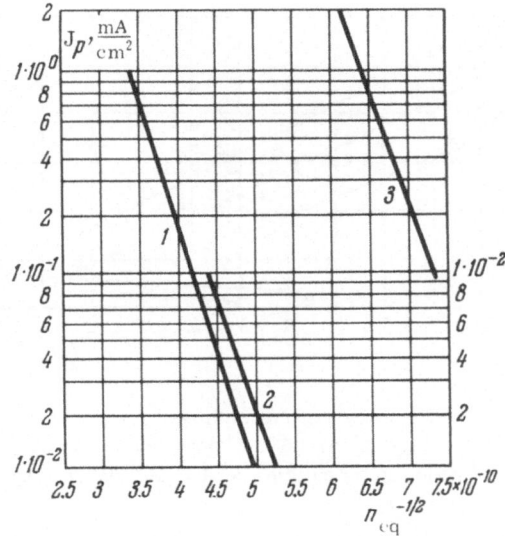

Fig. 7.15. Dependence of the density of a current flowing through a tunnel diode on the equivalent doping. 1)
Ge + As; 2) Ge + Sb (scale on the right); 3) Ge + p.

The properties of the tunnel diode, which determine its technical applications, can be described well by these equivalent circuit parameters [23, 29, 31, 32].

Peak Current. The dependence of the peak current J_p
on the impurity concentration follows from the concentration dependence of the tunnel effect probability. In fact, a calculation reported in [33] gives the following expression for J_p, which is similar to Eq. (7.1.3):

$$J_p = J_0 S \exp\left[- BE_g \left(\frac{\mathscr{E}m^*}{n_{eq}}\right)^{1/2}\right], \qquad (7.1.22)$$

where J_0 is a coefficient of proportionality, in units of the current
density; B is a constant which depends on the system of units; S
is the p−n junction area.

The expression (7.1.22) shows that the peak current for a
given semiconductor depends strongly on the equivalent doping.
This conclusion has been confirmed experimentally, as shown in

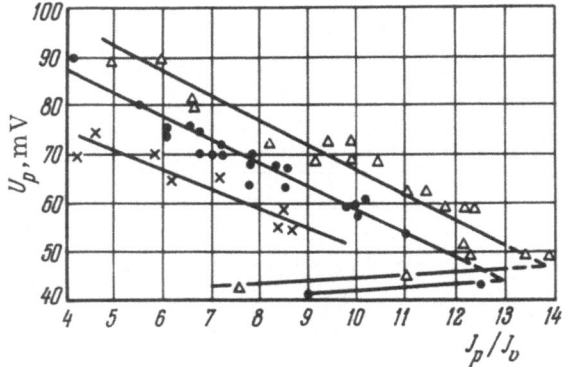

Fig. 7.16. Distribution of tunnel diodes in accordance with the values of J_p/J_v. $\triangle$) $3.5 \cdot 10^{19}$ donor atoms per cm^3; $\bullet$) $2.5 \cdot 10^{19}$ donor atoms per cm^3; $\times$) $1.8 \cdot 10^{19}$ donor atoms per cm^3.

Fig. 7.15, which is plotted on the basis of the results reported in [34, 35]. It is interesting to note that the doping of a crystal with different impurities gives rise (even for the same impurity concentration) to different values of J_p. Thus, Furukawa [35] has found that the tunnel current in p−n junctions made of arsenic-doped germanium is higher than in junctions made of antimony-doped germanium. This is in agreement with the different values of the electron mobility in these two materials (cf. Chap. 3).

Since Eq. (7.1.22) includes the quantity n_{eq}, we should expect a strong dependence of J_p (and consequently of J_p/J_v) on the impurity concentration in the alloyed part of the crystal. However, it is impossible to follow this dependence directly because the impurity concentration in the alloyed region cannot be found. Indirectly, this dependence has been observed by many workers [20, 28, 34, 36].

In Fig. 7.16, taken from [36], the unknown impurity concentration occurs implicitly in the value of U_p, plotted along the ordinate. This figure should be regarded as a diagram giving the distribution curve. It follows from it that, using the same donor concentration N_d, we can obtain devices with different values of U_p and these devices will have different values of J_p/J_v. This diagram is very useful in the selection of germanium crystals when a given value of the ratio J_p/J_v is required.

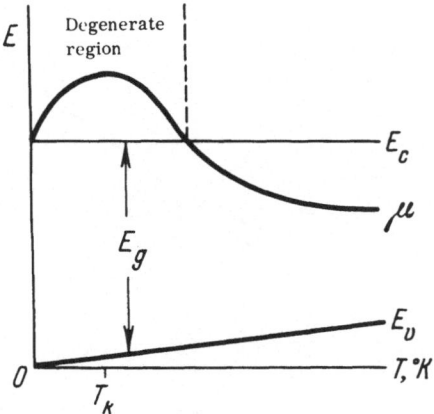

Fig. 7.17. Temperature dependence of the Fermi level and of the forbidden band width in a degenerate semiconductor.

Unfortunately, such diagrams are not yet available for other materials.

It is easy to show that the tunnel current should depend on the p – n junction temperature. The current should vary due to the temperature dependence of the Fermi levels in the p- and n-type regions and due to the temperature dependence of the forbidden band width. These two dependences are shown schematically in Fig. 7.17 for n-type germanium. Above a certain fairly low temperature T_k, the position of the Fermi level decreases. Therefore, the degree of degeneracy also decreases. However, a decrease in the forbidden band width causes the degree of degeneracy to increase. Since a material with a relatively low impurity concentration is degenerate in a narrow temperature range, the temperature dependence of μ is the dominant factor. For a heavily doped material, the second temperature dependence is also important.

In the first investigations of the tunnel diode [4, 37, 38] the experimenters used semiconductors which were relatively lightly doped ($N_d \leq 1 \cdot 10^{19}$ cm^{-3}) and therefore they observed a drop in the tunnel current when temperature was increased. The second type of temperature dependence was discovered later [34, 39].

The temperature dependence of the tunnel current is observed most clearly in the value of J_p. Figure 7.18 shows the relative

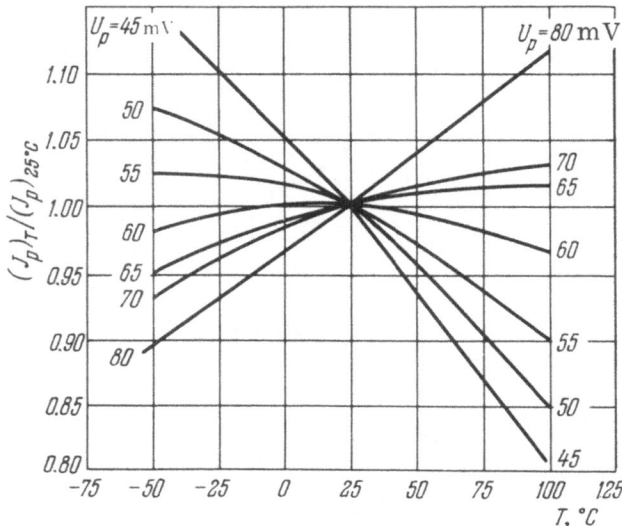

Fig. 7.18. Temperature dependences of the current J_p.

change in J_p with temperature for a germanium tunnel diode, according to Cady [36]. The voltage U_p is a parameter for curves in that figure.

According to Eq. (7.1.7), the probability of the tunnel effect in the case of indirect tunnel transitions should increase with rising temperature. Therefore, even in the case of relatively lightly doped crystals the current may increase with rising temperature. This is observed, for example, in antimony-doped germanium. The rise of the current with temperature in diodes made of such germanium is observed already at an antimony concentration of $5 \cdot 10^{18}$ cm^{-3} [39]. This is in excellent agreement with the experimental observation of the electron −phonon interaction in the tunnel effect in similar p−n junctions [40].

TABLE 7.2

Semi-conduc-tor	J_p/J_v	U_p, mV	U_v, mV	U_f, mV	T_{max}, °C	RC, sec	J_p/C, mA/pF
Ge	10—15	40— 70	270—350	450	250	$0.5 \cdot 10^{-9}$	0.3—1
Si	3— 4	80—100	400—500	700	400	$0.2 \cdot 10^{-8}$	< 0.5
GaAs	40—70	90—120	450—650	1000	600	$0.1 \cdot 10^{-9}$	10—15
GaSb	15—20	30— 50	200—250	450	300	$0.1 \cdot 10^{-9}$	
InSb	7—10			200	25	$0.5 \cdot 10^{-11}$	

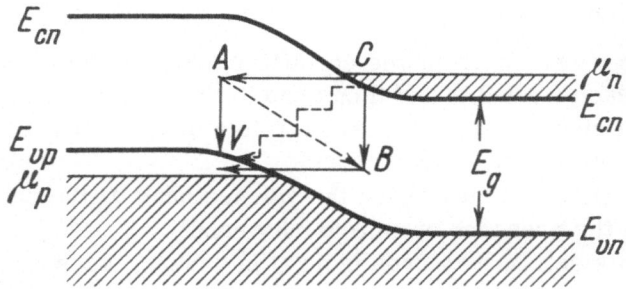

Fig. 7.19. Possible mechanisms of the excess current in tunnel diodes.

No systematic investigation of the temperature characteristics of $A^{III}B^V$ and Si tunnel diodes has yet been published. However, the band structure of these semiconductors suggests direct tunnel transitions in $A^{III}B^V$ compounds and consequently the temperature dependences of the tunnel current should be similar to those given in Fig. 7.18. In the case of Si diodes in which the tunnel transitions are indirect [40], the current should rise with increasing temperature at all impurity concentrations.

The working temperature range of tunnel diodes is governed primarily by the forbidden band width and therefore it is different for different semiconductors (Table 7.2).

Excess Current. According to our simplified treatment of the p−n junction band structure (Fig. 7.5e), the tunnel current should decrease to zero. This is not observed in real diodes and the current corresponding to the valley voltage U_V may be large. In one of the first investigations of tunnel diodes [41], Esaki suggested that the excess current is of "tunnel" origin. He assumed that deep levels in the forbidden band, with which the excess current is associated, are due to dislocations. Later, it was established experimentally that the current J_V is independent of pressure [42] and temperature [43] and depends very weakly on the donor and acceptor concentrations [44]. All these experimental observations support the tunnel nature of the excess current. Additional data obtained more recently [43, 45] have been used to develop a model [44] which explains the tunnel effect through deep levels. This model is shown schematically in Fig. 7.19. Several possible tunnel transition mechanisms can be seen in that figure:

1) an electron from the conduction band on the right can reach a level A and then drop to the valence band;
2) an electron from C may drop to a deep level B and leak from it to the valence band;
3) an electron may reach the valence band by the path CABV in the presence of an impurity band;
4) an electron may drop from C in several steps through local levels and this process may be accompanied by the emission of photons or phonons.

The last mechanism has been considered theoretically by Kane [45], who has shown that such transitions are practically impossible in germanium. In fact, photon emission has not been observed in any experimental investigation although the methods used have been more than sufficiently sensitive (with a margin of one order of magnitude [44]).

The most likely transition is that along CBV. A tunnel transition from B to V represents simply the field ionization. This has been considered theoretically by Franz [46], who has shown that the transition probability is the same as for interband transitions provided E_g is replaced with the impurity ionization energy. Using this approach, Chynoweth et al. [44] have obtained an expression for the excess current J_{ex} in the form

$$J_{ex} = \ln A + \ln D_{ex} - \frac{\alpha_{ex} L^{1/2}}{2} [E_g - eU_{ex} + e(\mu_v + \mu_c)], \qquad (7.1.23)$$

where D_{ex} is the density of states in the forbidden band (which represents the density of the B levels); U_{ex} is the bias voltage (in volts) in the excess current range; μ_v and μ_c are the positions of the Fermi levels in the valence and conduction bands;

$$\alpha_{ex} = \theta \left[\frac{8}{3} \cdot \frac{(m_{\parallel}^*)^{1/2}}{e\hbar} \right], \qquad (7.1.24)$$

where θ is a factor numerically close to unity.

The expression (7.1.23) can be easily checked experimentally since D_{ex}, L, E_g, and U_{ex} can be varied independently and their influence on J_{ex} can be investigated. A detailed investigation [44] has proved conclusively the validity of Eq. (7.1.23) for Si diodes, and

consequently it has established the tunnel nature of the excess current. The validity of the model has been finally confirmed by the
discovery – in the excess current region at liquid nitrogen temperature – of an additional negative conductance region [47]. Moreover, special doping of a crystal with gold impurities [48] has produced two maxima in the excess current range, corresponding to
two levels of gold, 0.35 eV and 0.54 eV below the bottom of the conduction band. Similarly, Chynoweth [48] has recorded five maxima
and has attributed them to the levels of phosphorus vacancies.

Esaki [48] has carried out an interesting investigation of the
impedance Z of a GaAs tunnel diode at microwave frequencies. In
the excess current range, he has detected oscillations of Z with a
period of about 10^{-8} sec. It is difficult to physically interpret this
period, but, bearing in mind that the equivalent circuit in this case
is of the delay-line type, Esaki has suggested that he has measured
the residence time of electrons at the B levels in the energy scheme
shown in Fig. 7.19.

Determination of the impurity levels in the forbidden band
from the additional maxima in the current – voltage characteristic
of tunnel diodes is known as tunnel spectroscopy. This method has
been applied successfully to the determination of the positions of
deep acceptor levels of Cu, Fe, Ni, and Co in heavily doped gallium
arsenide [49, 50].

The negative conductance region in the tunnel diode characteristic may also have some technical application.

Characteristic Voltages. The peak and valley volt-
ages U_p, U_v, associated with the corresponding currents J_p, J_v are
independent (in the first approximation) of the method of preparation of the diode. A more careful examination shows that both U_p
and U_v are governed by the value of n_{eq}, i.e., by the impurity concentrations on both sides of the p–n junction.

An explicit analytic dependence of U_p and U_v on the degree of
degeneracy on both sides of the junction is not known. It has been
shown experimentally that U_p increases somewhat when the impurity concentration is increased [28, 34]. In such investigations, it
is necessary to make allowance for a change in U_p due to the voltage drop across the resistance r_s. The value of U_v also increases
when the impurity concentrations in the n- and p-type regions are

increased [34]. This is in agreement with the treatment given in [19].

Negative Resistance. Negative resistance (or negative damping) raises the energy of a system and is possible only in the presence of a source of energy. In the case of tunnel diodes, such a source is the battery providing a bias voltage corresponding to that part of the current–voltage characteristic which has a negative slope, i.e., where the differential resistance R is negative.

The minimum negative resistance is inversely proportional to the probability of the tunnel effect in a given material, given by Eq. (7.1.3). Using Eq. (7.1.1), we can show that the higher the density of free carriers (i.e., the higher the impurity concentration) on both sides of the junction, the lower is the minimum negative resistance. In practice, the lower limit of R is set by the maximum solubility of impurities and by the possibility of obtaining a steep concentration gradient near the junction. Moreover, R is inversely proportional to the junction area.

Capacitance of p – n Junctions is given by the well-known expression [51]

$$C = 1.05S \left[\frac{\varkappa e N_d N_a}{8\pi (U_c - U)(N_d + N_a)} \right]^{1/2}, \qquad (7.1.25)$$

where S is the p–n junction area; the rest of the notation is the same as in Eq. (7.1.1).

This dependence has been verified on many occasions [52–54] in a wide range of voltages and its validity has been demonstrated up to bias voltage values at which the normal diode current begins to be observed [54]. Over this range of voltages, the absolute value of the capacitance varies only by 20–30%.

It is interesting to note that the quality factor $k = J_p/C$ is independent of the area and is only a function of n_{eq}.

This circumstance has allowed Cady [36] to plot curves shown in Fig. 7.20 for germanium tunnel diodes. It follows from this figure that, when n-type germanium is used, it is in practice difficult to make tunnel diodes with a quality factor $J_p/C > 1$. However, this can be done using other materials, for example GaAs, for which the factor k is approximately 10 mA/pF [55]. Typical values of J_p/C for diodes made of different semiconductors are listed in Table 7.2.

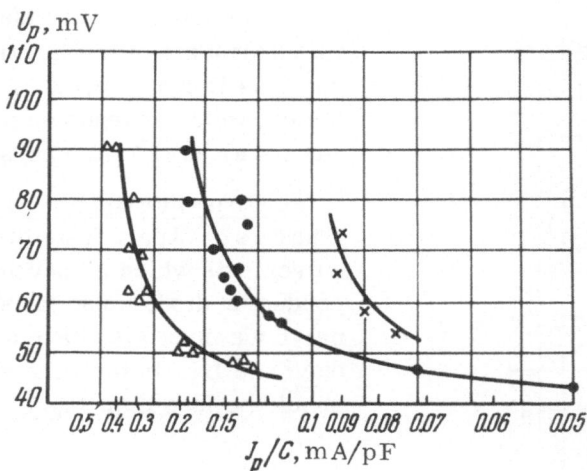

Fig. 7.20. Distribution of tunnel diodes in accordance with the values of J_p/C. △) $3.5 \cdot 10^{19}$ donor atoms per cm^3; ●) $2.5 \cdot 10^{19}$ donor atoms per cm^3; ×) $1.8 \cdot 10^{19}$ donor atoms per cm^3.

Time Constant. Another parameter which is independent of the junction area and depends only on n_{eq} is the time constant of the diode, RC, which increases almost exponentially when the free-carrier density is increased (Fig. 7.21) [23, 54].

Typical values of RC for diodes made of various semiconductors are listed in Table 7.2.

Loss Resistance. The use of low-resistivity semiconductors in tunnel diodes means that the loss resistance r_s is small. Thus, for germanium tunnel diodes the value of r_s under dc conditions is of the order of 1 Ω or even less.

The value of r_s for tunnel diodes cannot be estimated using the formulas for conventional diodes. This is because the resistivity in a tunnel diode is approximately the same on both sides of the junction. In conventional diodes, the main part of the semiconductor always has a higher resistivity and it is that part which governs the loss resistance. Moreover, calculations show [56] that if a tunnel diode is made by alloying an acceptor impurity to n-type germanium, then almost the whole loss resistance is concentrated in the alloyed region. In this case, the value of r_s should be inversely proportional to the p−n junction area. This is indeed observed experimentally [30].

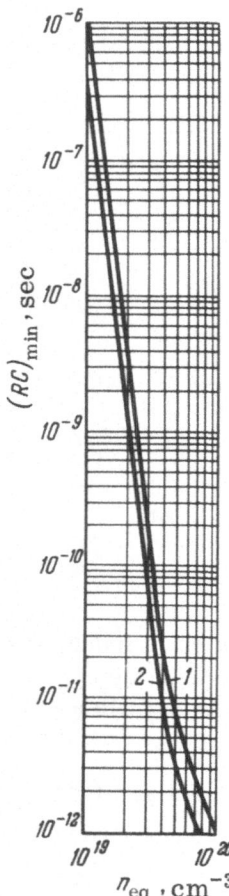

Fig. 7.21. Dependence of the tunnel-diode time constant on the equivalent doping for diodes made of Ge (curve 1) and GaAs (curve 2).

We note that the value of r_s/R is independent of the junction area and thus it can be used as a quality factor, which depends on n_{eq} in the same way as the time constant RC.

Construction of Tunnel Diodes. Apart from the usual requirements which a conventional semiconductor diode must satisfy, a tunnel diode must also have a low inductance L_S. In fact, the value of L_S should be less than L_{max}, given by

$$L_{max} = RCr_s. \qquad (7.1.26)$$

Typical values of the parameters of germanium tunnel diodes ($RC \approx 10^{-9}$ sec, $r_s \approx 1\ \Omega$) show that such diodes should have inductances less than 10^{-9} H in order to satisfy Eq. (7.1.26). Conventional diodes, with which electrical contact with the p—n junction is made • by means of thin wires, have inductances not less than $3 \cdot 10^{-9}$ H. Therefore, electrical contact with tunnel diodes is made either through a membrane [9] (Fig. 7.22a), a massive pressure electrode [57] (Fig. 7.22b) or a tag [58] (Fig. 7.22c). The latter construction ensures a value of L_S equal to $4.0 \cdot 10^{-10}$ H.

The best low-inductance electrical contact is provided by the direct soldering of a flat plate to an alloyed metal drop (Fig. 7.22d). In this case, the tunnel diode inductance is governed by the geometry of the crystal and the metal drop.

Circuits with Tunnel Diodes. Tunnel diodes can be used in amplifiers, oscillators, converters, and switches. The necessary relationships between the diode and circuit parameters

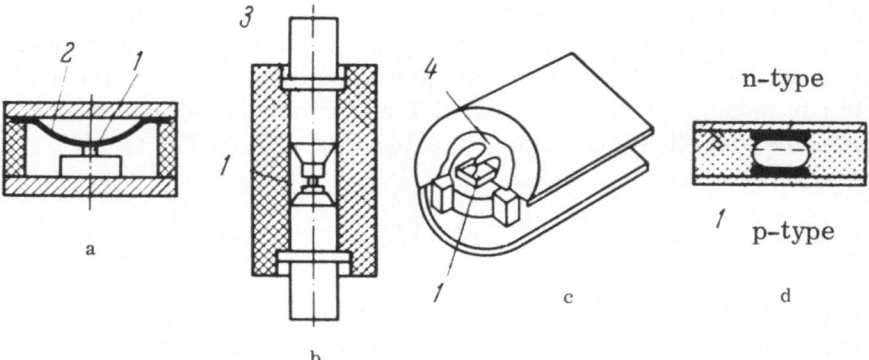

Fig. 7.22. Various constructions of tunnel diodes. 1) Crystal; 2) membrane; 3) pressure electrode; 4) contact tab.

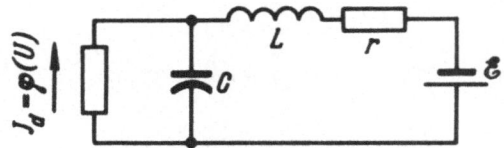

Fig. 7.23. Equivalent circuit of a tunnel diode connected to a bias voltage source; $r = r_s + r_1 + r_l$; $L = L_s + L_1$ ("s" refers to the diode; "1" to the circuit, and "l" to the load).

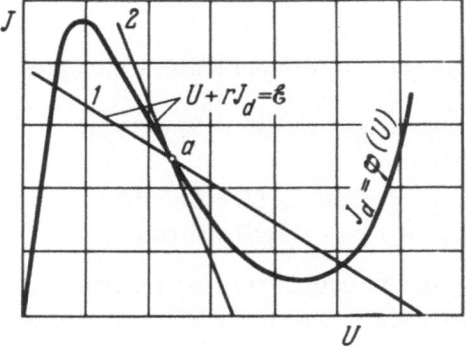

Fig. 7.24. Equilibrium states of the circuit shown in Fig. 7.23. 1) $r > 1/G$, equilibrium at point a always unstable; 2) $r < 1/G$, stability of equilibrium at point a governed by an additional condition $r > LG/C$.

are found by analyzing the stability. These relationships can be found easily by considering a very simple circuit containing a tunnel diode, connected to a bias voltage source through a resistance r and an inductance L (the values of L and r include the loss resistance r_s and the diode inductance L_s) (Fig. 7.23). The stable operation condition

$$U + rJ_d = \mathscr{E} \tag{7.1.27}$$

(a graphical solution of this equation, together with the current–voltage characteristic of the tunnel diode, is given in Fig. 7.24) determines the equilibrium state of this circuit.

The behavior of this simple circuit in the negative part of the current–voltage characteristic is given by a second-order differential equation [59]:

$$\frac{d^2U}{dt^2} + \left(\frac{r}{L} - \frac{G}{C}\right)\frac{dU}{dt} + \frac{U}{LC} + \frac{r}{LC}\varphi(U) + \frac{1}{C}\frac{dC}{dU}\left(\frac{dU}{dt}\right)^2 = \mathscr{E}\frac{1}{LC}. \tag{7.1.28}$$

According to Lyapunov, Eq. (7.1.28) can be linearized to determine the stability conditions of equilibrium points when the circuit is subjected to small perturbations. The linearized equation has the form [60]:

$$\frac{d^2U}{dt^2} + \left(\frac{r}{L} - \frac{G}{C}\right)\frac{dU}{dt} + \frac{1-rG}{LC}U = 0, \tag{7.1.29}$$

or, if expressed in terms of dimensionless parameters:

$$\ddot{U} + (p - s)\dot{U} + (1 - ps)U = 0, \tag{7.1.30}$$

where $\omega_0 = 1/\sqrt{LC}$; $p = \omega_0 rC$; $s = \omega_0 LG$; the differentiation with respect to the parameter $t_1 = \omega_0 t$ is denoted by $\dot{U}$ and $\ddot{U}$.

The stability of the equilibrium conditions can be checked by analysis of the roots of the characteristic equations corresponding to the differential equations (7.1.29) or (7.1.30). The equilibrium is stable only if the real parts of both roots of the characteristic equation are negative. The same stability conditions are obtained from the requirement that there should be no zeros of the impedance on the right-hand side of the complex frequency plane [52, 61].

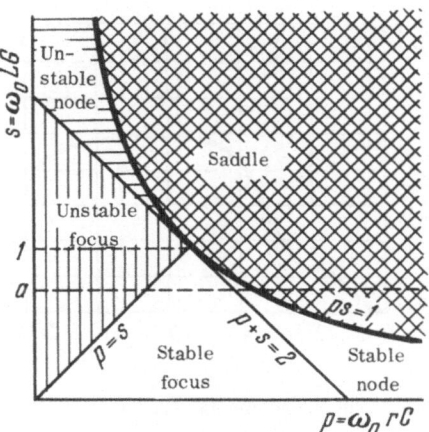

Fig. 7.25. Stability diagram
of the tunnel diode.

When

$$ps > 1 \qquad \left(r > \frac{1}{G}\right) \tag{7.1.31}$$

the roots of the equation are real and have different signs; the equation describes a system with a repulsive force, characterized by an unstable equilibrium of the saddle type, irrespective of the relationships between the other parameters of the system. Under these conditions, a tunnel diode can be used for switching.

When

$$ps < 1 \qquad \left(r < \frac{1}{G}\right) \tag{7.1.32}$$

the load line intersects the current−voltage characteristic at one point only and the dc stability conditions are satisfied.

Complex roots are obtained when

$$p + s < 2, \tag{7.1.33}$$

i.e., when the integral curves in the phase plane are in the form of two interlocking spirals. The asymptotic point of all these curves is a focus; periodic oscillations are possible.

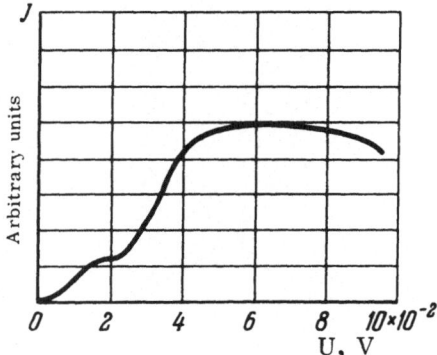

Fig. 7.26. Experimental current—voltage characteristics indicating the participation of phonons
in the tunnel effect.

The stability of the focus is determined by the Lyapunov
method. When

$$p > s \quad \left(r > \frac{LG}{C} \right) \tag{7.1.34}$$

the equilibrium is stable. Damped sinusoidal oscillations are possible in the stable focus region, bounded by the abscissa and the
straight lines $p = s$ and $(p + s) = 2$.

Such oscillations may appear if the following conditions are
obeyed:

$$p > s, \quad p + s < 2. \tag{7.1.35}$$

These conditions are employed in amplifiers, when

$$p < s \quad \left(r < \frac{LG}{C} \right), \quad p + s < 2 \tag{7.1.36}$$

the focus is unstable.

In the region enclosed by the ordinate and the straight lines
$p = s$, $(p + s) = 2$, periodic oscillations may be generated. In the region between the hyperbola $ps = 1$ and the straight line $(p + s) = 2$ the
equilibrium state is a node and the process is aperiodic. The stability is again governed by the criterion $p > s$. All these relationships are shown in Fig. 7.25 [60].

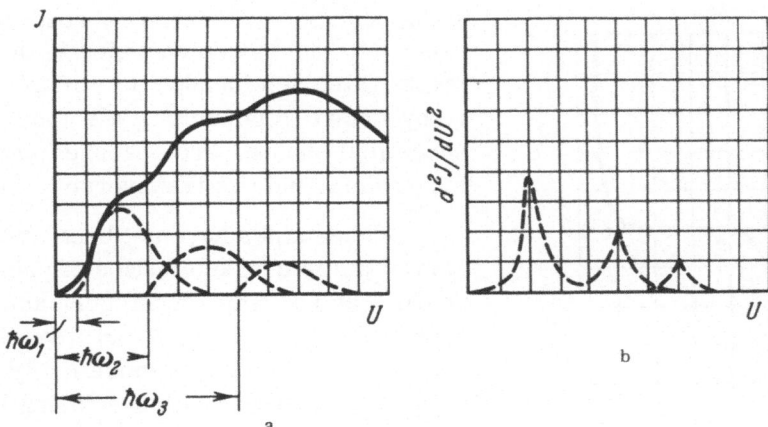

Fig. 7.27. Curves illustrating the participation of phonons in the tunnel effect (according to Hall [48]).

We must mention that at all frequencies stable operation is possible if the condition (7.1.34) is satisfied.

In a case of frequency-dependent $L(\omega)$ and $r(\omega)$, the expression (7.1.34) becomes:

$$r(\omega) > \frac{L(\omega)\,G}{C}. \tag{7.1.37}$$

In general, a circuit is more complex than that shown in Fig. 7.23. Such circuits are described by differential equations of third and higher orders. It is difficult to calculate directly the roots of such equation. The stability conditions in this case can be determined from the Hurwitz−Routh criterion (cf., for example, [62]).

For actual circuits of oscillators, amplifiers, switches, and many other devices in which tunnel diodes are used, the reader is referred to [32, 71] and to a bibliography on tunnel diodes [64].

Use of Tunnel Diodes in Physical Investigations. The quantum-mechanical nature of the tunnel diode can be used to study directly the electron−phonon interaction.

This interaction was first investigated by Holonyak et al. [40], who investigated the current−voltage characteristic of tunnel diodes at helium temperatures and have discovered characteristic kinks in the forward branch (Fig. 7.26). The voltages at which these

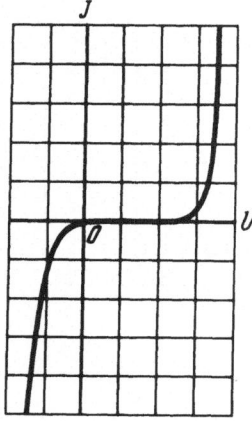

Fig. 7.28. Current—voltage characteristic of the backward diode.

kinks have been observed correspond exactly to the phonon energies in the investigated semiconductor. In antimony-doped germanium, all four acoustical phonons are observed: two longitudinal and two transverse.

These kinks have not been observed in the characteristics of phosphorus- and arsenic-doped germanium. This indicates either direct tunnel transitions or, which is more likely, indirect transitions with scattering by impurities. This result is confirmed also by the nature of the temperature dependence of the tunnel current and by the reverse branch of the current—voltage characteristic [65].

The electron—phonon interaction has also been observed in silicon tunnel diodes [40]. Such investigations can even be quantitative. Hall [48] has shown that a current—voltage characteristic with kinks can be regarded as a sum of several curves, which are shown dotted in Fig. 7.27a. These curves are shifted along the voltage axis by amounts proportional to $\hbar\omega_i$ of the corresponding phonons. Their amplitudes are different because of the different degree of interaction between phonons and electrons. The picture becomes clearer when the second derivative of the tunnel current with respect to the voltage is plotted (Fig. 7.27b). A careful investigation has shown that each of the dashed curves in Fig. 7.27a decreases asymptotically toward the voltage axis. This is regarded in [48] as a consequence of the broadening of the lower edge of the conduction band.

Low-temperature investigations of the tunnel current in diodes prepared from semiconducting compounds ($A^{III}B^V$ and others) have led to the discovery of an interaction between electrons and polarons [66].

Investigations of the current—voltage characteristics of tunnel diodes in strong magnetic fields [67] can be used to determine the effective carrier mass. Such a determination is based on the well-known de Haas—van Alphen effect. In tunnel diodes, this effect is manifested by oscillations of the current passing through the diode. Chynoweth [48] has investigated PbTe tunnel diodes in fields stronger than $4 \cdot 10^4$ G and has recorded at least nine such oscillations. By measuring the period of these oscillations we can find the value of the Bohr magneton, which is

$$\mu = \frac{e\hbar}{m^* c},$$ (7.1.38)

and, consequently, the effective mass m*. Investigations of tunnel diodes in magnetic fields are of obvious practical interest because the value of J_p/J_v decreases when the field intensity is increased [67, 68].

Measurements of the forward and reverse currents in a tunnel diode can yield information on the probability of the tunnel effect and consequently on all the quantities which occur in the expressions for this probability [Eqs. (7.1.4) and (7.1.5)].

To obtain such information, the dependence of the forbidden band width E_g on the hydrostatic pressure has been determined in [42, 69]. In [70], the value of D_{ex} has been found from Eq. (7.1.23), i.e., the density of the allowed states in the forbidden band has been determined. Finally, we have already mentioned tunnel spectroscopy, which allows us to study deep levels in the forbidden band of a semiconductor.

These examples do not exhaust all the possibilities of the use of tunnel diodes in physics investigations. Intensive studies of these diodes will undoubtedly bring more valuable information on physical phenomena in solids.

Other Devices with p − n Tunnel Junctions. Tunnel junctions are also used in backward diodes and tunnel transistors. The backward diode [73-78] differs from the tunnel diode by having one side of the p−n junction heavily doped (strongly degenerate) and the other side either nondegenerate or slightly degenerate. The current−voltage characteristic of the backward diode is shown in Fig. 7.28.

A theory of the current−voltage characteristic of the backward diode has been given in [79]; it is based on the theory of the tunnel diode characteristic. The backward diode parameters depend on the properties of the semiconducting material in exactly the same way as the tunnel diode parameters.

The most promising applications of the backward diode are in high-quality detectors and mixers working at low signal levels (in particular, in mixers with a low intermediate frequency) and in video signal detectors, where the backward diode has the advantage of much lower low-frequency noise than the point-contact diodes used for this purpose [77]. Backward diodes can be employed also as unidirectional elements in computers, where conventional diodes cannot be used because of the large forward voltage drop and slow response.

In the tunnel transistor [80, 81], the collector is an ordinary p−n junction and the emitter is a tunnel p−n junction. Such a structure opens up new possibilities in electronic circuits, particularly in computers.

§7.2. Semiconductor Lasers

Principle of Operation. Let us consider an assembly of particles which are not in thermodynamic equilibrium. For simplicity, we shall assume that particles can be in only two possible states: n and m, where n > m. If this assembly is subjected to directional radiation of frequency ν_{nm}, the number of upward transitions from the level m to the level n will be:

$$N_{mn} = B_{mn} u_\nu N_m, \tag{7.2.1}$$

and the number of downward transitions to the level m will be:

$$N_{nm} = (A_{nm} + u_\nu B_{nm}) N_n. \tag{7.2.2}$$

In these expressions, A and B are some constants representing the system of particles; N_n and N_m are the numbers of particles in the states n and m; u_ν is the density of the directional radiation, which is related to the transition probability P_{mn} by

$$P_{mn} = B_{mn} u_\nu.$$

If $N_n < N_m$, the losses in the incident beam of light will be $B_{nm} u_\nu (N_m - N_n)$ quanta per second. The spontaneous emission of $A_{nm} N_n$ quanta can be regarded as scattered radiation. Consequently, in this case the system of particles has a positive absorption coefficient and represents a normal system.

However, if in the assembly of particles considered, the higher energy state n is occupied by a larger number of particles than the lower energy state m $(N_n > N_m)$, then the absorption coefficient is negative. An assembly with such a distribution of particles is said to have an inverse population.

A negative absorption coefficient means that such an assembly can emit spontaneously. This assembly will also amplify radiation of the frequency

$$ v = \frac{E_n - E_m}{h}, $$

i.e., it will exhibit the laser effect.

To construct a laser, it is necessary to place a material with negative absorption in a resonator from which electromagnetic radiation can be extracted.

The simplest resonator is a system of two mirrors, known in classical optics as the Fabry–Perot interferometer.

To obtain population inversion in semiconductors, we can use various electron transitions: band–band, band–impurity level, impurity–impurity.

To obtain an inverse population, the total absorption for a given transition must be negative. Following the discussion given in [82], this condition can be written in the form

$$ n_r P_1 \left[f_c (1 - f_v) - f_v (1 - f_c) \right] > 0, \qquad (7.2.3) $$

where n_r is the number of photons in a given state; P_1 is the probability of forward and reverse transitions (two probabilities are assumed to be equal); f_c and f_v are the electron distribution functions in the conduction and valence bands of the semiconductor.

It follows from Eq. (7.2.3) that $f_c > f_v$ and therefore

$$ \mu_c - \mu_v > E_g, \qquad (7.2.4) $$

if the distribution functions are of the Fermi type.

The expression (7.2.4) is valid for a direct band—band transition (cf. Fig. 7.8a). If the transition is indirect (Fig. 7.8b), instead of Eq. (7.2.4) we have

$$\mu_c - \mu_v > E_g - \hbar v, \qquad (7.2.5)$$

where hν is the energy of the phonon participating in the indirect transition.

It is also shown in [82] that for a transition from an exciton state, the condition for an inverse population is

$$\mu_c - \mu_v > E_g - \hbar v - | E_{np} , \qquad (7.2.6)$$

where E_{np} is the binding energy of an exciton.

If the transition takes place from one of the main bands to impurity levels with an ionization energy E_i, lying close to the edge of the other band, the following condition must be satisfied

$$\mu_c - \mu_v > E_g - | E_i |. \qquad (7.2.7)$$

Semiconductor Laser Materials. In semiconductors, the absorption of light by direct transitions is more likely than the absorption by free carriers. The absorption due to indirect transitions is comparable with the free-carrier absorption. This makes it difficult to obtain a population inversion using indirect interband transitions.

Therefore, the laser effect in a semiconductor is most likely to be associated with direct interband transitions. It follows from Eq. (7.2.4) that the carrier gas in at least one of the bands should be degenerate. Thus, lasers can be made only from sufficiently heavily doped crystals.

The second requirement that semiconductor laser materials must satisfy is a band structure with a high probability of direct transitions. Comparison of the band structures of various semiconductors (Figs. 1.4-1.6) shows that the most suitable material is gallium arsenide. In fact, the first semiconductor lasers were made of this material [83-85]. We can also use other semiconductors, which may have a more complex energy spectrum or may not be heavily doped [86-89].

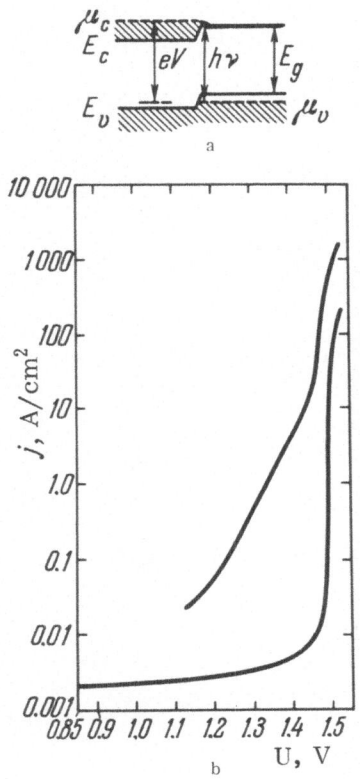

Fig. 7.29. a) Energy level diagram near a p−n junction in a degenerate semiconductor, subjected to a bias $U \approx E_g/e$; b) current−voltage characteristics of diffused GaAs diodes with $8 \cdot 10^{16}$ cm^{-3} (lower curve) and $1.6 \cdot 10^{18}$ cm^{-3} (upper curve) carrier densities (T = 4.2°K).

Experimental Results. Radiative recombination in a GaAs p−n junction during the passage of a current in the forward direction has been reported in [90, 91]. At 77°K, the emission has a maximum at about 1.47 eV. The energy band diagram near a p−n junction subjected to a bias $U \approx E_g/e$ is shown in Fig. 7.29a and the current−voltage characteristics are given in Fig. 7.29b. These characteristics show a rapid increase of the forward current at a voltage of about 1.5 V (~Eg/e). At higher values of the current the conduction band will have many electrons and the valence band many holes with which electrons can recombine; a population inversion will be established near the p−n junction. Such an injection method of producing a population inversion in a p−n junction in a degenerate semiconductor was first predicted theoretically [92].

If the energy losses due to other forms of absorption of light in a crystal can be made good by the emitted quanta, then a p−n junction will emit light in the inverse population state.

Soon after the observation of strong and narrow recombination lines emitted by GaAs diodes, reports were published of similar experiments on $A^{III}B^{V}$ compounds with direct interband transitions: InP [93], InAs [94], InSb [95], and GaSb [96].

When the forward current is increased, a recombination radiation line becomes narrower above a certain threshold value of

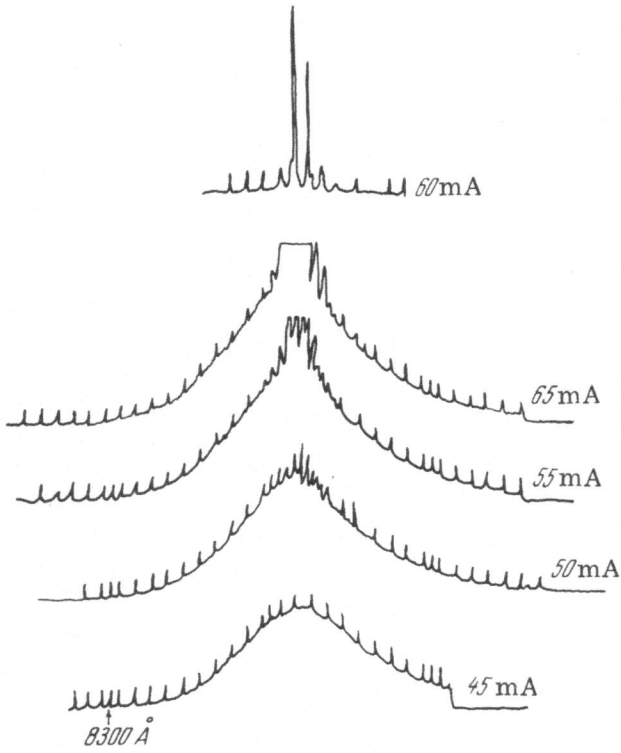

Fig. 7.30. Spectrum of a typical GaAs laser diode near the
laser threshold (55 mA) at 4°K. The separation between the
peaks is 11.3 Å.

TABLE 7.3

Semi-conductor	T, °K	λ, Å
GaAs	77	8400
	300	9000
InP	4	9000
	77	9100
InAs	2	51,000
	4	53,000
Ga $(As_{1-x}P_x)$		Up to 6600

the current, narrow modes (modes are
different types of electromagnetic field
oscillation) appear in the spectrum, and
finally the light emitted by the diode becomes directional (Fig. 7.30). If the temperature is increased to liquid nitrogen
temperature, the threshold current increases by a factor larger than 15. Initially, the laser effect in GaAs diodes
has been observed only under pulse conditions. However, the improvement in
the manufacturing technology of such diodes has rapidly led to the development
of devices generating continuously [97].

Observations of the laser effect have also been reported for $Ga(As_{1-x}P_x)$ [98] and $(In_{1-x}Ga_x)As$ [99] solid solutions and for InP [100] and InAs [101].

Table 7.3 lists the wavelengths of light generated by diodes made of different semiconductors.

It is hardly possible to discuss the numerous experimental papers on injection lasers; the reader is referred to suitable reviews [102-105]. The development of laser technology is so rapid that it would be premature to attempt a definitive review.

Preparation Methods. An injection laser is an ordinary diode in the form of a resonator, whose purpose is to select geometrically certain modes. The preparation of a GaAs laser with cleaved faces is described in [102]. For this purpose, one uses (110) natural cleavage faces. The GaAs is cut into plates with sides perpendicular to one or several (110) planes. The crystals are doped with Si, Ge, Se, or Te to obtain carrier densities of $3 \cdot 10^{18}$-$6 \cdot 10^{18}$ cm^{-3}. A p−n junction is prepared by the diffusion method The diffusant is Zn in the form of $ZnAs_2$.

Contacts with the p- and n-type regions are made using standard methods, which are reviewed, for example, in [106].

§7.3. Thermoelectric Devices

Principle of Operation [107, 108]. We shall consider a thermocouple (Fig. 7.31) with branches represented by the parameters α_1, σ_1, $\varkappa_1$ and α_2, σ_2, and $\varkappa_2$, respectively. The cross-sectional areas of the branches are S_1 and S_2. Assuming that heat transfer takes place only along the x axis, we shall calculate the coefficient representing the ratio of the amount of heat taken from a source to the amount of heat transformed into electric energy; this coefficient is known as the efficiency. Another interesting quantity is the maximum difference between temperatures that can be reached by extracting heat from a source.

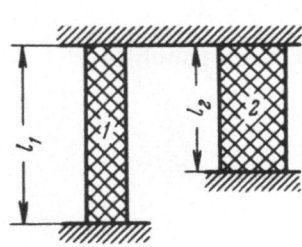

Fig. 7.31. Thermocouple consisting of two branches.

The flux of heat in one of the branches of the thermocouple i (i = 1, 2) along the x axis is

$$q_i = \mp a_i J T - \varkappa_i S_i \frac{dT}{dx}$$

(7.3.1)

(the notation is the same as in Chap. 3).

The generation of heat per unit length due to the Joule effect is

$$\frac{J^2}{\sigma_i A_i} = - \varkappa_i S_i \frac{d^2T}{dx^2}.$$

(7.3.2)

Solving this equation for the boundary conditions

$$(T)_{x=0} = T_h; \quad (T)_{x=l} = T_c,$$

we find

$$q_i = -\frac{J^2 \left[x - \left(\frac{l}{2} \right) \right]}{\sigma_i S_i} - \frac{\varkappa_i S_i (T_c - T_h)}{l_i}.$$

(7.3.3)

The heat pumping rate q is equal to the sum of q_1 and q_2 when $x = 0$. Assuming that the currents in the thermocouple branches flow in the opposite directions, we obtain

$$q = (a_2 - a_1) J T_h - \frac{J^2 R}{2} - K(T_c - T_h),$$

(7.3.4)

where $R = l_1/S_1\sigma_1 + l_2/S_2\sigma_2$ is the total electrical resistance of the branches connected in series; $K = S_1\varkappa_1/l_1 + S_2\varkappa_2/l_2$ is the total thermal conductivity of the branches connected in parallel.

The total electrical energy consumed in one branch is:

$$w_i = \pm \int_{T_h}^{T_c} a_1 J dT + \int_0^{l_i} \frac{J^2 dx}{\sigma_i S_i} = \mp a_i J (T_c - T_h) + \frac{J^2 l_i}{\sigma_i S_i},$$

and in both branches:

$$w = (a_2 - a_1) J (T_c - T_h) + J^2 R.$$

(7.3.5)

The thermocouple efficiency φ, equal to the ratio q/w becomes

$$\varphi = \frac{(a_2 - a_1) J T_h - \dfrac{J^2 R}{2} - K(T_c - T_h)}{(a_2 - a_1) J (T_c - T_h) + J^2 R}.$$
(7.3.6)

We shall find the value of the current J_φ, corresponding to the maximum efficiency, φ_{max}, for a given temperature gradient. For this purpose, we shall assume that $d\varphi/dT = 0$ and then

$$J_\varphi = \frac{(a_2 - a_1)(T_c - T_h)}{R(\sqrt{1 + ZT_a} - 1)},$$
(7.3.7)

where T_a is the average temperature given by $(T_c + T_h)/2$;

$$Z = \frac{(a_2 - a_1)^2}{KR}.$$

The maximum efficiency φ_{max} is

$$\varphi_{max} = \frac{T_h}{T_c - T_h} \cdot \frac{\sqrt{1 + ZT_a} - \dfrac{T_c}{T_h}}{\sqrt{1 + ZT_a} + 1}.$$
(7.3.8)

From the condition $dq/dJ = 0$, we find the value of J_q. corresponding to the maximum cooling rate:

$$J_q = \frac{(a_2 - a_1) T_h}{R},$$
(7.3.9)

and the efficiency expressed in terms of the power

$$\varphi_q = \frac{\dfrac{1}{2} Z T^2_h - (T_c - T_h)}{Z T_c T_h}.$$
(7.3.10)

From Eq. (7.3.1), we can easily find $(T_c - T_h)_{max}$:

$$-(T_c - T_h)_{max} = \frac{1}{2} Z T^2_h.$$
(7.3.11)

Since the figure of merit Z governs the maximum temperature drop and the efficiency of a thermocouple subjected to a small

temperature gradient, it can be regarded as a general characteristic of a thermocouple.

For any two given thermocouple materials, there is a maximum value of Z, which can be obtained by selecting an optimum
relationship between the dimensions of the branches. For this
purpose, it is necessary that the product RK be minimal. This
happens when

$$\left(\frac{S_1/l_1}{S_2/l_2}\right) = \sqrt{\frac{\sigma_2 \varkappa_2}{\sigma_1 \varkappa_1}}. \tag{7.3.12}$$

If this condition is satisfied, then

$$Z = \frac{(\alpha_2 - \alpha_1)^2}{\left[\left(\frac{\varkappa_1}{\sigma_1}\right)^{1/2} + \left(\frac{\varkappa_2}{\sigma_2}\right)^{1/2}\right]^2} \tag{7.3.13}$$

Thus, investigations of thermoelectric materials represent
a search for those semiconductor pairs for which Z, defined by Eq.
(7.3.13), has its maximum value.

To represent a thermocouple, one introduces separate figures of merit z_i for each of the branches:

$$z_i = \frac{\alpha_i^2 \sigma_i}{\varkappa_i}. \tag{7.3.14}$$

This expression can be rewritten in a somewhat different form:

$$zT = \frac{\alpha^2/L}{1 + \dfrac{\varkappa_p}{\varkappa_e}},$$

or, since for the majority of semiconductors $\varkappa_p/\varkappa_e \gg 1$, it follows
that

$$zT \approx \frac{\alpha^2}{L \dfrac{\varkappa_p}{\varkappa_e}}. \tag{7.3.15}$$

If we analyze the quantities α and L (cf. Chap. 3) we can easily find the relationship between zT with the degree of degeneracy
μ^* and, consequently, with the degree of doping of the semiconductor. High values of z are obtained for μ^* close to degeneracy.

Construction of semiconducting thermoelectric devices is described in [108].

§7.4. Hall Probes and Magnetoresistors Stable under Nuclear Radiation

It is known that the irradiation of semiconducting crystals with high-energy particles produces lattice defects, because atoms at lattice sites may be displaced into interstitial positions by elastic collisions with the incident particles. Numerous experiments on deuteron, α-particle, fast electron, fast and slow neutron irradiation of lightly doped germanium, silicon, and other single crystals have established that the effects of irradiation with different particles are similar and that the free-carrier density and the conductivity of semiconducting crystals usually decrease after irradiation [109]. Moreover, it has been found that radiation-induced changes in the electrical properties can be reversed (to a considerable degree) by high-temperature heating or even prolonged storage at low temperatures (below 170°K). The only exception to this rule is the effect of slow neutrons on Ge and Si. In this case, an appreciable residual and stable effect remains in these crystals because of certain nuclear reactions. For example, in Ge this effect is due to the transformation of $_{32}Ge^{70}$ and $_{32}Ge^{74}$ into stable elements, gallium and arsenic:

$$_{32}Ge^{70}(n\gamma)\,_{32}Ge^{71}, \quad _{32}Ge^{71} \xrightarrow{K} _{31}Ga^{71},$$

$$_{32}Ge^{74}(n\gamma)\,_{32}Ge^{75}, \quad _{32}Ge^{75} \xrightarrow{\beta^-} _{33}As^{75} + e\ .$$

The arsenic and gallium formed in these reactions act as donor and acceptor impurities, which affect the electrical properties of germanium crystals.

The number of captures of slow neutrons by a given isotope is

$$N = nvt\theta_i N_0 P_i, \tag{7.4.1}$$

where nvt is the integral slow-neutron flux; θ_i is the effective neutron cross section of a given isotope; P_i is the relative concentration of this isotope; N_0 is the concentration of germanium atoms per 1 cm³, which is $4.52 \cdot 10^{22}$ cm^{-3}.

TABLE 7.4

Isotope	θ_i, barn	P_i, %	$T_{1/2}$
$_{32}Ge^{70}$	3.4 ± 0.3	20.55	12 days
$_{32}Ge^{74}$	0.62 ± 0.6	36.74	82 min
$_{32}Ge^{76}$	0.36 ± 0.07	7.69	12 h

We must mention that natural germanium contains $\approx 8\%$ of the $_{32}Ge^{76}$ isotope which gives rise to selenium (after a suitable nuclear reaction). Since P_i and θ_i for this isotope are small, the formation of Se in the Ge lattice can be ignored.

The values of the effective cross sections, the corresponding isotopic abundances, and the half-lives of the three Ge isotopes are listed in Table 7.4.

Knowing the integral neutron flux nvt, as well as θ_i and P_i, we can calculate the concentrations of arsenic and gallium which are formed by slow-neutron irradiation.

The stability of semiconducting devices, in particular of the Hall probes, under the action of nuclear radiation depends strongly on the number of impurity centers formed by such radiation.

It is recommended that Hall probes be made of heavily doped germanium, silicon, or other degenerate semiconductors, since in this case the influence of irradiation on the electrical properties (in particular, on the carrier density) is negligibly small even for very large integral nuclear radiation fluxes, i.e., such a Hall probe should be stable under nuclear radiation.

According to Eq. (7.4.1), the concentrations of As and Ga atoms formed by the irradiation of germanium with slow neutrons are, respectively:

$$N_{As} = 1.03 \cdot 10^{-2} nvt; \quad N_{Ga} = 3.16 \cdot 10^{-2} nvt.$$

Hence, it is clear that if a Hall probe is made of heavily doped germanium with a free-carrier density of the order of 10^{19} cm^{-3}, such a probe should be stable up to integral slow-neutron fluxes of the order of nvt $\approx 10^{20}$

The sensitivity of a Hall probe made of heavily doped germanium would be somewhat lower than that made of lightly doped material.

One of the main requirements which a Hall probe must satisfy is the thermal stability [110]. A very important property of heavily doped semiconductors in general and of germanium in particular is the lack of a dependence of the Hall coefficient on temperature over a wide range of temperatures.

As mentioned in Chap. 3, the Hall coefficient of germanium doped with $N > 2 \cdot 10^{18}$ cm^{-3} of impurities is independent of temperature between 4.2 and 600°K. It follows that a Hall probe made of heavily doped germanium should be stable under nuclear radiation and should be unaffected by temperature variation in the range 4.2-600°K and possibly even at higher temperatures.

Irradiation may, in principle, affect also the mobility because of the additional scattering of carriers on defects. However, in the case of heavily doped semiconductors of the Ge type, the carrier mobility is governed mainly by the scattering on ionized impurities so that the additional scattering on defects may be ignored. Moreover, even if this additional scattering is important, it does not alter the Hall coefficient.

Fistul' [111] carried out direct experiments in which he irradiated germanium single crystals with slow neutrons, and in which he investigated the magnetoresistance and the Hall effect of the irradiated samples. These experiments have shown that germanium crystals with a carrier density higher than 10^{18} cm^{-3} can be used to prepare stable Hall and magnetoresistance probes.

§7.5. Strain Gauges

It is known that the electrical resistivity of semiconductors and metals is affected by deformation. The relative change in the resistivity per unit strain is known as the strain sensitivity. This quantity is dimensionless, and gives the change in the electrical resistivity due to deformation. However, it is more convenient to measure a dimensional quantity known as the piezoresistance. If X is a mechanical stress and $\Delta\rho/\rho$ is the change in the resistivity of a sample due to the stress X, then the piezoresistance is

$$\Pi = \frac{\Delta\rho}{\rho X}. \qquad (7.5.1)$$

Usually, X is measured in dyn/cm^2 and, therefore Π has the dimensions of cm^2/dyn. Both the strain sensitivity and the piezo-

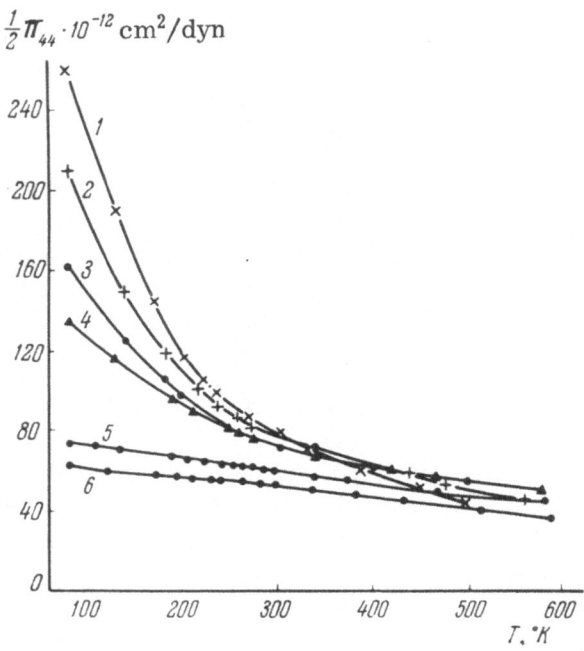

Fig. 7.32. Temperature dependence of the piezoresistance of germanium samples with the following electron densities: 1) $4.5 \cdot 10^{17}$ cm^{-3}; 2) $2.1 \cdot 10^{18}$ cm^{-3}; 3) $6.6 \cdot 10^{18}$ cm^{-3}; 4) $8.0 \cdot 10^{18}$ cm^{-3}; 5) $2.2 \cdot 10^{19}$ cm^{-3}; 6) $3.3 \cdot 10^{19}$ cm^{-3}.

resistance are tensors of the fourth rank, which generally have 21 independent components. For cubic crystals, the number of independent components reduces to three. Combinations of three components Π_{11}, Π_{12}, and Π_{44} give the whole piezoresistance for any direction in a crystal. To find the values of Π_{11}, Π_{12}, and Π_{44}, it is necessary to carry out at least three measurements in two samples with different orientations [112]. Of these three measurements, two should be "longitudinal" and one "transverse," or the converse. In the "longitudinal" measurements, the current through a sample and the stretching or compressing force are parallel; in the "transverse" measurements, they are perpendicular.

If we use l, m, and n to denote the cosines of the angles between the long dimension of a sample and the crystallographic axes of a cube x, y, and z, the longitudinal piezoresistance for any orientation of a sample becomes [113]

$$\frac{\Delta\rho}{\rho X} = \Pi_{11} - 2\,(l^2 m^2 + m^2 n^2 + n^2 l^2)\,(\Pi_{11} - \Pi_{12} - \Pi_{44}). \qquad (7.5.2)$$

When all the three components Π_{11}, Π_{12}, and Π_{44} are small, it is necessary to introduce a correction for the change in the sample geometry due to deformation. If a sample is cut along the [110] direction and the "longitudinal" measurements are carried out, Eq. (7.5.2) becomes

$$\left.\frac{\Delta\rho}{\rho X}\right|_{[110]} = \frac{1}{2}\,(\Pi_{11} + \Pi_{12} + \Pi_{44}). \qquad (7.5.3)$$

The theory given in [114, 115] predicts that semiconductors with energy minima along the [111] directions in the k-space and the ellipsoidal constant-energy surfaces at the minima k_i have components Π_{11} and Π_{12} whose contribution to the total piezoresistance is two orders of magnitude smaller than the contribution of Π_{44}. This makes it possible to write Eq. (7.5.3) for n-type germanium in the following form:

$$\left.\frac{\Delta\rho}{\rho X}\right|_{[110]} = \frac{1}{2}\,\Pi_{44}. \qquad (7.5.4)$$

The theory predicts both the magnitude and temperature dependence of the piezoresistance of n-type germanium and silicon. For degenerate n-type germanium, we should have

$$\Pi_{44} = -\frac{\Sigma_U}{3c_{44}k_0 T}\,\frac{u_\perp - u_\parallel}{2u_\perp + u_\parallel}\,r\,\frac{F_{r+1}(\mu^*)}{F_r(\mu^*)}, \qquad (7.5.5)$$

where Σ_U is the deformation potential constant; c_{44} is a component of the rigidity tensor; $u_\perp$ and $u_\parallel$ are the carrier mobilities at right-angles and parallel to the major axes of the constant-energy ellipsoid in the k-space.

Experimental investigations of the piezoresistance of heavily doped n-type germanium are reported in [116, 117] and of heavily doped silicon, in [118].

Figure 7.32 shows the temperature dependence of Π_{44} for germanium samples doped to various degrees. The optimum impurity concentration for temperature-stable germanium strain gauges is $2 \cdot 10^{19}$ cm^{-3}. The strain sensitivity of a gauge made from such germanium is approximately 40, if the crystal is oriented along the [110] direction, and 60 if the crystal is cut along the [111] direction].

These results, as well as investigations of the piezoresistance of silicon [118], show that heavily doped semiconductors are promising materials for temperature-stable strain gauges.

Appendices

A. 1. NUMERICAL VALUES OF FERMI – DIRAC DISTRIBUTION FUNCTION $f = (1 + e^{-\mu^*})^{-1}$

μ^*	f	μ^*	f	μ^*	f
−4	0.9820	0.8	0.3100	3.2	0.0392
−3.5	0.9707	1.0	0.2689	3.4	0.0323
−3.0	0.9526	1.2	0.2315	3.6	0.0266
−2.5	0.9241	1.4	0.1978	3.8	0.0219
−2.0	0.8808	1.6	0.1680	4.0	0.0180
−1.5	0.8176	1.8	0.1418	4.5	0.0110
−1.0	0.7310	2.0	0.1192	5.0	0.0067
−0.5	0.6225	2.2	0.0997	6.0	0.0025
0.0	0.5000	2.4	0.0832	7.0	0.0009
0.2	0.4502	2.6	0.0691	8.0	0.0003
0.4	0.4013	2.8	0.0573	9.0	0.0001
0.6	0.3543	3.0	0.0474	10.0	0.00004

367

A.2. TABLE OF FERMI INTEGRALS
WITH INTEGRAL INDICES

μ^*	F_0	F_1	F_2	F_3
—4	0,01815	0,01820	0,03660	0,1098
—3	0,04859	0,04920	0,09900	0,2978
—2,8	0,05904	0,05990	0,12076	0,3635
—2,6	0,07164	0,07290	0,14719	0,4436
—2,4	0,08684	0,08870	0,17943	0,5413
—2,2	0,10510	0,10788	0,21863	0,6603
—2	0,12690	0,13101	0,26626	0,8053
—1,8	0,15300	0,15893	0,32408	0,9819
—1,6	0,18390	0,19253	0,39416	1,1967
—1,4	0,22040	0,23286	0,47900	1,4578
—1,2	0,26330	0,28112	0,58151	1,7748
—1	0,31327	0,33870	0,70500	2,1598
—0,9	0,34118	0,37135	0,77608	2,3818
—0,8	0,37111	0,40695	0,85386	2,6262
—0,7	0,40320	0,44564	0,93807	2,8949
—0,6	0,43750	0,48766	1,03234	3,1904
—0,5	0,47403	0,53322	1,13437	3,5153
—0,4	0,51298	0,58255	1,24588	3,8720
—0,3	0,55435	0,63590	1,36766	4,2631
—0,2	0,60155	0,69350	1,50050	4,6937
—0,1	0,64439	0,75561	1,64536	5,1653
0	0,69320	0,82247	1,80300	5,6822
0,2	0,79813	0,97150	2,16110	6,8601
0,4	0,91302	1,14240	2,58520	8,2824
0,6	1,03745	1,33730	3,04600	9,9790
0,8	1,17110	1,55800	3,56930	11,9951
1	1,31322	1,80630	4,12000	14,3950
1,2	1,4633	2,0837	5,213	17,213
1,4	1,61044	2,3920	6,003	20,472
1,6	1,78391	2,7324	7,021	24,493
1,8	1,95294	3,1060	8,189	28,918
2	2,12689	3,5135	9,445	34,304
2,2	2,30510	3,9570	11,004	40,3697
2,4	2,48681	4,4462	12,679	47,5187
2,6	2,67168	4,9520	14,547	55,676
2,8	2,85929	5,5050	16,641	64,936
3	3,04862	6,0957	18,870	75,722
4	4,0182	9,6267	34,592	154,21
5	5,0067	14,138	58,120	290,94
6	6,0025	19,642	91,744	513,00
7	7,0009	26,144	137,36	853,42
8	8,0003	33,645	196,99	1351,2
9	9,0001	42,145	272,61	2051,3
10	10,000	51,645	366,23	3004,8
11	11,000	62,145	479,86	4268,7
12	12,000	73,645	615,48	5906,0
13	13,000	86,145	775,10	7985,0
14	14,000	99,645	960,72	10592
15	15,000	114,15	1174,3	13778
16	16,000	129,65	1418	17659
17	17,000	146,15	1693,6	22318
18	18,000	163,65	2003,2	27854
19	19,000	182,15	2348,8	34375
20	20,000	201,65	2732,5	41985

A.3. TABLE OF FERMI INTEGRALS
WITH FRACTIONAL INDICES IN THE RANGE
$0 \leq \mu* \leq 20$

$\mu*$	$F_{-1/2}$	$F_{1/2}$	$F_{3/2}$	$F_{5/2}$	$F_{7/2}$	$F_{9/2}$	$F_{11/2}$
0	1,0722	0,678094	1,1528	3,08259	11.1837	51.2904	284.903
0,1	1,1405	0,733403	1,2586	3,38384	12.3146	56.5735	314.542
0,2	1,2109	0,792181	1,3730	3,71261	13.5560	62.3900	347.232
0,3	1,2830	0,854521	1,49646	4,07110	14.9168	68.7916	383.279
0,4	1,3566	0,920505	1,62954	4,46164	16.4091	75.8348	423.020
0,5	1,4317	0,990209	1,77279	4,88671	18,0440	83.5812	466.826
0,6	1,5079	1,06369	1,92679	5,34893	19.8341	92.0977	515.100
0,7	1,5851	1,14102	2,09209	5,35105	21.7929	101.457	568.288
0,8	1,6630	1,22222	2,26929	6,39597	23.9349	111.739	626,872
0,9	1,7415	1,30733	2,45895	6,98673	26.2754	123.028	691.385
1,0	1,8204	1,39638	2,66168	7,62653	28.8313	135.419	762.405
1,1	1,8995	1,48937	2,87806	8,31871	31.6201	149.011	840.566
1,2	1,9786	1,58632	3,10869	9,06675	34.6609	163.915	926,558
1,3	2,0575	1,68723	3,35416	9,87429	37.9738	180.247	1021.13
1,4	2,1361	1,79207	3,61506	10,7451	41.5803	198.135	1125.12
1,5	2,2144	1,90083	3,89198	11,6832	45.5032	217.717	1239.39
1,6	2,2920	2,01350	4,18550	12,6925	49.7668	239.139	1364.94
1,7	2,3691	2,13003	4,49622	13,7773	54.3968	262.562	1502.81
1,8	2,4453	2,25039	4,82470	14,9421	59.4203	288.155	1654.16
1,9	2,5208	2,37455	5,17152	16,1912	64.8661	316.103	1820.22
2,0	2,5954	2,50246	5,53725	17,5294	70.7645	346.603	2002.34
2,1	2,6690	2,63407	5,92245	18,9615	77.1476	379.864	2201.98
2,2	2,7417	2,76934	6,32766	20,4923	84.0491	416.113	2420.74
2,3	2,8133	2,90822	6,75343	21,2270	91.5043	455.591	2660.30
2,4	2,8839	3,05066	7,20030	23,8708	99,5507	498.556	2922.53
2,5	2,9535	3,19660	7,66880	25,7290	108,227	545.281	3209.40
2,6	3,0219	3,34599	8,15946	27,7070	117,575	596,061	3523.08
2,7	3,0893	3,49878	8,67277	29,8106	127,637	651.206	3865.87
2,8	3,1557	3,65491	9,20925	32,0453	138,458	711.048	4240.27
2,9	3,2210	3,81433	9,76941	34,4172	150,084	775.939	4648,95
3,0	3,2852	3,97699	10,3537	36,9321	162,566	846.252	5094,79
3,1	3,3484	4,14283	10,9627	39,5961	175,954	922.384	5580.89
3,2	3,4106	4,31181	11,5967	42,4155	190,302	1004.75	6110.56
3,3	3,4718	4,48388	12,2564	45,3966	205,664	1093.81	6687.35
3,4	3,5320	4,65898	12,9420	48,5458	222,099	1190.01	7315.06
3,5	3,5913	4,83707	13,6542	51,8698	239,666	1293.87	7997.76
3,6	3,6497	5,01810	14,3933	55,3752	258,429	1405.89	8739.81
3,7	3,7071	5,20202	15,1598	59,0687	278,451	1526.64	9545.84
3,8	3,7637	5,38880	15,9540	62,9574	299,800	1656,70	10420.8
3,9	3,8194	5,57838	16,7766	67,0481	322,545	1796,67	11370.0
4,0	3,8743	5,77073	17,6277	71,3480	346,758	1947.21	12399.1
4,1	3,9284	5,96580	18,5079	75,8644	372,514	2108.98	13514.0
4,2	3,9818	6,16356	19,4176	80,6044	399,889	2282.71	14721.2
4,3	4,0344	6,36396	20,3571	85,5756	428,964	2469.14	16027.3
4,4	4,0862	6,56698	21,3269	90,7855	459,890	2669.05	17439.7
4,5	4,1374	6,77257	22,3273	96,2416	492,542	2883.26	18965,9
4,6	4,1878	6,98070	23,3588	101,952	527,219	3112.63	20614.1
4,7	4,2376	7,19134	24,4217	107,924	563,939	3358.06	22392.7
4,8	4,2868	7,40445	25,5163	141,165	602,797	3620.49	24311.0
4,9	4,3352	7,62001	26,6431	120,684	643,887	3900.91	26378,6
5,0	4,3832	7,83797	27,8024	127,489	687,309	4200.34	28605,5
5,1	4,4306	8,05832	28,9946	134,588	733,164	4519.86	31002.6
5,2	4,4774	8,28103	30,2201	141,990	781,556	4860.57	33581.3

μ^*	$F_{-1/2}$	$F_{1/2}$	$F_{3/2}$	$F_{5/2}$	$F_{7/2}$	$F_{9/2}$	$F_{11/2}$
5,3	4,5236	8,50606	31,4791	149,701	832,593	5223,65	36353,4
5,4	4,5694	8,73339	32,7720	157,732	886,384	5610,32	39331,6
5,5	4,6146	8,96299	34,0992	166,090	943,044	6021,83	42529,3
5,6	4,6594	9,19485	35,4610	174,784	1002,69	6459,50	45960,4
5,7	4,7036	9,42893	36,8578	183,823	1065,43	6924,71	49639,8
5,8	4,7474	9,66521	38,2898	193,216	1131,40	7418,88	53582,9
5,9	4,7908	9,90367	39,7574	202,971	1200,73	7943,48	57806,1
6,0	4,8338	10,1443	41,2610	213,098	1273,53	8500,05	62326,6
6,1	4,8762	10,3870	42,8008	223,605	1349,94	9090,20	67162,3
6,2	4,9182	10,6319	44,3772	234,501	1430,10	9715,56	72332,3
6,3	4,9600	10,8789	45,9905	245,797	1514,14	10379,9	77856,2
6,4	5,0012	11,1279	47,6410	257,500	1602,20	11078,9	83755,0
6,5	5,0422	11,3790	49,3290	269,620	1694,44	11820,5	90050,4
6,6	5,0828	11,6321	51,0548	282,167	1790,99	12604,5	96765,3
6,7	5,1230	11,8873	52,8187	295,151	1892,00	13433,0	103924
6,8	5,1628	12,1444	54,6211	308,580	1997,64	14308,0	111550
6,9	5,2024	12,4035	56,4621	322,464	2108,06	15231,6	119671
7,0	5,2416	12,6646	58,3422	336,814	2223,42	16206,0	128314
7,1	5,2804	12,9277	60,2616	351,639	2343,89	17233,5	137508
7,2	5,3190	13,1927	62,2206	366,948	2469,63	18316,3	147281
7,3	5,3572	13,4596	64,2195	382,752	2600,81	19456,9	157666
7,4	5,3952	13,7284	66,2586	399,061	2737,61	20657,9	168695
7,5	5,4328	13,9991	68,3381	415,885	2880,21	21921,7	180401
7,6	5,4702	14,2717	70,4584	433,234	3028,79	23251,0	192821
7,7	5,5074	14,5461	72,6197	451,118	3183,54	24648,5	205990
7,8	5,5442	14,8224	74,8223	469,547	3344,64	26117,1	219947
7,9	5,5808	15,1005	77,0665	488,532	3512,29	27659,7	234732
8,0	5,6170	15,3805	79,3526	508,084	3686,68	29279,2	250387
8,1	5,6532	15,6622	81,6808	528,212	3868,01	30978,7	266954
8,2	5,6890	15,9458	84,0514	548,928	4056,49	32761,5	284479
8,3	5,7240	16,2311	86,4646	570,241	4252,33	34630,7	303008
8,4	5,7600	16,5183	88,9208	592,164	4455,73	36589,7	322589
8,5	5,7950	16,8071	91,4202	614,705	4666,92	38642,0	343273
8,6	5,8300	17,0978	93,9630	637,877	4886,10	40791,1	365113
8,7	5,8646	17,3901	96,5496	661,690	5113,51	43040,7	388162
8,8	5,8990	17,6842	99,1801	686,156	5349,36	45394,5	412477
8,9	5,9334	17,9800	101,855	711,284	5593,89	47856,4	438116
9,0	5,9674	18,2776	104,574	737,087	5847,34	50430,4	465139
9,1	6,0012	18,5768	107,338	763,575	6109,93	53120,4	493610
9,2	6,0348	18,8777	110,147	790,760	6381,92	55930,7	523594
9,3	6,0682	19,1803	113,002	818,652	6663,55	58865,6	555157
9,4	6,1016	19,4845	115,902	847,264	6955,06	61929,4	588370
9,5	6,1346	19,7904	118,847	876,607	7256,72	65126,7	623304
9,6	6,1674	20,0980	121,839	906,692	7568,77	68462,0	660034
9,7	6,2002	20,4072	124,877	937,530	7891,49	71940,2	698638
9,8	6,2326	20,7180	127,961	969,134	8225,14	75566,0	739196
9,9	6,2650	21,0304	131,092	1001,51	8569,98	79344,5	781789
10,0	6,2972	21,3445	134,270	1034,68	8926,29	83280,7	826504
10,1	6,3290	21,6601	137,495	1068,65	9294,35	87379,9	873428
10,2	6,3608	21,9774	140,768	1103,44	9674,44	91647,4	922652
10,3	6,3926	22,2962	144,089	1139,04	10066,8	96088,7	974272
10,4	6,4240	22,6166	147,457	1175,48	10471,9	100709	1028380
10,5	6,4554	22,9386	150,874	1212,77	10889,8	105515	1085090
10,6	6,4866	23,2622	154,339	1250,92	11320,9	110512	1144480
10,7	6,5176	23,5873	157,853	1289,95	11765,5	115706	1206690
10,8	6,5484	23,9139	161,415	1329,85	12224,0	121103	1271800
10,9	6,5792	24,2421	165,027	1370,66	12696,5	126710	1339940
11,0	6,6096	24,5718	168,688	1412,37	13183,5	132532	1411220
11,1	6,6402	24,9031	172,398	1455,01	13685,3	138577	1485760

TABLE A.3 (continued)

μ^*	$F_{-1/2}$	$F_{1/2}$	$F_{3/2}$	$F_{5/2}$	$F_{7/2}$	$F_{9/2}$	$F_{11/2}$
11,2	6,6704	25,2359	176,159	1498,58	14202,2	144851	1563700
11,3	6,7006	25,5701	179,969	1543,09	14734,4	151362	1645140
11,4	6,7306	25,9059	183,830	1588,56	15282,4	158115	1730240
11,5	6,7604	26,2432	187,741	1635,01	15846,5	165118	1819120
11,6	6,7902	26,5820	191,703	1682,44	16427,1	172379	1911920
11,7	6,8196	26,9222	195,716	1730,87	17024,3	179905	2008780
11,8	6,8492	27,2639	199,780	1780,30	17638,8	187704	2109860
11,9	6,8784	27,6071	203.895	1830,76	18270,7	195782	2215310
12,0	6,9076	27,9518	208,062	1882,25	18920,4	204150	2325270
12,1	6,9368	28,2979	212,281	1934,79	19588,4	212814	2439930
12,2	6,9658	28,6455	216,551	1988,40	20274,9	221782	2559430
12,3	6,9946	28,9945	220,874	2043,07	20980,4	231064	2683940
12,4	7,0232	29,3449	225,250	2098,84	21705,2	240667	2813650
12,5	7,0518	29,6968	229,678	2155,70	22449,7	250601	2948740
12,6	7,0802	30,0501	234,159	2213,68	23214,3	260875	3089380
12,7	7,1086	30,4048	238,693	2272,79	23999,4	271497	3235760
12,8	7,1368	30,7610	243,280	2333,03	24805,4	282478	3388090
12,9	7,1650	31,1185	247,921	2394,43	25632,7	293825	3546560
13,0	7,1930	31,4775	252,616	2457,00	26481,6	305550	3711370
13,1	7,2210	31,8378	257,365	2520,74	27352,7	317662	3882730
13,2	7,2486	32,1996	262,167	2585,68	28246,3	330171	4060870
13,3	7,2764	32,5627	267,024	2651,83	29162,8	343087	4246000
13,4	7,3040	32,9272	271,936	2719,20	30102,7	356421	4438340
13,5	7,3314	33,2931	276,903	2787,81	31066,4	370183	4638140
13,6	7,3588	33,6603	281,924	2857,66	32054,3	384385	4845630
13,7	7,3860	34,0290	287,001	2928,77	33066,9	399036	5061050
13,8	7,4132	34,3989	292,133	3001,16	34104,6	414149	5284650
13,9	7,4402	34,7703	297,321	3074,84	35167,9	429734	5516700
14,0	7,4672	35,1430	302,564	3149,83	36257,2	445804	5757450
14,1	7,4940	35,5170	307,864	3226,13	37372,9	462369	6007170
14,2	7,5208	35,8924	313,219	3303,76	38515,6	479443	6266150
14,3	7,5474	36,2691	318,631	3382,74	39685,7	497038	6534650
14,4	7,5740	36,6471	324,100	3463,08	40883,7	515165	6812990
14,5	7,6006	37,0265	329,626	3544,80	42110,0	533837	7101440
14,6	7,6268	37,4072	335,208	3627,90	43365,2	553068	7400310
14,7	7,6532	37,7892	340,848	3712,41	44649,7	572870	7709920
14,8	7,6794	38,1725	346,545	3798,33	45964,1	593257	8030570
14,9	7,7054	38,5571	352,300	3885,68	47308,7	614242	8362610
15,0	7,7314	38,9430	358,112	3974,48	48684,2	635840	8706350
15,1	7,7574	39,3303	363,983	4064,75	50091,0	658063	9062150
15,2	7,7832	39,7188	369,911	4156,48	51529,7	680926	9430340
15,3	7,8090	40,1086	375,898	4249,71	53000,7	704445	9811290
15,4	7,8346	40,4997	381,944	4344,43	54504,7	728632	10205400
15,5	7,8602	40,8921	388,048	4440,68	56042,0	753504	10612900
15,6	7,8858	41,2857	394,212	4538,46	57613,3	779075	11034300
15,7	7,9112	41,6806	400,434	4637,79	59219,1	805361	11470000
15,8	7,9364	42,0768	406,716	4738,69	60860,0	832377	11920400
15,9	7,9618	42,4743	413,057	4841,16	62536,4	860140	12385800
16,0	7,9868	42,8730	419,458	4945,22	64249,0	888666	12866700
16,1	8,0120	43,2730	425,919	5050,89	65998,2	917970	13363400
16,2	8,0370	43,6742	432,440	5158,18	67784,8	948070	13876600
16,3	8,0618	44,0767	439,021	5267,12	69609,2	978982	14406500
16,4	8,0868	44,4804	445,663	5377,70	71472,0	1010720	14953600
16,5	8,1116	44,8854	452,366	5489,95	73373,7	1043310	15518400
16,6	8,1362	45,2916	459,129	5603,89	75315,1	1076770	16101400
16,7	8,1608	45,6990	465,953	5719,52	77296,7	1111100	16703000
16,8	8,1854	46,1076	472,839	5836,87	79319,0	1146340	17323800
16,9	8,2098	46,5175	479,785	5955,95	81382,7	1182500	17964200

μ*	$F_{-1/2}$	$F_{1/2}$	$F_{3/2}$	$F_{5/2}$	$F_{7/2}$	$F_{9/2}$	$F_{11/2}$
17,0	8,2342	46,9286	486,794	6076,77	83488,4	1219590	18624700
17,1	8,2586	47,3410	493,864	6199,35	85636,6	1257640	19305900
17,2	8,2828	47,7545	500,996	6323,70	87828,1	1296670	20008300
17,3	8,3070	48,1692	508,190	6449,85	90063,4	1336690	20732400
17,4	8,3312	48,5852	515,447	6577,80	92343,2	1377730	21478800
17,5	8,3552	49,0024	522,766	6707,58	94668,1	1419810	22248100
17,6	8,3792	49,4207	530,148	6839,19	97038,7	1462940	23040800
17,7	8,4030	49,8403	537,592	6972,66	99455,8	1507150	23857500
17,8	8,4268	50,2610	545,100	7107,99	101920	1524600	24698900
17,9	8,4506	50,6830	552,671	7245,21	104432	1598890	25565400
18,0	8,4744	51,1061	560,305	7384,33	106992	1646450	26457900
18,1	8,4980	51,5304	568,003	7525,37	109601	1695190	27376800
18,2	8,5216	51,9559	575,764	7668,34	112260	1745100	28322800
18,3	8,5450	52,3825	583,589	7813,26	114969	1796230	29296600
18,4	8,5684	52,8104	591,479	7960,14	117729	1848580	30298900
18,5	8,5918	53,2394	599,433	8109,00	120541	1902190	31330300
18,6	8,6152	53,6696	607,451	8259,86	123406	1957080	32391500
18,7	8,6384	54,1009	615,533	8412,73	126323	2013260	33483300
18,8	8,6616	54,5334	623,681	8567,64	129295	2070780	34606300
18,9	8,6846	54,9671	631,894	8724,58	132321	2129640	35761400
19,0	8,7076	55,4019	640,171	8883,59	135402	2189870	36949200
19,1	8,7306	55,8378	648,514	9044,67	138540	2251510	38170500
19,2	8,7536	56,2749	656,923	9207,85	141734	2314570	39426100
19,3	8,7764	56,7132	665,397	9373,14	144985	2379080	40716800
19,4	8,7994	57,1526	673,937	9540,55	148295	2445060	42043400
19,5	8,8220	57,5931	682,543	9710,11	151664	2512550	43406600
19,6	8,8448	58,0348	691,215	9881,83	155093	2581570	44807400
19,7	8,8674	58,4776	699,953	10055,7	158582	2652140	46246600
19,8	8,8900	58,9215	708,758	10231,8	162132	2724300	47725100
19,9	8,9124	59,3666	717,630	10410,1	165744	2798070	49243700
20,0	8,9350	59,8128	726,568	10590,6	169419	2873480	50803300

A.4. TABLE OF FERMI INTEGRALS
WITH FRACTIONAL INDICES IN THE RANGE
$-4 \leq \mu* \leq 0$

$\mu*$	$F_{-1/2}$	$F_{1/2}$	$F_{3/2}$	$F_{5/2}$	$F_{7/2}$	$F_{9/2}$	$F_{11/2}$
—4,0	0,03204	0,016128	0,0242685	0,060773	0,212877	0,958331	5,27190
—3,9	0,03536	0,017812	0,0268125	0,067153	0,235245	1,05908	5,82621
—3,8	0,03904	0,019670	0,0296220	0,074201	0,259961	1,17040	6,43879
—3,7	0,04306	0,021721	0,0327240	0,081988	0,287271	1,29342	7,11577
—3,6	0,04752	0,023984	0,0361485	0,090590	0,317447	1,42937	7,86390
—3,5	0,05240	0,026480	0,0399300	0,100092	0,350788	1,57960	8,69065
—3,4	0,05778	0,029233	0,0441060	0,110588	0,387627	1,74560	9,60431
—3,3	0,06372	0,032296	0,0487140	0,122181	0,428327	1,92903	10,6140
—3,2	0,07022	0,035615	0,0538020	0,134985	0,473294	2,13173	11,7297
—3,1	0,07740	0,039303	0,0594165	0,149126	0,522971	2,35570	12,9627
—3,0	0,08526	0,043366	0,0656115	0,164742	0,577852	2,60317	14,3253
—2,9	0,09390	0,047842	0,0724470	0,181985	0,638479	2,87662	15,8309
—2,8	0,10336	0,052770	0,0799860	0,201024	0,705450	3,17875	17,4948
—2,7	0,11374	0,058194	0,0883020	0,222043	0,779426	3,51257	19,3333
—2,6	0,12512	0,064161	0,0974715	0,245246	0,861134	3,88139	21,3650
—2,5	0,13758	0,070724	0,107580	0,270857	0,951377	4,28886	23,6099
—2,4	0,15118	0,077938	0,118722	0,299122	1,05104	4,73903	26,0905
—2,3	0,16606	0,085864	0,130998	0,330312	1,16110	5,23635	28,8314
—2,2	0,18228	0,094566	0,144521	0,364725	1,28263	5,78574	31,8600
—2,1	0,19994	0,104116	0,159410	0,402686	1,41682	6,39261	35,2062
—2,0	0,21918	0,114588	0,175800	0,444554	1,56497	7,06296	38,9035
—1,9	0,24010	0,126063	0,193836	0,490723	1,72851	7,80339	42,9883
—1,8	0,26278	0,138627	0,213674	0,541623	1,90902	8,62116	47,5013
—1,7	0,28736	0,152373	0,235484	0,597724	2,10825	9,52431	52,4872
—1,6	0,31394	0,167397	0,259451	0,659544	2,32810	10,5217	57,9953
—1,5	0,34262	0,183802	0,285773	0,727646	2,57066	11,6230	64,0800
—1,4	0,37352	0,201696	0,314666	0,802644	2,83825	12,8390	70,8016
—1,3	0,40674	0,221193	0,346361	0,885212	3,13339	14,1815	78,2261
—1,2	0,44236	0,242410	0,381110	0,976079	3,45886	15,6636	86,4268
—1,1	0,48048	0,265471	0,419177	1,07604	3,81771	17,2995	95,4842
—1,0	0,52114	0,290501	0,460848	1,18597	4,21325	19,1050	105,487
—0,9	0,56444	0,317630	0,506432	1,30679	4,64916	21,0975	116,534
—0,8	0,61040	0,346989	0,556250	1,43954	5,12940	23,2959	128,732
—0,7	0,65902	0,378714	0,610647	1,58530	5,65835	25,7212	142,201
—0,6	0,71034	0,412937	0,662989	1,74527	6,24076	28,3964	157,071
—0,5	0,76434	0,449793	0,734660	1,92074	6,88184	31,3467	173,488
—0,4	0,82096	0,489414	0,805065	2,11308	7,58725	34,5997	191,608
—0,3	0,88014	0,531931	0,881628	2,32378	8,36313	38,1858	211,608
—0,2	0,94182	0,577470	0,964796	2,55444	9,21622	42,1381	233,680
—0,1	1,0059	0,626152	1,05503	2,80677	10,1538	46,4931	258,034

A.5. TRANSPORT INTEGRALS Φ_n

Recurrence Formulas for the Calculation of the Integrals Φ_n. The integrals Φ_n, mentioned in Chap. 3 in connection with the transport phenomena, are more general than the Fermi integrals, since they allow for the mixed nature of the scattering process. Integrals of the type Φ_n, describing the main effects, will be written in the form

$$\Phi_n = \int_0^\infty \frac{(\varepsilon^*)^n}{(\varepsilon^*)^2 + a^2} \frac{\partial f}{\partial \varepsilon^*} d\varepsilon^*, \qquad (A.5.1)$$

where f is the Fermi−Dirac distribution function. Equation (A.5.1) is valid if the index n is an integer (even or odd).

For fractional indices, the transport integral becomes

$$\Phi_{k+\frac{1}{2}} = \int_0^\infty \frac{(\varepsilon^*)^{k+1/2}}{((\varepsilon^*)^2 + a^2)^2} \frac{\partial f}{\partial \varepsilon^*} d\varepsilon^*, \qquad (A.5.2)$$

where k is an integer.

We can show that for any odd integer n we have

$$\Phi_n = (-1)^{n+1} a^{n-1} \Phi_1 + \sum_{j=1}^{\frac{n-1}{2}} (-1)^{j+1} S_{n-2j} a^{2j-2}, \qquad (A.5.3)$$

where

$$\Phi_1 = \int_0^\infty \frac{\varepsilon^*}{(\varepsilon^*)^2 + a^2} \frac{\partial f}{\partial \varepsilon^*} d\varepsilon^*, \qquad (A.5.4)$$

$$S_n = \int_0^\infty (\varepsilon^*)^n \frac{df}{d\varepsilon^*} d\varepsilon^* = -n F_{n-1}(\mu^*), \qquad (A.5.5)$$

and

$$S_0 = -(1 + e^{-\mu^*})^{-1}, \qquad (A.5.6)$$

and $F_{n-1}(\mu^*)$ is the usual Fermi integral.

For example, the integral Φ_3, which occurs in all the transport phenomena, is

$$\Phi_3 = S_1 + a^2\Phi_1 = -F_0(\mu^*) + a^2\Phi_1. \tag{A.5.7}$$

For even integers n (n = 2k), Φ_n can be represented in the form

$$\Phi_{2k} = a^{2k}\Phi_0 + \sum_{j=1}^{k} S_{2k-2j}(-1)^j a^{2j}, \tag{A.5.8}$$

where

$$\Phi_0 = \int_0^\infty \frac{1}{(\varepsilon^*)^2 + a^2} \frac{\partial f}{\partial \varepsilon^*} d\varepsilon^*. \tag{A.5.9}$$

If, for example, n = 4, then k = 2 and

$$\Phi_4 = -a^2 S_2 + a^4 S_0 + a^4 \Phi_0, \tag{A.5.10}$$

or, using Eqs. (A.5.5) and (A.5.6), we can rewrite Eq. (A.5.10) in the form

$$\Phi_4 = 2a^2 F_1(\mu^*) - a^4 \frac{1}{1 + e^{-\mu^*}} + a^4 \Phi_0. \tag{A.5.11}$$

Similarly for k ≥ 4, we can express the integrals (A.5.2) with fractional indices in the following form:

$$\Phi_{k+1/2} = S_{k-4+1/2} - 2a^2 T_{k-4} + a^2 R_{k-4} \quad (k \geq 4), \tag{A 5.12}$$

where

$$S_{k-4+1/2} = -(k-4+1/2) F_{k-4+1/2-1}(\mu^*), \tag{A.5.13}$$

$$T_{k-4} = \int_0^\infty \frac{(\varepsilon^*)^{k-4+1/2}}{(\varepsilon^*)^2 + a^2} \frac{\partial f}{\partial \varepsilon^*} d\varepsilon^*. \tag{A.5.14}$$

$$R_{k-4} = \int_0^\infty \frac{(\varepsilon^*)^{k-4+1/2}}{((\varepsilon^*)^2 + a^2)^2} \cdot \frac{\partial f}{\partial \varepsilon^*} d\varepsilon^*. \tag{A.5.15}$$

APPENDIX 5

TABLE A.5.1. Numerical Values of the Integral $\Phi_3(\mu^*, a)$

μ^* \ a	0	1	2	3	4	5
—4	0,018150	0,011956	0,007691	0,005237	0,003745	0,002786
—3	0,048587	0,032180	0,020752	0,014146	0,010123	0,007535
—2	0,126928	0,085222	0,055299	0,037816	0,027110	0,020202
—1	0,313262	0,217005	0,142957	0,098526	0,070955	0,053029
0	0,693147	0,508869	0,346122	0,242684	0,176578	0,132845
1	1,313262	1,043028	0,748440	0,541380	0,401620	0,306041
2	2,126928	1,821806	1,398347	1,057170	0,807562	0,627857
3	3,048587	2,760661	2,262657	1,797801	1,423407	1,135688
4	4,018150	3,769654	3,259067	2,713997	2,230106	1,830907
5	5,006715	4,798274	4,314198	3,737460	3,178263	2,683968
6	6,002476	5,827182	5,385240	4,812585	4,214602	3,652053
7	7,000912	6,851306	6,453448	5,905082	5,297331	4,694226
8	8,000336	7,870417	7,512839	6,997542	6,399254	5,778642
9	9,000124	8,885471	8,562789	8,082662	7,504879	6,883546
10	10,000046	9,897468	9,604454	9,158271	8,706371	7,995408

μ^* \ a	6	7	8	9	10
—4	0,002141	0,001689	0,001362	0,001119	0,000934
—3	0,005791	0,004570	0,003686	0,003029	0,002529
—2	0,015539	0,012269	0,009901	0,008138	0,006796
—1	0,040869	0,032313	0,026102	0,021472	0,017941
0	0,102845	0,081575	0,066052	0,054431	0,045541
1	0,239033	0,190804	0,155221	0,128371	0,107700
2	0,497419	0,401207	0,328938	0,273667	0,230672
3	0,917032	0,750275	0,621858	0,521760	0,442733
4	1,511195	1,257555	1,056237	0,895574	0,766346
5	2,266216	1,920546	1,636912	1,404479	1,213489
6	3,151653	2,719951	2,353471	2,044678	1,785072
7	4,132387	3,628289	3,185977	2,802804	2,473064
8	5,176766	4,617221	4,110857	3,660353	3,263692
9	6,260016	5,661899	5,105214	4,597467	4,140536
10	7,364484	6,742637	6,149193	5,595484	5,086918

TABLE A.5.1 (concluded)

a μ*	11	12	13	14	15
—4	0,000791	0,000677	0,000586	0,000511	0,000450
—3	0,002140	0,001833	0,001586	0,001384	0,001219
—2	0,005752	0,004926	0,004263	0,003723	0,003277
—1	0,015192	0,013016	0,011267	0,009841	0,008665
0	0,038607	0,033106	0,028676	0,025062	0,022078
1	0,091500	0,078599	0,068179	0,059656	0,052603
2	0,196700	0,169471	0,147362	0,129198	0,114115
3	0,379552	0,328431	0,286602	0,252018	0,223148
4	0,661475	0,575586	0,504603	0,445426	0,395682
5	1,055779	0,924756	0,815180	0,722916	0,644706
6	1,566603	1,382213	1,225948	1,092889	0,979016
7	2,190066	1,947206	1,738455	1,558543	1,402964
8	2,916498	2,613465	2,349169	2,118496	1,916829
9	3,732865	3,371051	3,050848	2,767785	2,517529
10	4,624857	4,208095	3,833946	3,498977	3,199483

a μ*	16	17	18	19	20
—4	0,000399	0,000356	0,000320	0,000288	0,000262
—3	0,001081	0,000964	0,000866	0,000781	0,000708
—2	0,002906	0,002593	0,002328	0,002101	0,001905
—1	0,007685	0,006859	0,006158	0,005558	0,005040
0	0,019587	0,017489	0,015706	0,014179	0,012861
1	0,046708	0,041733	0,037500	0,033871	0,030737
2	0,101469	0,090771	0,081648	0,073811	0,067032
3	0,198833	0,178187	0,160523	0,145306	0,132113
4	0,353540	0,317577	0,286676	0,259957	0,236716
5	0,577974	0,520677	0,471187	0,428196	0,390652
6	0,881057	0,796354	0,722743	0,658459	0,602056
7	1,267222	1,150241	1,047275	0,956819	0,877037
8	1,740108	1,584819	1,447950	1,326934	1,219589
9	2,296059	2,099751	1,925397	1,770185	1,631672
10	2,931784	2,692388	2,478081	2,285954	2,113414

TABLE A.5.2. Numerical Values of the Integral $\Phi_4(\mu^*, a)$

μ^*	a = 1	2	3	4	5
—4	0,029594	0,021778	0,016135	0,012218	0,009468
—3	0,080001	0,058947	0,043702	0,033108	0,025665
—2	0,214211	0,158389	0,117662	0,089241	0,069232
—1	0,560398	0,417961	0,312027	0,237368	0,184512
0	1,393232	1,058776	0,799277	0,612286	0,478145
1	3,165742	2,485531	1,916149	1,488146	1,172954
2	6,383243	5,232208	4,161888	3,303380	2,644169
3	11,39937	9,763251	8,056550	6,570073	5,372951
4	18,37165	16,31619	13,97130	11,74723	9,829425
5	27,34464	25,01544	22,07452	19,09228	16,36184
6	38,32798	35,80269	32,39774	28,72225	25,16803
7	51,31580	48,66126	44,90388	40,63836	36,31552
8	66,30984	63,56951	59,55053	54,80349	49,80147
9	83,30475	80,50606	76,29334	71,16239	65,57893

μ^*	a = 6	7	8	9	10
—4	0,007498	0,006052	0,004967	0,004138	0,003493
—3	0,020326	0,016409	0,013471	0,011225	0,009477
—2	0,054860	0,044305	0,036386	0,030323	0,025605
—1	0,146401	0,118351	0,097262	0,081106	0,068512
0	0,380615	0,308387	0,253877	0,211976	0,179237
1	0,938984	0,765040	0,631897	0,529224	0,447373
2	2,143000	1,758325	1,461080	1,229485	1,045936
3	4,421410	3,672371	3,081070	2,609395	2,231502
4	8,239760	6,944620	5,894305	5,041132	4,344507
5	13,98854	11,97852	10,29798	8,904459	7,739421
6	21,93517	19,09310	16,64023	14,54442	12,76962
7	32,21629	28,48091	25,16179	22,25014	19,70237
8	44,87878	40,25024	36,01337	32,12056	28,83144
9	59,91427	54,42591	49,27226	44,53330	40,24684

TABLE A.5.2 (concluded)

a / μ*	11	12	13	14	15
—4	0,002985	0,002574	0,002242	0,001965	0,001738
—3	0,008092	0,006983	0,006079	0,005332	0,004716
—2	0,021867	0,018866	0,016427	0,014420	0,012748
—1	0,058531	0,050514	0,043992	0,038619	0,034151
0	0,153257	0,132353	0,115318	0,101283	0,089598
1	0,383218	0,331583	0,289323	0,254175	0,220065
2	0,897349	0,778309	0,680849	0,599130	0,516834
3	1,927411	1,677178	1,470670	1,296232	1,152119
4	3,771988	3,297555	2,903560	2,565399	2,288074
5	6,765491	5,962456	5,268770	4,687936	4,191646
6	11,26012	9,971705	8,871560	7,933522	7,126962
7	17,53500	15,64525	14,00680	12,58713	11,35773
8	25,81757	23,23302	20,94188	18,92685	17,15958
9	36,38961	32,94355	29,83122	27,17054	24,75706

a / μ*	16	17	18	19	20
—4	0,001547	0,001384	0,001248	0,001126	0,001026
—3	0,004199	0,003756	0,003384	0,003059	0,002779
—2	0,011347	0,010157	0,009144	0,008272	0,007516
—1	0,030399	0,027217	0,024502	0,022167	0,020142
0	0,079772	0,071446	0,064332	0,058211	0,052906
1	0,196775	0,176962	0,159774	0,145230	0,132581
2	0,462340	0,415764	0,375914	0,341091	0,294396
3	1,037174	0,925401	0,816671	0,741662	0,675059
4	2,052098	1,849396	1,674557	1,521764	1,378944
5	3,757397	3,393829	3,080945	2,807741	2,569822
6	6,428711	5,815685	5,281964	4,813842	4,407941
7	10,28417	9,354292	8,524022	7,812245	7,147676
8	15,60102	14,23463	13,03098	11,94831	11,00122
9	22,61493	20,54904	19,00546	17,49935	16,14461

TABLE A.5.3. Numerical Values of the Integral $\Phi_{9/2}(\mu^*, a)$

μ^* \ a	0	1	2	3	4	5
—4	0,016031	0,006097	0,002678	0,001340	0,000734	0,000432
—3	0,042647	0,016416	0,007236	0,003626	0,001990	0,001170
—2	0,109636	0,043508	0,019359	0,009741	0,005358	0,001961
—1	0,260650	0,110921	0,050508	0,025690	0,014216	0,008401
0	0,536172	0,260074	0,124540	0,064922	0,036431	0,021719
1	0,910280	0,528589	0,276526	0,151084	0,087179	0,052912
2	1,297737	0,899089	0,529350	0,310501	0,187432	0,117245
3	1,642625	1,299844	0,864732	0,552254	0,353565	0,230561
4	1,937183	1,670110	1,233553	0,857460	0,585458	0,400734
5	2,191631	1,989737	1,591746	1,191963	0,865961	0,623050
6	2,416873	2,263070	1,917385	1,525428	1,170919	0,883129
7	2,620778	2,500759	2,206068	1,839854	1,479175	1,163554
8	2,808524	2,712253	2,461576	2,128225	1,776825	1,449179
9	2,983701	2,904373	2,689854	2,390100	2,056824	1,729341
10	3,148529	3,081759	2,896458	2,627985	2,316855	1,997765

μ^* \ a	6	7	8	9	10
—4	0,000268	0,000174	0,000117	0,000081	0,000058
—3	0,000726	0,000471	0,000318	0,000221	0,000158
—2	0,001961	0,001274	0,000858	0,000597	0,000426
—1	0,005234	0,003406	0,002298	0,001600	0,001144
0	0,013610	0,008892	0,006019	0,004199	0,003007
1	0,033565	0,022121	0,015068	0,010563	0,007593
2	0,075962	0,050834	0,035025	0,024769	0,017926
3	0,153958	0,105374	0,073857	0,052929	0,038712
4	0,277614	0,195456	0,140052	0,102134	0,075755
5	0,449107	0,326360	0,239858	0,178544	0,134664
6	0,662137	0,497091	0,375292	0,285643	0,219465
7	0,905095	0,701295	0,543879	0,423498	0,331730
8	1,165153	0,929721	0,739832	0,589106	0,470473
9	1,431162	1,172712	0,955783	0,777430	0,632651
10	1,694865	1,421845	1,184306	0,982560	0,813959

TABLE A.5.3 (concluded)

α / μ^*	11	12	13	14	15
—4	0,000042	0,000032	0,000024	0,000019	0,000015
—3	0,000115	0,000086	0,000065	0,000050	0,000039
—2	0,000312	0,000233	0,000177	0,000137	0,000107
—1	0,000837	0,000625	0,000475	0,000367	0,000288
0	0,002203	0,001647	0,001253	0,000969	0,000760
1	0,005579	0,004181	0,003188	0,002469	0,001939
2	0,013245	0,009969	0,007628	0,005925	0,004664
3	0,028845	0,021860	0,016822	0,013128	0,010376
4	0,057093	0,043673	0,033871	0,0266605	0,021144
5	0,102899	0,079619	0,062344	0,049369	0,039508
6	0,170311	0,133506	0,105700	0,084492	0,068161
7	0,261696	0,208039	0,166699	0,134635	0,109589
8	0,377413	0,304409	0,247004	0,201688	0,165742
9	0,515975	0,422259	0,347031	0,286563	0,237830
10	0,674519	0,559913	0,466022	0,389178	0,326246

α / μ^*	16	17	18	19	20
—4	0,000012	0,000009	0,000007	0,000006	0,000005
—3	0,000031	0,000025	0,000020	0,000017	0,000014
—2	0,000085	0,000068	0,000055	0,000045	0,000037
—1	0,000228	0,000183	0,000149	0,000122	0,000101
0	0,000604	0,000485	0,000394	0,000367	0,000323
1	0,001542	0,001240	0,001008	0,000827	0,000685
2	0,003715	0,002996	0,002439	0,002004	0,001661
3	0,008297	0,006706	0,005473	0,004507	0,003742
4	0,016987	0,013784	0,011289	0,009324	0,007762
5	0,031928	0,026040	0,021420	0,017761	0,014836
6	0,055469	0,045513	0,037636	0,031350	0,026294
7	0,089877	0,074247	0,061762	0,051716	0,043576
8	0,137070	0,114068	0,095503	0,080428	0,068115
9	0,198413	0,166399	0,140276	0,118858	0,101213
10	0,274615	0,232146	0,197100	0,168074	0,143943

TABLE A.5.4. Numerical Values of the Integral $\Phi_{11/2}(\mu^*, a)$

μ^* \ a	0	1	2	3	4
— 4	0,024192	0,014730	0,008049	0,004598	0,002760
— 3	0,065050	0,039801	0,021797	0,012464	0,007488
— 2	0,171880	0,106490	0,058657	0,033635	0,020239
—1	0,435750	0,277940	0,155340	0,089715	0,054203
0	1,017140	0,686580	0,396300	0,232740	0,141980
1	2,094560	1,535430	0,939640	0,570420	0,355040
2	3,753690	3,000970	1,992590	1,273960	0,820430
3	5,965480	5,115710	3,702390	2,527990	1,705610
4	8,656090	7,792490	6,087200	4,444630	3,161190
5	11,75696	10,92326	9,059200	7,025910	5,264690
6	15,21643	14,42771	12,50706	10,19723	8,009600
7	18,99696	18,25450	16,33971	13,86076	11,33379
8	23,07073	22,37073	20,40451	17,92896	15,15527
9	27,41634	26,75382	24,93082	22,33525	19,39598
10	32,01671	31,38691	29,62201	27,03254	23,99163

μ^* \ a	5	6	7	8	9
—4	0,001733	0,001131	0,000764	0,000531	0,000378
—3	0,004703	0,003071	0,002073	0,001441	0,001027
—2	0,012725	0,008313	0,005615	0,003904	0,002785
—1	0,034165	0,022357	0,015118	0,010521	0,007508
0	0,090038	0,059162	0,040121	0,027979	0,019999
1	0,228100	0,151210	0,103200	0,072301	0,051864
2	0,539400	0,363430	0,250970	0,177390	0,128110
3	1,159460	0,800550	0,563030	0,403560	0,294640
4	2,237650	1,593660	1,148070	0,838550	0,621450
5	3,889550	2,867350	2,123680	1,586210	1,197080
6	6,166480	4,707310	3,589500	2,746760	2,115120
7	9,057320	7,146650	5,608090	4,398160	3,458470
8	12,51045	10,17219	8,200390	6,585970	5,287590
9	16,45795	13,74086	11,35267	9,323030	7,635710
10	20,83223	17,79623	15,02868	12,59590	10,51038

TABLE A.5.4 (concluded)

a / μ^*	10	11	12	13	14
—4	0,000276	0,000205	0,000155	0,000120	0,000093
—3	0,000749	0,000557	0,000422	0,000325	0,000253
—2	0,002031	0,001511	0,001144	0,000881	0,000687
—1	0,005479	0,004078	0,003090	0,002378	0,001856
0	0,014612	0,010887	0,008254	0,006357	0,004965
1	0,037997	0,028370	0,021547	0,016617	0,012995
2	0,094358	0,070749	0,053915	0,041697	0,032684
3	0,218890	0,165260	0,126650	0,098410	0,077438
4	0,467260	0,356260	0,275240	0,215300	0,170370
5	0,913560	0,705150	0,550400	0,434250	0,346150
6	1,641560	1,285100	1,015090	0,809031	0,650490
7	2,732490	2,171960	1,738140	1,400910	1,137320
8	4,253840	3,434710	2,786380	2,272560	1,864130
9	6,252490	5,127970	4,217690	3,481920	2,886840
10	8,752800	7,288190	6,076250	5,077290	4,255200

a / μ^*	15	16	17	18	19	20
—4	0,000074	0,000059	0,000048	0,000039	0,000032	0,000027
—3	0,000200	0,000160	0,000130	0,000106	0,000087	0,000072
—2	0,000543	0,000435	0,000352	0,000287	0,000237	0,000197
—1	0,001468	0,001175	0,000950	0,000776	0,000639	0,000531
0	0,003928	0,003144	0,002544	0,002078	0,001713	0,001423
1	0,010292	0,008245	0,006675	0,005456	0,004499	0,003741
2	0,025935	0,020811	0,016871	0,013807	0,011398	0,009485
3	0,061648	0,049605	0,040311	0,033057	0,027338	0,022785
4	0,136280	0,111110	0,089791	0,073857	0,061240	0,051159
5	0,278610	0,226300	0,185390	0,153100	0,127390	0,106750
6	0,527460	0,431190	0,355220	0,294790	0,246330	0,207190
7	0,929980	0,765800	0,634910	0,529850	0,444940	0,375880
8	1,538130	1,276680	1,065890	0,895010	0,755710	0,641530
9	2,404590	2,012600	1,692810	1,430820	1,215250	1,037060
10	3,578620	3,021070	2,560600	2,179230	1,862340	1,598100

It follows, for example from Eq. (A.5.12), that the integral $\Phi_{9/2}$, which governs the mobility, can be found from the formula

$$\Phi_{9/2} = -\frac{1}{2} F_{-1/2}(\mu^*) - 2a^2 \int_0^\infty \frac{(\varepsilon^*)^{1/2}}{(\varepsilon^*)^2 + a^2} \frac{\partial f}{\partial \varepsilon^*} d\varepsilon^* + a^2 \int_0^\infty \frac{(\varepsilon^*)^{1/2}}{((\varepsilon^*)^2 + a^2)^2} \frac{\partial f}{\partial \varepsilon^*} d\varepsilon^*.$$

(A.5.16)

In exactly the same way, we find the integral $\Phi_{11/2}$, which occurs in the expression for the Nernst—Ettingshausen transverse effect coefficient $Q^\perp$:

$$\Phi_{11/2} = -\frac{3}{2} F_{1/2}(\mu^*) - 2a^2 \int_0^\infty \frac{(\varepsilon^*)^{3/2}}{(\varepsilon^*)^2 + a^2} \frac{\partial f}{\partial \varepsilon^*} d\varepsilon^* + a^2 \int_0^\infty \frac{(\varepsilon^*)^{3/2}}{((\varepsilon^*)^2 + a^2)^2} \frac{\partial f}{\partial \varepsilon^*} d\varepsilon^*.$$

(A.5.17)

Thus, Eqs. (A.53), (A.58), (A.512) can be used to calculate the transport integrals describing the mixed scattering using the tabulated Fermi integrals and certain other integrals Φ_1, Φ_0, T_{k-4}, and R_{k-4}, which can be calculated more accurately than the integrals Φ_n, and $\Phi_{k+1/2}$ because the latter converge less rapidly at infinity.

The recurrence formulas given here have been used to calculate the integrals Φ_3, Φ_4, $\Phi_{9/2}$, and $\Phi_{11/2}$; the integrals Φ_0, Φ_1, T_{k-4}, and R_{k-4} used in these calculations have been computed using a "Strela" computer. The numerical values of the integrals Φ_n are given in Tables A.5.1-A.5.4.

A.6. TRANSPORT INTEGRALS φ_n

It is shown in §3.3 that integrals of the φ_n type are used to describe the transport phenomena in the case of scattering by ions

TABLE A.6.1. Values of D
for n-Type Ge, Si, and GaAs
at Various Temperatures

T, °K	4,2	78	300	500
D_{Ge}	0.397	1.7	3.35	4.36
D_{Si}	0.216	0.925	1.82	2.37
D_{GaAs}	1.36	5.96	11.8	15.23

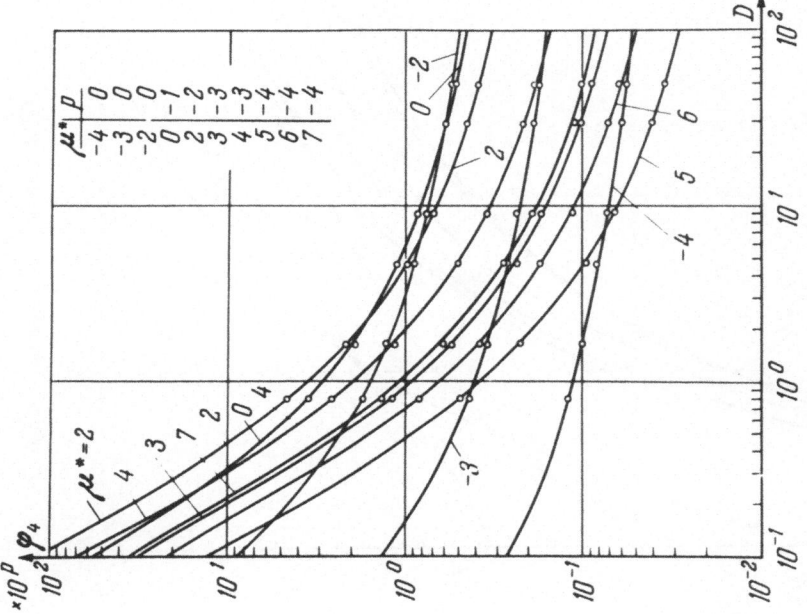

Fig. A.6.2. Values of the integral φ_4.

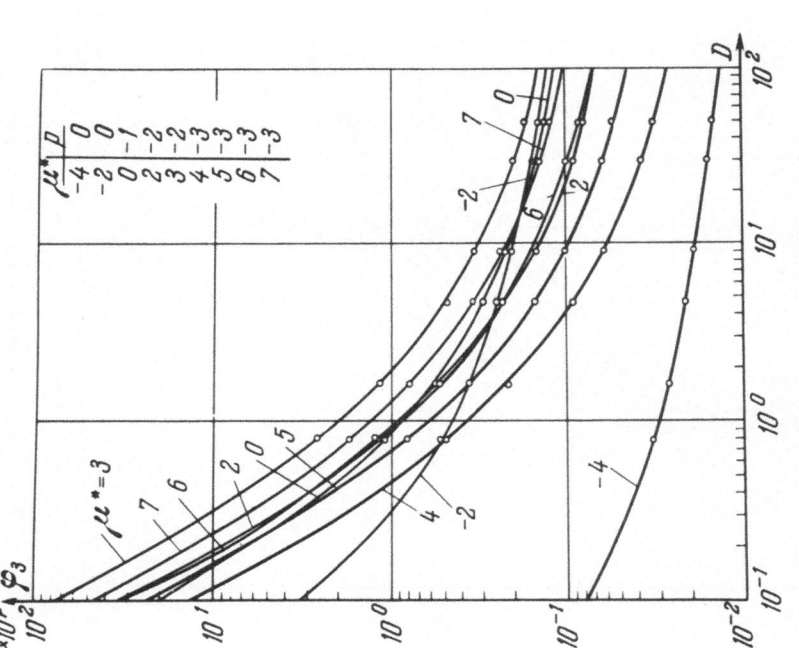

Fig. A.6.1. Values of the integral φ_3.

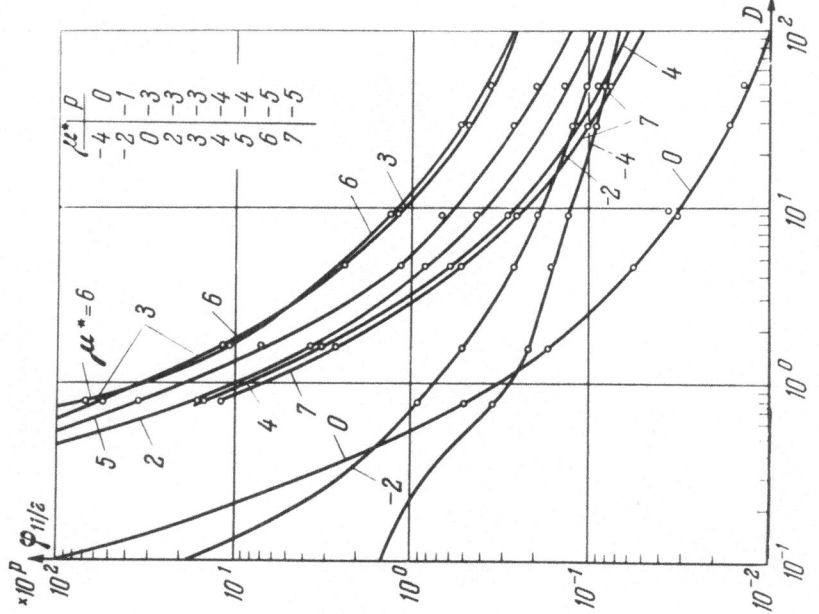

Fig. A.6.4. Values of the integral $\varphi_{11/2}$.

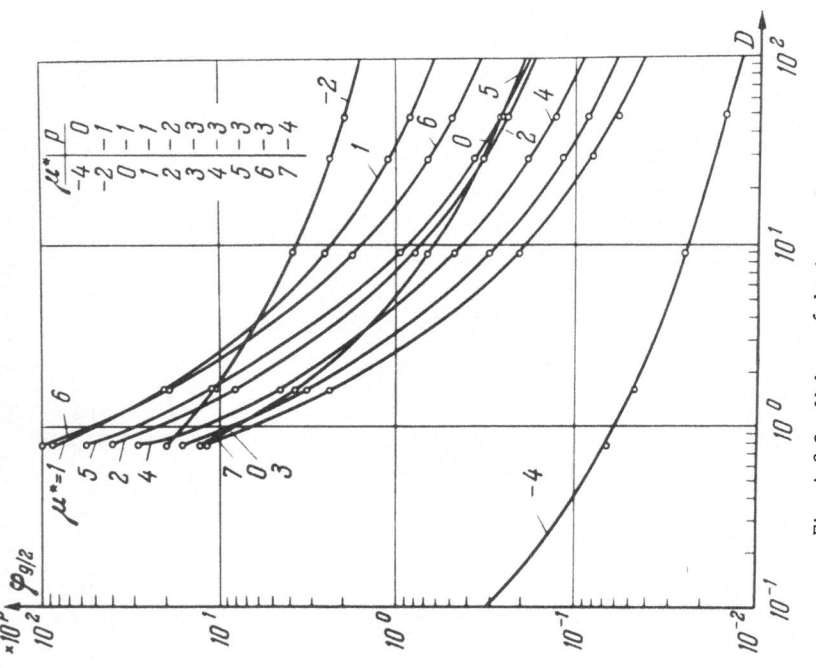

Fig. A.6.3. Values of the integral $\varphi_{9/2}$.

when the screening factor g(b) is allowed for. In this appendix, we shall give the numerical values of the integrals φ_n, calculated on an M-20 computer. Since the values of the integrals φ_n have been calculated less accurately than the integrals Φ_n, we shall not tabulate them but we shall give them in the form of curves (cf. Figs. A.6.1-A.6.4).

The abscissas in these figures give the quantity D,

$$D = \frac{\varkappa h}{e^2} \left(\frac{2k_0 T}{m_N^*} \right)^{1/2}.$$

(A.6.1)

This quantity determines the function g(b), since

$$b = D \frac{\varepsilon^*}{F_{-1/2}(\mu^*)},$$

(A.6.2)

which can be shown quite easily by comparing Eq. (A.6.1) with the expression for b given in Table 3.2. The values of D for n-type germanium, silicon, and gallium arsenide, calculated by Fistul', are given in Table A.6.1.

A.7. TABLE OF VALUES OF u_H/u_{H_0}

μ^* \ a	0	1	2	3	4	5	6
—4	0,9981	0,5762	0,3934	0,2891	0,2215	0,1752	0,1414
—3	0,9919	0,5764	0,3940	0,2830	0,2211	0,1754	0,1417
—2	0,9761	0,5768	0,3956	0,2911	0,2233	0,1765	0,1426
—1	0,9402	0,5776	0,3992	0,2946	0,2264	0,1790	0,1447
0	0,8741	0,5775	0,4066	0,3023	0,2331	0,1847	0,1495
1	0,7833	0,5727	0,4175	0,3170	0,2453	0,1954	0,1587
2	0,6895	0,5576	0,4274	0,3319	0,2623	0,2110	0,1726
3	0,6089	0,5321	0,4319	0,3471	0,2807	0,2294	0,1897
4	0,5448	0,5009	0,4277	0,3570	0,2967	0,2473	0,2076
5	0,4946	0,4686	0,4169	0,3604	0,3604	0,2623	0,2239
6	0,4545	0,4384	0,4018	0,3585	0,3127	0,2736	0,2370
7	0,4231	0,4121	0,3866	0,3517	0,3152	0,2787	0,2472
8	0,3968	0,3891	0,3704	0,3432	0,3135	0,2838	0,2540
9	0,3747	0,3696	0,3551	0,3339	0,3093	0,2838	0,2583
10	0,3568	0,3517	0,3398	0,3246	0,3042	0,2821	0,2600

TABLE A.7 (concluded)

a / μ^*	7	8	9	10	11	12	13
—4	0,1164	0,0971	0,0818	0,0702	0,0600	0,0534	0,0462
—3	0,1165	0,0975	0,0824	0,0706	0,0607	0,0530	0,0463
—2	0,1173	0,0979	0,0829	0,0708	0,0613	0,0535	0,0469
—1	0,1191	0,0992	0,0842	0,0721	0,0623	0,0543	0,0477
0	0,1231	0,1030	0,0872	0,0746	0,0645	0,0562	0,0494
1	0,1310	0,1097	0,0930	0,0797	0,0689	0,0601	0,0528
2	0,1432	0,1203	0,1023	0,0878	0,0761	0,0665	0,0585
3	0,1580	0,1342	0,1139	0,0986	0,0858	0,0747	0,0663
4	0,1756	0,1498	0,1289	0,1117	0,0975	0,0857	0,0758
5	0,1920	0,1656	0,1437	0,1254	0,1104	0,0967	0,0867
6	0,2065	0,1810	0,1580	0,1385	0,1223	0,1088	0,0977
7	0,2184	0,1929	0,1708	0,1504	0,1351	0,1206	0,1079
8	0,2268	0,2031	0,1820	0,1589	0,1453	0,1317	0,1189
9	0,2336	0,2081	0,1912	0,1725	0,1555	0,1410	0,1291
10	0,2379	0,2167	0,1979	0,1809	0,1648	0,1504	0,1376

a / μ^*	14	15	16	17	18	19	20
—4	0,0420	0,0377	0,0340	0,0286	0,0247	0,0235	0,0216
—3	0,0408	0,0320	0,0324	0,0293	0,0261	0,0246	0,0223
—2	0,0416	0,0369	0,0331	0,0296	0,0267	0,0242	0,0219
—1	0,0421	0,0376	0,0335	0,0301	0,0273	0,0248	0,0226
0	0,0437	0,0389	0,0348	0,0313	0,0283	0,0257	0,0234
1	0,0468	0,0416	0,0373	0,0336	0,0304	0,0276	0,0252
2	0,0518	0,0462	0,0414	0,0373	0,0338	0,0307	0,0280
3	0,0586	0,0527	0,0467	0,0425	0,0382	0,0348	0,0323
4	0,0675	0,0604	0,0541	0,0490	0,0445	0,0405	0,0371
5	0,0773	0,0681	0,0620	0,0561	0,0509	0,0467	0,0425
6	0,0875	0,0782	0,0714	0,0646	0,0586	0,0535	0,0493
7	0,0977	0,0884	0,0799	0,0731	0,0662	0,0612	0,0561
8	0,1079	0,0977	0,0892	0,0816	0,0748	0,0688	0,0629
9	0,1172	0,1070	0,0977	0,0892	0,0816	0,0756	0,0697
10	0,1257	0,1155	0,1062	0,0977	0,0901	0,0833	0,0765

A.8. TABLE OF VALUES OF THE HALL FACTOR

a μ*	0	1	2	3	4	5	6
—4	1,1774	1,031977	1,09539	1,1821	1,2662	1,3466	1,4146
—3	1,1752	1,0313	1,0931	1,1788	1,2633	1,3406	1,4083
—2	1,1698	1,0298	1,0882	1,1709	1,2532	1,3289	1,3961
—1	1,1574	1,0264	1,0769	1,1532	1,2304	1,3018	1,3654
0	1,1351	1,0216	1,0574	1,0942	1,1885	1,2518	1,3088
1	1,1055	1,0177	1,0340	1,0797	1,1321	1,1833	1,23047
2	1,0768	1,0178	1,0162	1,0429	1,0788	1,1164	1,11524
3	1,0544	1,0174	1,0076	1,0193	1,0410	1,0664	1,0921
4	1,0386	1,0173	1,0053	1,0080	1,0190	1,0348	1,0523
5	1,0279	1,0161	1,0055	1,0032	1,0079	1,0169	1,0281
6	1,0207	1,0141	1,0060	1,0021	1,0031	1,0075	1,0143
7	1,0158	1,0121	1,0063	1,0023	1,0014	1,0031	1,0069
8	1,0123	1,0102	1,0062	1,0027	1,0010	1,0012	1,0031
9	1,0099	1,0086	1,0058	1,0030	1,0012	1,0006	1,0013
10	1,0080	1,0072	1,0053	1,0032	1,0015	1,0006	1,0005

a μ*	7	8	9	10	11	12	13
—4	1,4758	1,5260	1,5651	1,6121	1,6431	1,6739	1,6977
—3	1,4671	1,5226	1,5670	1,6071	1,6377	1,6659	1,6910
—2	1,4549	1,5045	1,5361	1,5855	1,6210	1,6506	1,6742
—1	1,4214	1,4697	1,5122	1,5487	1,5803	1,6075	1,6305
0	1,3592	1,4033	1,4416	1,4747	1,5034	1,5285	1,5499
1	1,2727	1,3100	1,3426	1,3711	1,3958	1,4176	1,4365
2	1,1854	1,2151	1,2414	1,2646	1,2850	1,3029	1,3186
3	1,1167	1,1393	1,1598	1,1782	1,1945	1,2090	1,2217
4	1,0698	1,0866	1,1023	1,1166	1,1295	1,1411	1,1515
5	1,0403	1,0524	1,0642	1,0792	1,0853	1,0946	1,1030
6	1,0224	1,0310	1,0396	1,0480	1,0560	1,0633	1,0701
7	1,0120	1,0179	1,0241	1,0304	1,0365	1,0423	1,0478
8	1,0061	1,0100	1,0144	1,0190	1,0236	1,0282	1,0326
9	1,0029	1,0054	1,0084	1,0117	1,0152	1,0187	1,0222
10	1,0013	1,0028	1,0048	1,0071	1,0097	1,0123	1,0151

TABLE A.8 (concluded)

μ^* \ a	14	15	16	17	18	19	20
—4	1,7241	1,7418	1,7584	1,7740	1,7822	1,8033	1,8063
—3	1,7149	1,7320	1,7482	1,7654	1,7757	1,7894	1,8006
—2	1,6990	1,7128	1,7303	1,7385	1,7445	1,7523	1,7790
—1	1,6513	1,6714	1,6822	1,6950	1,7122	1,7210	1,7326
0	1,5692	1,5859	1,6013	1,6130	1,6246	1,6342	1,6420
1	1,4532	1,4678	1,4805	1,4913	1,5014	1,5100	1,5187
2	1,3324	1,3444	1,3544	1,3650	1,3734	1,3808	1,4043
3	1,2331	1,2431	1,2520	1,2600	1,2671	1,2734	1,2790
4	1,1607	1,1690	1,1729	1,1830	1,1890	1,1943	1,1991
5	1,1106	1,1175	1,1237	1,1293	1,1343	1,1389	1,1430
6	1,0764	1,0821	1,0873	1,0920	1,0963	1,1003	1,1038
7	1,0529	1,0577	1,0632	1,0661	1,0698	1,0731	1,0762
8	1,0368	1,0407	1,0444	1,0478	1,0509	1,0538	1,0565
9	1,0256	1,0288	1,0318	1,0347	1,0374	1,0399	1,0423
10	1,0178	1,0204	1,0229	1,0253	1,0276	1,0298	1,0318

A.9. TABLES OF VALUES OF THE RATIO OF INTEGRALS $\Phi_4(\mu^*, a)/\Phi_3(\mu^*, a)$

μ^* \ a	1	2	3	4	5
—4	2,47530	2,83158	3,08103	3,26257	3,39828
—3	2,48604	2,84055	3,08939	3,27057	3,40618
—2	2,51357	2,86423	3,11145	3,29183	3,42701
—1	2,58242	2,92369	3,16695	3,34533	3,47945
0	2,73790	3,05897	3,29349	3,46751	3,59927
1	3,03514	3,32095	3,53938	3,70536	3,83267
2	3,50380	3,74171	3,93682	4,09056	4,21142
3	4,12922	4,31495	4,48134	4,61920	4,73101
4	4,87356	5,0064	5,14787	5,26758	5,36863
5	5,69885	5,79840	5,90629	6,00715	6,09614
6	6,57745	6,64830	6,73188	6,81494	6,89148
7	7,48998	7,54035	7,60428	7,67148	7,73622
8	8,42520	8,46145	8,51021	8,56405	8,61820
9	9,37539	9,40185	9,43914	9,48215	9,52692

TABLE A.9 (continued)

a μ*	6	7	8	9	10
—4	3,50209	3,58303	3,64687	3,69808	3,73962
—3	3,50991	3,59070	3,65458	3,70578	3,74732
—2	3,53049	3,61115	3,67495	3,72610	3,76760
—1	3,58234	3,66264	3,72624	3,77728	3,81872
0	3,70086	3,78041	3,84359	3,89439	3,93572
1	3,92826	4,00956	4,07095	4,12261	4,15388
2	4,30824	4,38259	4,44181	4,49263	4,53430
3	4,82143	4,89470	4,95462	5,00114	5,04029
4	5,45250	5,52232	5,58046	5,62894	5,66912
5	6,17266	6,23704	6,29111	6,34004	6,37782
6	6,95990	7,01965	7,07051	7,11330	7,15357
7	7,79605	7,84968	7,89767	7,93854	7,96680
8	8,66927	8,71742	8,76055	8,77537	8,83400
9	9,57097	9,61266	9,65136	9,68648	9,72020

a μ*	11	12	13	14	15
—4	3,77368	3,80189	3,82546	3,84531	3,86218
—3	3,78137	3,80957	3,83314	3,85300	3,86987
—2	3,80164	3,80984	3,85341	3,87327	3,89014
—1	3,85273	3,88091	3,90448	3,92435	3,94123
0	3,96968	3,99785	4,02143	4,04133	4,05824
1	4,18818	4,21867	4,24358	4,26068	4,18351
2	4,56202	4,59258	4,62025	4,63730	4,52907
3	5,07812	5,10664	5,13141	5,14341	5,16303
4	5,70239	5,72904	5,75415	5,75943	5,78261
5	6,40805	6,44760	6,46332	6,48476	6,50164
6	7,18762	7,21432	7,23648	7,25922	7,27972
7	8,00661	8,03472	8,05704	8,07623	8,09555
8	8,85225	8,88974	8,91459	8,93410	8,95207
9	9,74844	9,77249	9,77801	9,81671	9,88387

TABLE A.9 (concluded)

a μ*	16	17	18	19	20
—4	3,87661	3,88902	3,89977	3,90915	3,91735
—3	3,88429	3,89671	3,90746	3,91683	3,92503
—2	3,90457	3,91699	3,92775	3,93712	3,94533
—1	3,95567	3,96811	3,97888	3,98827	3,99649
0	4,07272	4,08520	4,09601	4,10544	4,11370
1	4,21288	4,24036	4,26066	4,28774	4,31342
2	4,55647	4,58037	4,60409	4,62114	4,35292
3	5,21631	5,19343	5,08757	5,10414	5,10971
4	5,80443	5,82346	5,84129	5,85391	5,82531
5	6,50098	6,51811	6,53869	6,55714	6,57829
6	7,29659	7,30289	7,30822	7,31077	7,32148
7	8,11554	8,13247	8,13924	8,16481	8,14980
8	8,96555	8,98186	8,99961	9,00448	9,02043
9	9,84945	9,86078	9,87093	9,88561	9,89453

A.10. TABLE OF VALUES OF $(\sqrt{\pi}/4)Q^{\perp}/Q^{\perp}_{L_0}$

a μ*	1	2	3	4	5
—4	0,0316	—0,0594	—0,0887	—0,0970	—0,0955
—3	0,0328	—0,0588	—0,0880	—0,0990	—0,0948
—2	0,0350	—0,0568	—0,0869	—0,0953	—0,0940
—1	0,0408	—0,0525	—0,0839	—0,0929	—0,0924
0	0,0517	—0,0430	—0,0769	—0,0879	—0,0888
1	0,0680	—0,0350	—0,0596	—0,0787	—0,0818
2	0,0837	—0,0205	—0,0474	—0,0653	—0,0718
3	0,0936	0,0149	—0,0280	—0,0495	—0,0594
4	0,0957	0,0294	—0,0094	—0,0328	—0,0458
5	0,0895	0,0419	0,0061	—0,0262	—0,0360
6	0,0789	0,0443	0,0140	—0,0123	—0,0217
7	0,0704	0,0508	0,0223	0,0029	—0,0125
8	0,0621	0,0480	0,0256	0,0105	0,0021
9	0,0591	0,0465	0,0289	0,0140	0,0046

TABLE A.10 (continued)

a / μ*	6	7	8	9	10
—4	—0,0895	—0,0828	—0,0761	—0,0698	—·0,0630
—3	—0,0896	—0,0830	—0,0752	—0,0688	—0,0598
—2	—0,0899	—0,0824	—0,0755	—0,0685	—0,0625
—1	—0,0877	—0,0814	—0,0756	—0,0679	—0,0614
0	—0,0850	—0,0795	—0,0728	—0,0665	—0,0607
1	—0,0802	—0,0755	—0,0701	—0,0643	—0,0595
2	—0,0718	—0,0695	—0,0658	—0,0609	—0,0563
3	—0,0626	—0,0620	—0,0598	—0,0575	—0,0541
4	—0,0517	—0,0538	—0,0532	—0,0511	—0,0487
5	—0,0407	—0,0448	—0,0460	—0,0451	—0,0442
6	—0,0321	—0,0354	—0,0380	—0,0387	—0,0388
7	—0,0219	—0,0273	—0,0307	—0,0326	—0,0335
8	—0,0140	—0,0243	—0,0270	—·0,0290	—0,0295
9	—0,0047	—0,0130	—0,0175	—0,0199	—0,0225

a / μ*	11	12	13	14	15
—4	—0,0575	—0,0491	—0,0440	—0,0389	—0,0354
—3	—0,0569	—0,0513	—0,0460	—0,0390	—0,0360
—2	—0,0562	—0,0518	—0,0465	—0,0418	—0,0370
—1	—0,0558	—0,0507	—0,0460	—0,0423	—0,0381
0	—0,0551	—0,0502	—0,0457	—0,0416	—0,0380
1	—0,0544	—0,0495	—0,0452	—0,0412	—0,0375
2	—0,0521	—0,0475	—0,0435	—0,0401	—0,0370
3	—0,0494	—0,0460	—0,0419	—0,0392	—0,0359
4	—0,0460	—0,0432	—0,0402	—0,0380	—0,0350
5	—0,0416	—0,0390	—0,0362	—0,0347	—0,0312
6	—0,0380	—0,0360	—0,0335	—0,0324	—0,0300
7	—0,0333	—0,0341	—0,0315	—0,0306	—0,0287
8	—0,0298	—0,0293	—0,0283	—0,0265	—0,0257
9	—0,0230	—0,0232	—0,0234	—0,0236	—0,0237

TABLE A.10 (concluded)

a μ^*	16	17	18	19	20
—4	—0,0317	—0,0300	—0,0287	—0,0282	—0,0281
—3	—0,0330	—0,0310	—0,0295	—0,0270	—0,0270
—2	—0,0340	—0,0320	—0,0304	—0,0274	—0,0268
—1	—0,0350	—0,0327	—0,0298	—0,0275	—0,0253
0	—0,0350	—0,0322	—0,0296	— 0,0275	—0,0254
1	—0,0346	—0,0320	—0,0307	—0,0280	—·0,0254
2	—0,0344	—0,0322	—0,0302	—0,0287	—·0,0281
3	—0,0336	— 0,0307	—0,0290	—0,0270	—0,0262
4	—0,0325	—0,0299	—0,0273	—0,0262	—0,0255
5	—0,0292	—0,0280	—0,0262	—0,0255	—0,0246
6	—0,0278	—0,0265	—0,0252	—0,0243	—0,0238
7	—0,0270	—0,0253	—0,0250	—0,0241	—0,0239
8	—0,0251	—0,0250	—0,0245	←0,0240	—0,0240
9	—0,0239	—0,0239	—0,0240	—0,0242	—0,0245

Literature

The following short-form abbreviations have been used in the reference lists:

DAN — Doklady Akademii Nauk
FTT — Fizika Tverdogo Tela
UFN — Uspekhi Fizicheskikh Nauk
ZhÉTF — Zhurnal Éksperimental'noi i Teoreticheskoi Fiziki
ZhTF — Zhurnal Tekhnicheskoi Fiziki

CHAPTER 1

1. F. Bloch, Z. Physik, 52:555 (1928).
2. É. I. Adirovich, Some Problems in the Theory of Luminescence of Crystals [in Russian], Gostekhizdat, 1956.
3. N. B. Hannay (ed.), Semiconductors, 1959.
4. T. O. Woodruff, Solid State Physics, Vol. 4 (1957).
5. F. Herman, Phys. Rev., 93:1214 (1954).
6. F. Herman, Phys. Rev., 95:847 (1954).
7. F. Herman, Proc. IRE, 43:1703 (1955).
8. B. Lax and J. Mavroides, Solid State Phys., 11:261 (1960).
9. B. Lax, UFN, 70(1):111 (1960).
10. R. K. Willardson, T. C. Harman, and A. C. Beer, Phys. Rev., 96:1512 (1954).
11. G. Della Pergola and D. Sette, Phys. Rev., 104:598 (1956).
12. H. P. Furth and R. W. Waniek, Phys. Rev., 104:343 (1956).
13. G. G. Macfarlane et al., Phys. Rev., 108:1377 (1957).
14. G. G. Macfarlane et al., Phys. Rev., 111:1245 (1958).
15. F. Herman, J. Electronics, 1:103 (1955).
16. F. Herman, J. Phys. Chem. Solids, 8:380 (1959).
17. C. Hilsum and A. C. Rose-Innes, Semiconducting III—V Compounds, Pergamon Press, Oxford, 1961.
18. Selected Constants Relative to Semiconductors, Pergamon Press, New York, 1961.

19.   A. I. Ansel'm, Introduction to the Theory of Semiconductors [in Russian],
      Fizmatgiz, 1963.
20.   D. I. Blokhintsev, Fundamentals of Quantum Mechanics [in Russian], "Vysshaya
      shkola," 1961.
21.   C. H. Kittel and A. H. Mitchell, Phys. Rev., 96:1488 (1954).
22.   J. M. Luttinger and W. Kohn, Phys. Rev., 97:863 (1955).
23.   J. M. Luttinger and W. Kohn, Phys. Rev., 96:802 (1954).
24.   W. Kohn and J. M. Luttinger, Phys. Rev., 97:1721 (1955).
25.   W. Kohn and J. M. Luttinger, Phys. Rev., 98:915 (1955).
26.   W. Kohn and D. Schechter, Phys. Rev., 99:1903 (1955).
27.   E. Burstein, G. S. Picus, and N. Sclar, Report on the Photoconductivity Conference,
      New York, November, 1954.
28.   G. Busch and H. Labhart, Helv. Phys. Acta, 14:463 (1946).
29.   N. F. Mott and W. D. Twose, UFN, 79(4):691 (1963).
30.   P. Debye and E. M. Conwell, Phys. Rev., 93:693 (1954).
31.   G. L. Pearson and J. Bardeen, Phys. Rev., 75:865 (1949).
32.   W. Baltensperger, Phil. Mag., 44:1355 (1953).
33.   M. Lax and J. C. Phillips, Phys. Rev., 110:41 (1958).
34.   V. L. Bonch-Bruevich, FTT, 4:2660 (1962).
35.   V. L. Bonch-Bruevich, FTT, 5:1852 (1963).
36.   T. P. Brody, J. Appl. Phys., 33:100 (1962).
37.   H. M. James and A. S. Ginzberg, J. Phys. Chem., 57:840 (1953).
38.   E. M. Conwell, Phys. Rev., 103:51 (1956).
39.   H. L. Frisch and S. P. Lloyd, Phys. Rev., 120:1175 (1960).
40.   V. L. Bonch-Bruevich and A. G. Mironov, FTT, 3:3009 (1961).
41.   R. H. Parmenter, Phys. Rev., 97:587 (1955).
42.   V. L. Bonch-Bruevich, FTT (collection of papers), 2:177 (1959).
43.   P. Aigrain, Physica, 20:978 (1954).
44.   V. L. Bonch-Bruevich and Sh. M. Kogan, Ann. Phys., 9:125 (1960).
45.   L. V. Keldysh and G. P. Proshko, FTT, 5:3378 (1963).
46.   F. N. Horn, Phys. Rev., 97:1521 (1955).
47.   G. B. Dubrovskii, Dissertation for Candidate's Degree [in Russian], Leningrad,
      1963.
48.   J. I. Pankove, Phys. Rev. Letters, 4:31 (1960).
49.   J. I. Pankove, Phys. Rev. Letters, 4:454 (1960).
50.   C. Haas, Phys. Rev., 125:1965 (1962).
51.   J. I. Pankove and P. Aigrain, Phys. Rev., 126:956 (1962).
52.   G. B. Dubrovskii and B. K. Subashiev, FTT, 4:3018 (1962).
53.   H. S. Sommers, Phys. Rev., 124:1101 (1961).
54.   F. Stern and R. M. Talley, Phys. Rev., 100:1638 (1955).
55.   F. Stern and J. R. Dixon, J. Appl. Phys., 30:268 (1959).
56.   V. Roberts and J. E. Quarrington, J. Electronics, 1:152 (1955).
57.   R. Newman, Phys. Rev., 111:1518 (1958).
58.   S. W. Kurnick and J. M. Powell, Phys. Rev., 116:597 (1959).
59.   D. Long, Phys. Rev., 99:388 (1955).
60.   V. I. Fistul' and A. M. Agaev, FTT, 7:3042 (1965).

CHAPTER 2

1. L. D. Landau and E. M. Lifshits, Statistical Physics [in Russian], Fizmatgiz, 1964.
2. M. A. Leontovich, Statistical Physics [in Russian], Gostekhizdat, 1944.
3. J. E. Mayer and M. G. Mayer, Statistical Mechanics, Wiley, New York, 1940.
4. J. W. Gibbs, Elementary Principles in Statistical Mechanics, Dover, New York, 1902.
5. A. G. Samoilovich, Thermodynamics and Statistical Physics [in Russian], Gostekhizdat, 1953.
6. V. G. Levich, Introduction to Statistical Physics [in Russian], Gostekhizdat, 1950.
7. C. Hilsum and A. C. Rose-Innes, Semiconducting III−V Compounds, Pergamon Press, Oxford, 1961.
8. R. A. Smith, Semiconductors, Cambridge University Press, 1961.
9. P. T. Landsberg, Proc. Phys. Soc. (London), A65:604 (1952).
10. A. G. Samoilovich, DAN UkrSSR, No. 3, p. 174 (1954).
11. P. Debye and E. M. Conwell, Phys. Rev., 93:693 (1954).
12. G. L. Pearson and J. Bardeen, Phys. Rev., 75:865 (1949).
13. K. Lehovec and H. Kedesday, J. Appl. Phys., 22:65 (1961).
14. W. Shockley, Electrons and Holes in Semiconductors, Van Nostrand, New York, 1950.
15. J. S. Blakemore, Electr. Commun., 29:2 (1952).
16. E. Moser, Zeits. Angew. Math. Phys., Vol. 4, No. 6 (1953).
17. L. L. Korenblit and T. G. Shraifel'd, ZhTF, 25:1182 (1955).
18. V. L. Bonch-Bruevich, FTT (collection of papers), 2:177 (1959).
19. I. P. Zvyagin, FTT, 5:581 (1963).
20. V. L. Bonch-Bruevich, FTT, 4:2660 (1962).
21. R. B. Dingle, Appl. Sci. Research, B6:225 (1957).
22. P. Rhodes, Proc. Phys. Soc. (London), A204:396 (1950).
23. J. McDougall and E. S. Stoner, Trans. Roy. Soc. (London), A237:67 (1939).
24. A. C. Beer, M. N. Chase, and R. E. Choquard, Helv. Phys. Acta, 28:529 (1955).
25. J. S. Blakemore, Semiconductor Statistics, Pergamon Press, Oxford, 1962.
26. E. Jahnke and F. Emde, Tables of Functions, Dover, New York, 1945.

CHAPTER 3

1. A. H. Wilson, Proc. Roy. Soc. (London), 133A:458 (1931).
2. F. F. Vol'kenshtein, Electrical Conductivity of Semiconductors [in Russian], Gostekhizdat, 1947.
3. A. I. Ansel'm, Introduction to the Theory of Semiconductors [in Russian], Fizmatgiz, 1963.
4. H. Brooks, Collection: Problems in Modern Physics [Russian translation], Vol. 8, p. 74 (1957).
5. R. A. Smith, Semiconductors, Cambridge University Press, 1961.
6. Collection: Problems in Quantum Theory of Irreversible Processes [Russian translation], IL, 1962.

7. V. L. Bonch-Bruevich and S. V. Tyablikov, Green's Function Method in Statistical Mechanics, Wiley, New York, 1962.
8. A. G. Samoilovich, M. I. Klinger, and L. L. Korenblit, FTT (collection of papers), 2:121 (1959).
9. C. Herring and E. Vogt, Collection: Problems in Physics of Semiconductors [Russian translation], IL, 1957, p. 167.
10. A. G. Samoilovich, I. Ya. Korenblit, I. V. Dakhovskii, and V. D. Iskra, FTT, 3:3285 (1961).
11. I. M. Tsidil'kovskii, Thermomagnetic Effects in Semiconductors, Infosearch, London, 1962.
12. R. E. Peierls, Quantum Theory of Solids, Oxford University Press, 1955.
13. V. Wieringen, Proc. Phys. Soc. (London), A67:206 (1954).
14. D. A. Greenwood, Proc. Phys. Soc. (London), 71:585 (1958).
15. H. Balesky, Physica, 27:693 (1961).
16. F. Seitz, Phys. Rev., 73:549 (1948).
17. J. Bardeen and W. Shockley, Phys. Rev., 80:72 (1950).
18. A. H. Wilson, Theory of Metals, Cambridge University Press, 1954.
19. W. Shockley and J. Bardeen, Phys. Rev., 77:407 (1950).
20. G. E. Pikus, ZhTF, 28:2390 (1958).
21. H. A Bethe and A. Sommerfeld, Electron Theory of Metals [Russian translation], ONTI, 1938.
22. P. J. W. Debye, Structure of Matter [Russian translation], 6, ITL, Kharkov, 1936.
23. B. I. Davydov and I. M. Shmushkevich, UFN, 24:21 (1940).
24. E. H. Sondheimer and D. J. Howarth, Proc. Roy. Soc. (London), A219:53 (1953).
25. K. S. Shifrin, ZhTF, 14:40 (1944).
26. N. F. Mott and H. S. W. Massey, Theory of Atomic Collisions, 2nd ed., Oxford University Press, 1949.
27. E. M. Conwell and V. F. Weisskopf, Phys. Rev., 77:388 (1950).
28. H. Brooks, Phys. Rev., 83:879 (A) (1951).
29. L. I. Schiff, Quantum Mechanics, 2nd ed., McGraw—Hill, New York, 1955.
30. P. Csavinszky, J. Phys. Soc. Japan, 16:1865 (1961).
31. G. L. Pearson and J. Bardeen, Phys. Rev., 75:865 (1949).
32. C. Erginsoy, Phys. Rev., 79:1013 (1950).
33. B. M. Askerov, Theory of Transport Phenomena in Semiconductors [in Russian], Izd. AN AzSSR, Baku, 1963.
34. L. Spitzer and R. Harm, Phys. Rev., 89:977 (1953).
35. A. I. Ansel'm and B. M. Askerov, FTT, 3:3672 (1961).
36. A. I. Ansel'm and V. I. Klyachkin, ZhÉTF, 22:297 (1952).
37. H. Jones, Phys. Rev., 81:149 (1951).
38. O. Madelung, Z. Naturforsch., 9a:667 (1954).
39. Yu. N. Obraztsov, ZhTF, 25:995 (1955).
40. R. Mansfield, Proc. Phys. Soc. (London), B69:76 (1956).
41. D. B. Agarwal, Z. Physik, 163:207 (1961).
42. V. I. Fistul', É. M. Omel'yanovskii, and Z. I. Tatarov, FTT, 6:974 (1964).
43. J. M. Ziman, Electrons and Phonons, Oxford University Press, 1960.
44. C. Herring, Bell System Tech. J., 34:237 (1955).

45.   A. G. Samoilovich, I. Ya. Korenblit, I. V. Dakhovskii, and V. D. Iskra, FTT, 3:2939 (1961).

46.   A. G. Samoilovich, I. Ya. Korenblit, and I. V. Dakhovskii, DAN SSSR, 139:355 (1961).

47.   I. Ya. Korenblit, FTT, 4:168 (1962).

48.   V. I. Fistul', É. M. Omel'yanovskii, D. G. Andrianov, and I. V. Dakhovskii, Proc. Intern. Conference on Semicond., Paris, 1964.

49.   D. G. Andrianov, I. V. Dakhovskii, É. M. Omel'yanovskii, and V. I. Fistul', FTT, 6:2825 (1964)

50.   É. M. Omel'yanovskii, Dissertation for Candidate's Degree [in Russian], 1963.

51.   É. M. Omel'yanovskii and V. I. Fistul', Zavodskaya laboratoriya, 30(5):559 (1964).

52.   G. Busch and U. Winkler, Determination of Characteristic Parameters of Semiconductors [Russian translation]. IL, 1959.

53   E. M Omel'yanovskii, V. I. Fistul', and M. G. Mil'vidskii, FTT, 5:921 (1963).

54   M. Glicksman, Progress in Semiconductors, 3:1 (1958)

55.   I. M. Ryzhik and I. S. Gradshtein, Tables of Integrals, Sums, Series, and Products [in Russian], Gostekhizdat, 1951

56.   I. V. Dakhovskii, FTT, 5:2332 (1963).

57   P. Debye and E. M. Conwell, Phys Rev., 93:693 (1954)

58.   Y. Furukawa, J. Phys. Soc. Japan, 16:687 (1961).

59.   W. G. Spitzer, F. A. Trumbore, and R. A. Logan, J. Appl. Phys., 32:1822 (1961).

60.   B. G. Zhurkin, V. S. Zemskov, and K. V. Yurkina, FTT, 3:3509 (1961).

61.   V. I. Fistul' and A. Ya. Gubenko, FTT, 3:1617 (1961).

62   W. Shockley, Electrons and Holes in Semiconductors, Van Nostrand, New York, 1950.

63.   V. I. Fistul', M. I. Iglitsyn, and É. M. Omel'yanovskii, FTT, 4:1065 (1962).

64.   E. Putley, The Hall Effect and Related Phenomena, Butterworths, London, 1960.

65.   Selected Constants Relative to Semiconductors, Pergamon Press, New York, 1961.

66.   A. G. Samoilovich and L. L. Korenblit, UFN, 57(4):629 (1955).

67   V. I. Fistul', Collection: Proc. Conf. on Impact Ionization and Tunnel Effects in Semiconductors [in Russian], Baku, 1962, p. 159.

68.   V. I. Fistul' and B. N. Khusainova, Trudy instituta Giredmet, 10:232 (1963).

69.   F. J. Morin and J. P. Maita, Phys. Rev., 96:28 (1954).

70.   R. A. Logan, J. F. Gilbert, and F. A. Trumbore, J Appl. Phys., 32:131 (1961).

71.   L. Esaki and Y. Miyahara, Solid State Electronics, 1:13 (1960).

72.   Y. Furukawa, J. Phys. Soc. Japan, 16:577 (1961).

73.   V. I. Fistul', Doctoral Dissertation [in Russian], Moscow, 1965.

74.   I. P. Zvyagin, Dissertation for Candidate's Degree [in Russian], Moscow, 1964; A. A. Abrikosov, ZhÉTF, 44:2039 (1963).

75   W. Kohn and J. M. Luttinger, Phys. Rev., 98:915 (1955).

76.   O. A. Golikova, B. Ya. Moizhes, and L. S. Stil'bans, FTT, 3:3105 (1961).

77.   O. A. Golikova, B. Ya. Moizhes, and A. G. Orlov, FTT, 4:3482 (1962).

78.   F. A. Trumbore and A. A. Tartaglia, J. Appl. Phys., 29:1511 (1958).

79   M. N. Vinogradova, O. A. Golikova, B. P. Mistrenin, and L. S. Stil'bans, FTT, 2:1429 (1960).

80. N. L. Pisarenko, Izv. AN SSSR, Ser. Fiz., No. 5-6, p. 631 (1938).

81. W. Ehrenberg, Electric Conduction in Semiconductors and Metals, Oxford University Press, 1958.

82   A. D. Middleton and W. Scanlon, Phys. Rev., 92:219 (1953).

83. T. H. Geballe and G. W. Hull, Phys. Rev., 94:1134 (1954).

84. P. I. Baranskii, FTT, 3:1786 (1961).

85. V. I. Fistul' and K. V. Cherkas, FTT, 4:3288 (1962).

86   K. V. Cherkas and V. I. Fistul', FTT, 6:16 (1964).

87. T. H. Geballe and G. W. Hull, Phys. Rev., 98:940 (1955).

88. I. A. Smirnov, B. Ya. Moizhes, and E. D. Nensberg, FTT, 2:1992 (1960).

89. C. Herring, Halbleiter und Phosphore, 1958, Vol. 5, p. 184.

90. M. N. Vinogradova, O. A. Golikova, I. N. Dubrovskaya, and B. Ya. Moizhes, FTT, 5:1657 (1963).

91. O. A. Golikova and A. G. Orlov, FTT, 5:1908 (1963).

92. M. Cardona and H. S. Sommers, Phys. Rev., 122:1382 (1961).

93. P. Debye, Vorträge über kinetische Theorie der Materie, Teubner, 1914.

94. A. Eucken, Z. Physik, 12:1005 (1911).

95. R. Peierls, Ann. Phys., 3:1055 (1929).

96. R. Peierls, Ann. Inst. Poincaré, 5:177 (1935).

97. J. Drabble and H. J. Goldsmid, Thermal Conduction in Semiconductors, Pergamon Press, Oxford, 1961.

98. I. A. Smirnov, Dissertation for Candidate's Degree [in Russian], Leningrad, 1961.

99. H. B. G. Casimir, Physica, 5:495 (1938).

100. V. Mielczarek and H. P. Frederikse, Phys. Rev., 115:888 (1959).

101. G. A. Slack, Phys. Rev., 105:832 (1957).

102. A. F. Ioffe, Physics of Semiconductors [in Russian], Izd. AN SSSR, 1957.

103. P. Klemens, Solid State Phys., Vol. 7, No. 4 (1958).

104. A. F. Ioffe, B. Ya. Moizhes, and D. S. Stil'bans, FTT, 2:2834 (1960).

105. R. W. Keyes, Phys. Rev., 115:564 (1959).

106. J. Tavernier, Abstracts Intern. Conf. on Semiconductor Physics, Prague, 1960

107. V. P. Zhuze, DAN SSSR, 99:711 (1954).

108. V. P. Zhuze and T. A. Kontorova, ZhTF, 28:1727 (1958).

109. T. A. Kontorova, Collection: Some Problems in the Strength of Solids [in Russian], Izd. AN SSSR, 1959.

110. A. V. Ioffe and A. F. Ioffe, DAN SSSR, 97:757 (1954).

111. A. F. Ioffe, Can. J. Phys., 34:1342 (1956).

112. P. J. Price, Phil. Mag., 46:1192 (1955).

113 G. E. Pikus, ZhTF, 26:49 (1956).

114. L. Genzel, Z. Physik, 135:177 (1953).

115. G. Slack and C. Glassbrenner, Phys. Rev., 120:782 (1960).

116 R. Berman, Advances in Physics, 2:103 (1953).

117. T. H. Geballe and G. W. Hull, Phys. Rev., 110:773 (1958).

118. P. G. Klemens, Proc. Roy. Soc. (London), 208:108 (1951).

119 J. Carruthers, T. H. Geballe, H. Rosenberg, and J. M. Ziman, Proc. Roy. Soc. (London), 238:202 (1957)

120. R. Berman, Proc. Roy. Soc. (London), 208:90 (1951).

121. R. Berman, E. Foster, and J. M. Ziman, Proc. Roy. Soc. (London), 231:130 (1955).

122. G. A. Slack, Phys. Rev., 105:829 (1957).

123. M. E. Straumanis, Collection: Germanium [Russian translation], IL, 1956.

124. M. I. Aliev, V. I. Fistul', and D. G. Arasly, FTT, 6:3700 (1964).

125. M. I. Aliev, D. G. Arasly, and V. I. Fistul', Izv. AN AzSSR, No. 5, 103 (1965).

126 D. G Arasly, Dissertation for Candidate's Degree [in Russian], Baku, 1967.

127. C. S. Fuller, Collection: Semiconductors (ed. by N. B. Hannay), Reinhold, New York, 1959.

128. C. Kittel, Introduction to Solid State Physics, 2nd ed., Wiley, New York, 1956.

129. E. D. Devyatkova, ZhTF, 27:461 (1957).

130. K. Hashimoto, Mem. Fac. Sci. Kyusyu Univ., B2:187 (1958).

131. M. Aliev, Thermal Conductivity [in Russian], Baku, 1964.

132. L. I. Domanskaya, É. M. Omel'yanovskii, V. I. Fistul', and I. M. Tsidil'kovskii, FTT, 5:3046 (1963).

133. L. I. Domanskaya, G. I. Kharus, and I. M. Tsidil'kovskii, FTT, 7:46 (1965).

134. O. V. Emel'yanenko and D. N. Nasledov, FTT, 1:985 (1959).

135. C. Herring and E. Vogt, Phys. Rev., 101:944 (1956).

136. W. Sasaki, C. Yamanoushi, and G. M. Hatoyama, Proc. Intern. Conf. Semiconductors, Prague, 1961, p. 159.

137. B. Abeles and S. Meiboom, Phys. Rev., 95:31 (1954).

138. D. G. Andrianov and V. I. Fistul', FTT, 5:1480 (1963).

139. Y. Furukawa, J. Phys. Soc. Japan, 18:737 (1963).

140. Y. Furukawa, J. Phys. Soc. Japan, 18:1270 (1963).

141. Y. Furukawa, J. Phys. Soc. Japan, 18:1374 (1963).

142. M. Mirzabaev, V. M. Tuchkevich, and Yu. V. Shmartsev, FTT, 5:1625 (1963).

143. J. Toyozawa, J. Phys. Soc. Japan, 17:986 (1962).

144. H. R. Frederikse and W. R. Hosler, Can. J. Phys., 34:1377 (1956).

145. H. Fritzsche and K. Lark-Horovitz, Phys. Rev., 99:400 (1955).

146. J. T. Edmond, R. F. Broom, and F. A. Cunnel, Report of the Meeting on Semiconductors, London, 1956, p. 109.

147. O. V. Emel'yanenko and D. N. Nasledov, ZhTF, 28:1177 (1958).

148. O. V. Emel'yanenko, T. S. Lagunova, and D. N. Nasledov, FTT, 3:198 (1961).

149. D. N. Nasledov, J. Appl. Phys. Suppl., 32:2140 (1961).

150. N. V. Zotova, T. S. Lagunova, and D. N. Nasledov, FTT, 5:3329 (1963).

151. D. G. Andrianov and V. I. Fistul', FTT, 6:470 (1964).

152. D. G. Andrianov, Dissertation for Candidate's Degree [in Russian], Moscow, 1965.

153. C. Hilsum and A. C. Rose-Innes, Semiconducting III−V Compounds, Pergamon Press, Oxford, 1961.

154. O. Madelung, Physics of III−V Compounds, Wiley, New York, 1964.

155. H. Rupprecht, R. Weber, and H. Weiss, Z. Naturforsch., 15a:783 (1960).

156. J. Kolodziejczak, Acta Phys. Polon., 21:637 (1961).

157. J. Kolodziejczak, L. Sosnowski, and W. Zawadzki, Proc. Intern. Conf. on Semicond. Phys., Exeter, 1962, p. 94.

158.   G. I. Guseva and I. M. Tsidil'kovskii, FTT, 4:2490 (1962).
159.   I. M. Tsidil'kovskii, FTT, 4:2539 (1962).

CHAPTER 4

1.   P. Drude, Theory of Optics, Dover, New York, 1902.
2.   H. Y. Fan, W. G. Spitzer, and R. J. Collins, Phys. Rev., 101:569 (1956).
3.   R. Rosenberg and M. Lax, Phys. Rev., 112:843 (1958).
4.   S. Visvanathan, Phys. Rev., 120:376 (1960).
5.   H. J. G. Meyer, Phys. Rev., 112:298 (1958).
6.   B. Donovan, Proc. Phys. Soc. (London), 76:574 (1960).
7.   V. A. Yakovlev, FTT, 2:1624 (1960).
8.   G. G. Macfarlane et al., Phys. Rev., 108:1137 (1957).
9.   G. G. Macfarlane et al., Phys. Rev., 108:1377 (1957).
10.   G. G. Macfarlane et al., Phys. Rev., 111:1245 (1958).
11.   D. L. Dexter, Proc. Photocond. Conf. at Atlantic City, New York, 1956, p. 155.
12.   T. P. McLean, Progress in Semiconductors, 5:53 (1960).
13.   G. B. Dubrovskii, Dissertation for Candidate's Degree [in Russian], Leningrad, 1964.
14.   G. B. Dubrovskii, FTT, 5:3361 (1963).
15.   J. Bardeen, F. J. Blatt, and L. H. Hall, Proc. Photocond. Conf. at Atlantic City, New York, 1956, p. 146.
16.   E. Burstein, Phys. Rev., 93:632 (1954).
17.   T. S. Moss, Optical Properties of Semiconductors, Butterworths, London, 1959.
18.   R. L. Hartman, Phys. Rev., 127:765 (1962).
19.   J. I. Pankove, Ann. Phys., 6:331 (1961).
20.   J. I. Pankove, Phys. Rev. Letters, 4:454 (1960).
21.   J. I. Pankove and P. Aigrain, Phys. Rev., 126:956 (1962).
22.   B. S.Bagaev, G. P. Proshko, and A. P. Shotov, FTT, 4:3229 (1962).
23.   C. Haas, Phys. Rev., 125:1965 (1962).
24.   M. G. Mil'vidskii, V. I. Fistul', and S. P. Grishina, FTT, 6:2762 (1964).
25.   W. G. Spitzer and H. Y. Fan, Phys. Rev., 106:882 (1957).
26.   K. B. Tolpygo, Trudy Instituta Fiziki AN UkrSSR, No. 3, p. 84 (1952).
27.   M. Cardona, W. Paul, and H. Brooks, Helv. Phys. Acta, 35:329 (1960).
28.   W. G. Spitzer, F. A. Trumbore, and H. A. Logan, J. Appl. Phys., 32:1822 (1961).
29.   E. P. Rashevskaya and V. I. Fistul', FTT, 4:2601 (1962).
30.   L. E. Howarth and J. F. Gilbert, J. Appl. Phys., 34:236 (1963).
31.   H. R. Philipp and E. A. Taft, Phys. Rev., 113:1002 (1959).
32.   H. R. Philipp and E. A. Taft, Phys. Rev., 120:37 (1960).
33.   J. C. Phillips, J. Phys. Chem. Solids, 12:208 (1960).
34.   J. Tauc and A. Abraham, Proc. Semicond. Conf. Prague, 1960, p. 375.
35.   E. F. Gross and V. S. Sobolev, Proc. Second All-Union Conf. on Optical and Photoelectric Effects in Semiconductors [in Russian], L'vov, 1961.
36.   M. Cardona, J. Appl. Phys., 32:958 (1961).
37.   M. Cardona, Z. Physik, 161:99 (1961).
38.   M. Cardona and H. S. Sommers, Phys. Rev., 122:1382 (1961).
39.   H. Ehrenreich, H. R. Philipp, and J. C. Phillips, Phys. Rev. Letters, 8:59 (1962).

40.   G. S. Landsberg, Optics [in Russian], Gostekhizdat, 1957.

41.   A. K. Walton and T. S. Moss, Proc. Phys. Soc. (London), 78:1393 (1961).

42.   E. W. J. Mitchell, Proc. Phys. Soc. (London), B68:973 (1955).

43.   T. S. Moss, Optical Properties of Semiconductors, Butterworths, London, 1959.

44.   M. J. Stephen and A. B. Lidiard, J. Phys. Chem. Solids, 9:43 (1959).

45.   B. Lax and S. Zwerdling, Progress in Semiconductors, 5:221 (1960).

46.   Yu. I. Ukhanov and Yu. V. Mal'tsev, FTT, 5:2926 (1963).

47.   Yu. I. Ukhanov and Yu. V. Mal'tsev, Izv. AN SSSR, Ser. Fiz., 28(6):989(1964).

48.   M. Tanenbaum and H. B. Briggs, Phys. Rev., 91:1561 (1953).

49.   H. J. Hrostowski, G. H. Wheatley, and W. F. Flood, Phys. Rev., 95:1683 (1954).

50.   R. G. Breckenridge, R. F. Blunt, W. R. Hosler, P. R. Frederikse, et al., Phys. Rev., 96:571 (1954).

51.   W. Kaiser and H. Y. Fan, Phys. Rev., 98:966 (1955).

52.   I. Kudman and L. J. Vieland, J. Phys. Chem. Solids, 24:967 (1963).

53.   V. Roberts and J. E. Quarrington, J. Electronics, 1:152 (1955).

54.   F. Stern and R. M. Talley, Phys. Rev., 100:1683 (1955).

55.   R. Newman, Phys. Rev., 111:1518 (1958).

56.   W. J. Turner and W. E. Reese, J. Appl. Phys., 35:350 (1964)

57.   D. E. Hill, Phys. Rev., 133:A866 (1964).

58.   E. O. Kane, J. Phys. Chem. Solids, 1:249 (1957).

59.   F. Stern, J. Appl. Phys., 32:2166 (1961).

60.   S. W. Kurnick and J. M. Powell, Phys. Rev., 116:597 (1959).

61.   J. R. Dixon and D. P. Enright, Bull. Am. Phys. Soc., 3:255 (1958).

62.   W. G. Spitzer and J. M. Whelan, Phys. Rev., 114:59 (1959).

63.   W. J. Turner and W. E. Reese, Phys. Rev, 117:1003 (1960).

64.   S. Visvanathan, Phys. Rev., 120:376 (1960).

65.   O. Madelung, Physics of III−V Compounds, Wiley, New York, 1964.

66.   H. Yoshinaga and R. Oetjen, Phys. Rev., 101:526 (1956).

67.   W. G. Spitzer and H. Y. Fan, Phys. Rev., 106:882 (1957).

68.   A. G. Chynoweth and R. A. Logan, Phys. Rev., 118:1470 (1961).

69.   A. R. Calawa, R. H. Rediker, B. Lax, and A. L. McWhorter, Phys. Rev. Letters, 5:55 (1961).

CHAPTER 5

1.   F. A. Trumbore, Bell System Tech. J., 39:205 (1960).

2.   B. I. Boltaks, Diffusion in Semiconductors, Infosearch, London, 1963.

3.   H. Reiss and C. S. Fuller, J. Metals, 8:276 (1956).

4.   H. Reiss, C. S. Fuller, and F. J. Morin, Bell System Tech. J., 35:535 (1956).

5.   J. Teltow, Ann. Phys., 5:63 (1949).

6.   N. B. Hannay (ed.), Semiconductors, Reinhold, New York, 1959, Chap. V.

7.   R. L. Longini and R. F. Greene, Phys. Rev., 102:992 (1956).

8.   W. M. Valenta and C. Ramasastry, Phys. Rev., 106:73 (1957).

9.   Y. Furukawa, J. Phys. Soc. Japan, 16:687 (1961).

10.   W. G. Spitzer, F. A. Trumbore, and R. A. Logan, J. Appl. Phys., 32:1822 (1961).

11.   V. I. Fistul' and É. M. Omel'yanovskii, FTT, 4:1370 (1962).

12. V. I. Fistul' and K. V. Cherkas, FTT, 4:3288 (1962).

13. É. M. Omel'yanovskii, V. I. Fistul', and M. G. Mil'vidskii, FTT, 5:921(1963).

14. É. M. Omel'yanovskii and V. I. Fistul', Zavodskaya laboratoriya, 30:559 (1964).

15. G. L. Pearson, J. D. Struthers, and H. C. Theuerer, Phys. Rev., 77:809 (1950).

16. G. L. Pearson and J. Bardeen, Phys. Rev., 75:865 (1949).

17. V. I. Fistul' and M. G. Mil'vidskii, Paper presented at Conf. on Degenerate Semiconductors [in Russian], Moscow, 1962.

18. V. I. Fistul', M. G. Mil'vidskii, É. M. Omel'yanovskii, and S. P. Grishina, DAN SSSR, 149(5):1119 (1963).

19. L. J. Vieland and I. Kudman, J. Phys. Chem. Solids, 24:437 (1965).

20. V. I. Fistul', É. M. Omel'yanovskii, O. V. Pelevin, and V. B. Ufimtsev, Izv. AN SSSR, Neorg. materialy, 2:657 (1966).

21. O. A. Golikova, B. Ya. Moizhes, and A. G. Orlov, FTT, 4:3482 (1962).

22. O. A. Golikova, Dissertation for Candidate's Degree [in Russian], LPI, Leningrad, 1964.

23. V. I. Fistul', Doctoral Dissertation [in Russian], Moscow, 1965.

24. N. G. Karpel', Zavodskaya laboratoriya, 30:1078 (1964).

25. J. Black, J. Electrochem. Soc., 8:924 (1964).

26. C. Elbaum, UFN, 79(3):545 (1963).

27. M. S. Chupakhin, G. G. Glavin, and V. I. Fistul', DAN SSSR, 150(5):1059 (1963).

28. M. S. Chupakhin, G. G. Glavin, and V. I. Fistul', Collection: Chemical Binding in Semiconductors and Solids [in Russian], "Nauka i tekhnika," Minsk, 1965.

29. R. Brown et al., Compound Semiconductors, Reinhold, New York, 1962, Vol. 1, p. 106.

30. E. Erhard and A. Adler, Z. Metall., 52:529 (1961).

31. V. M. Glazov and V. N. Vigdorovich, Microhardness of Metals [in Russian], Metallurgizdat, 1962.

32. H. Kodera, Japan J. Appl. Phys., 2:193 (1963).

33. J. Takashi and K. Makoto, Japan J. Appl. Phys., 2:194 (1963).

34. M. G. Mil'vidskii, O. G. Stolyarov, and A. V. Berkova, FTT, 6:3259 (1964).

35. T. Furuoga, Japan J. Appl. Phys., 1:135 (1962).

36. W. Bardsley, J. M. Callan, et al., Solid State Electronics, 3:142 (1961).

37. W. A. Tiller and J. W. Rutter, Can. J. Phys., 31:15 (1953).

38. V. V. Voronkov, FTT, 6:2984 (1964).

39. M. G. Mil'vidskii and V. V. Voronkov, FTT, 6:3736 (1964).

40. M. G. Mil'vidskii and S. P. Grishina, FTT, 6:483 (1964).

41. M. G. Mil'vidskii, V. I. Fistul', and S. P. Grishina, FTT, 6:2762 (1964).

42. W. C. Dash, J. Appl. Phys., 27:1153 (1956)

43. V. G. Fomin, M. G. Mil'vidskii, S. P. Grishina, N. S. Belyatskaya, and M. A. Gurevich, Kristallografiya, 9:219 (1964).

44. V. I. Fistul' and A. M. Agaev, FTT, 7:3042 (1965).

45. T. P. Brody, J. Appl. Phys., 33:100 (1962).

46. N. A. Belova, Radiotekhnika i élektronika, 8:2091 (1963).

47. N. A. Belova, Dissertation for Candidate's Degree [in Russian], Radio Engineering and Electronics Institute, Academy of Sciences of the USSR, Moscow, 1964.
48. Ya. I. Frenkel', Kinetic Theory of Liquids [in Russian], Izd. AN SSSR, 1945.
49. V. I. Danilov, Structure and Crystallization of Liquids [in Russian], Izd. AN UkrSSR, 1956.
50. E. G. Shvidkovskii, Some Problems in the Viscosity of Molten Metals [in Russian], Gostekhizdat, 1955.
51. N. S. Greengrich, Uspekhi khimii, Vol. 15, No. 3 (1946).
52. G. M. Martynkevich, Dissertation for Candidate's Degree [in Russian], Moscow State University, Moscow, 1961.
53. A. N. Nesmeyanov, Vapor Pressure of Chemical Elements [in Russian], Izd. AN SSSR, 1961.
54. I. B. Borovskii and N. P. Il'in, Zavodskaya laboratoriya, 23, No. 9 (1957).

CHAPTER 6

1. J. W. Rutter and B. Chalmers, Can. J. Phys., 31:15 (1953).
2. W. G. Cochran, Proc. Cambridge Phil. Soc., 30:365 (1934).
3. J. A. Burton et al., J. Chem. Phys., 21:1987 (1953).
4. D. T. J. Hurle, Solid State Electronics, 3:37 (1961).
5. H. Kodera, Japan J. Appl. Phys , 2(9):527 (1963).
6. W A. Tiller et al., Acta Metallurgica, 1:428 (1953).
7. W. A. Tiller and J. W. Rutter, Can. J. Phys., 34(1):96 (1956)
8. C. Elbaum, Progr. Metal. Phys., 8:203 (1959).
9. D. E. Temkin, Collection: Crystallization and Phase Transitions [in Russian], Izd. AN BSSR, Minsk, 1962.
10. V. V. Voronkov, FTT, 6:2984 (1964).
11. M. G. Mil'vidskii and V. V. Voronkov, FTT, 6:3736 (1964).
12. D. Walton et al., J. Metals, 7:1023 (1955).
13. T. S. Plaskett and W. C. Winegard, Can. J. Phys., 37:1555 (1959).
14. G. Mortimer, J. Electrochem. Soc., 105:739 (1958).
15. V. G. Alekseeva and P. Ya. Eliseev, FTT, 1:1135 (1960).
16. F. A. Trumbore, J. Electrochem. Soc., 103:597 (1956).
17. W. W. Tyler and H. H. Woodbury, Phys. Rev., 96:874 (1954).
18. W. W. Tyler et al., Phys. Rev. 97:669 (1955).
19. H. H. Woodbury and W. W. Tyler, Phys. Rev., 100:659 (1955).
20. S. G. Ellis, J. Appl. Phys., 26:1140 (1955).
21. F. D. Rosi, RCA Rev., 19:349 (1958).
22. G. F. Bolling et al., Can. J. Phys., 34:234 (1956).
23. W. Bardsley et al. Solid State Electronics, 3:142 (1961).
24. B. M. Turovskii, ZhFKh, 36:8 (1962).
25. H. Kodera, Japan J. Appl. Phys., 2:212 (1963).
26. M. G. Mil'vidskii and V. V. Eremeev, FTT, 6:1962 (1964).
27. J. C. Brice and P. A. C. Whiffin, Solid State Electronics, 7:183 (1964).
28. H. R. Shanks et al., Phys. Rev., 130:1743 (1963).
29. H. Kodera, Japan J. Appl. Phys., 2:527 (1963).

30. C. D. Thurmond and M. Kowalchik, Bell System Tech. J., 39:169 (1960).

31. F. A. Trumbore, Bell System Tech. J., 39:205 (1960).

32. V. I. Fistul', M. G. Mil'vidskii, S. P. Grishina, and É. M. Omel'yanovskii, DAN SSSR, 149:1119 (1963).

33. D. J. Turnbull and R. E. Cech, J. Appl. Phys., 21:840 (1950).

34. J. Leverton, J. Appl. Phys., 29:1241 (1958).

35. S. V. Airapetyants and G. I. Shmelev, FTT, 2:4 (1960).

36. G. R. Blackwell, J. Electron. and Control, 10:459 (1961).

37. V. N. Romanenko, Izd. AN SSSR, Metallurgiya i gornoe delo, 2:86 (1963).

38. D. A. Petrov and V. S. Zemskov, Growth of Crystals. Vol. 1, Consultants Bureau, New York, 1959, p. 209.

39. D. A. Petrov and B. L. Kalachev, Proc. Conf. on Experimental Techniques [in Russian], Moscow, 1959, p. 513.

40. W. Nelson, Transistors, RCA Laboratories, 1956, Vol. 1.

41. W. G. Pfann, Zone Melting, 2nd ed., Wiley, New York, 1966.

42. S. E. Bradshaw and A. J. Mlawsky, J. Electronics, 2:134 (1956).

43. A. Trainor and P. Harries, Proc. Phys. Soc. (London), 74:669 (1959).

44. Yu. M. Shashkov, Metallurgy of Semiconductors [in Russian], Metallurgizdat, 1960, p. 67.

45. D. A. Petrov et al., Proc. Third Conf. on Semiconducting Materials [in Russian], Izd. AN SSSR, 1961, p. 187.

46. J. Bridgers, J. Appl. Phys., 27:746 (1956).

47. D. A. Petrov, Zhurn. Vsesoyuzn. khimichesk. obshchestva im. D. I. Mendeleeva, 5:550 (1960).

48. B. M. Turovskii and M. G. Mil'vidskii, FTT, 3:2519 (1961).

49. M. G. Mil'vidskii, S. P. Grishina, and V. V. Eremeev, Izv. AN SSSR, Neorganicheskie materialy, No. 10 (1965).

50. J. A. M. Dikhoff, Philips Tech. Rev., 25:195 (1963/64).

51. H. Ueda, J. Phys. Soc. Japan, 16:61 (1961).

52. H. C. Gatos et al., J. Appl. Phys., 32:2057 (1961).

53. G. V. Komarov and A. R. Regel', FTT, 5:773 (1963).

54. V. V. Voronkov, FTT, 6:3736 (1964).

55. M. G. Mil'vidskii, Kristallografiya, 6:249 (1961).

56. F. Gutberlet-Vieweg, Z. Physik, 179:425 (1964).

57. W. R. Wilcox, Electronics News, No. 473, Vol. 10, 1, Sec. 1, p. 5 (1965).

58. R. N. Hall, J. Phys. Chem., 57:86 (1953).

59. K. F. Hulme and J. B. Mullin, Phil. Mag. 4:1286 (1959).

60. K. F. Hulme and J. B. Mullin, J. Phys. Chem.,Solids, 17:1 (1960).

61. W. P. Allred and R. T. Rate, J. Electrochem. Soc., 108:258 (1961).

62. M. G. Mil'vidskii and A. V. Berkova, FTT, 5:709 (1963); 5:513 (1963).

63. C. Z. Lemay, J. Appl. Phys. 34:439 (1963).

64. J. A. M. Dikhoff, Solid State Electronics, 1:202 (1960).

65. A. Trainor and B. E. Bartlett, Solid State Electronics, 2:106 (1961).

66. K. F. Hulme and J. B. Mullin, Solid State Electronics, 5:211 (1962).

67. W. A. Tiller, Acta Met., 10:681 (1962).

68. W. A. Tiller, Growth and Perfection of Crystals, New York, 1958, p. 332.
69. M. G. Mil'vidskii and S. P. Grishina, FTT, 6:483 (1964).
70. M. G. Mil'vidskii, V. G. Fomin, and S. P. Grishina, Kristallografiya, 3:9 (1964).
71. A. I. Goss et al., Acta Met., 4:332 (1956).
72. W. A. Tiller, J. Appl. Phys., 29:611 (1958).
73. W. T. Read, Dislocations in Crystals [Russian translation], Metallurgizdat, 1957, p. 199. [English edition: McGraw-Hill, New York, 1953.]
74. Ya. S. Umanskii et al., Physical Metallography [in Russian], Metallurgizdat, 1955, p. 138.
75. M. G. Mil'vidskii, V. I. Fistul', and S. P. Grishina, FTT, 6:2762 (1964).
76. P. E. Schmidt and R. Stickler, J. Electrochem. Soc., 111:1188 (1964).
77. H. Kodera, Japan J. Appl. Phys., 2:193 (1963).
78. T. Jizuka and M. Kikuchi, Japan J. Appl. Phys., 2:194 (1963).
79. F. A. Trumbore, J. Electrochem. Soc., 107:198C (1960).
80. C. D. Thurmond and M. Kowalchik, Bell System Tech. J., 39:169 (1960).
81. U. Merten and A. P. Hatcher, J. Phys. Chem. Solids, 23:533 (1962).
82. U. Merten and A. P. Hatcher, J. Phys. Chem. Solids, 24:1064 (1963).
83. M. G. Mil'vidskii and O. V. Pelevin, Izv. AN SSSR, Neorganicheskie materialy, 1:1454 (1965).
84. J. R. Patel et al., Metallurgy of Elemental and Compound Semiconductors, New York, 1960, p. 45.
85. M. G. Mil'vidskii, O. G. Stolyarov, and A. V. Berkova, FTT, 6:3259 (1964).
86. W. C. Dash, J. Appl. Phys., 30:459 (1959).
87. C. S. Fuller, Collection: Semiconductors (ed. by N. B. Hannay), Reinhold, New York, 1959.
88. R. L. Longini and R. F. Greene, Phys. Rev., 102:992 (1956).
89. M. G. Mil'vidskii, O. G. Stolyarov, and A. V. Berkova, FTT, 6:3170 (1964).
90. M. G. Mil'vidskii, V. B. Osvenskii, and O. G. Stolyarov, Izv. AN SSSR, Neorganicheskie materialy, 1, No. 9 (1965).
91. A. Steineman and U. Zimberlis, Solid State Electronics, 6:597 (1963).
92. G. E. Brock and C. F. Aliotta, IBM J. Res. Develop., 6:372 (1962).
93. V. G. Alekseeva and P. G. Eliseev, FTT, 1:1195 (1960).
94. S. Prussin, J. Appl. Phys., 32:1876 (1961).
95. J. Washburn and G. Thomas, J. Appl. Phys., 35:1909 (1964).
96. H. Ino et al., Japan J. Appl. Phys., 3:692 (1964).
97. M. L. Joshi and F. Welhelm, J. Electrochem. Soc., 112:185 (1965).
98. J. Black and P. Lublin, J. Appl. Phys., 35:2462 (1964).
99. F. A. Trumbore, Metallurgy of Elemental and Compound Semiconductors, New York, 1960, p. 15.
100. V. S. Zemskov et al., Proc. Conf. on Impact Ionization and Tunnel Effect in Semiconductors [in Russian], Izd. AN AzSSR, Baku, 1962, p. 130.
101. E. J. Mehalchick and P. K. Marshall, J. Electrochem. Soc., 109:267 (1962).
102. J. van der Boomgaard, Phillips Res. Rep., 10:319 (1955).
103. J. Coorissen and A. M. Van Run, The Institution of Electrical Engineers, Paper No. 3022F, March, 1960.

104. F. A. Trumbore and E. M. Porbansky, J. Appl. Phys., 31:2068 (1960).
105. W. Heinke, Z. Angew. Phys., 15:128 (1963).
106. F. A. Trumbore et al., J. Phys. Chem. Solids, 11:239 (1959).
107. Compound Semiconductors (ed. R. K. Willardson and H. L. Goering), Reinhold, New York, 1962, Vol. 1, p. 385.
108. M. G. Mil'vidskii and O. V. Pelevin, Izv. AN SSSR, Neorganicheskie materialy, Vol. 1, No. 9 (1965).
109. S. Nielsen et al., Semicond. Prod., 6:35 (1965).
110. V. J. Silvestri and F. Fang, J. Electrochem. Soc., 111:1164 (1964).
111. F. V. Willians, Solid State Electronics, 7:833 (1964).
112. H. S. Veloric et al., J. Appl. Phys., 27:895 (1956).
113. K. Lehovec and A. Levitas, J. Appl. Phys., 28:106 (1957).
114. J. Halpern and R. H. Rediker, Proc. IRE, 46:1068 (1958).
115. H. Rupprecht and C. Z. Lemay, J. Appl. Phys., 35:1970 (1964).

CHAPTER 7

1. W. Shockley, Electrons and Holes in Semiconductors, Van Nostrand, New York, 1950.
2. B. M. Vul, FTT, 2:2961 (1961).
3. A. G. Chynoweth, Progress in Semiconductors, 4:95 (1960).
4. L. Esaki, Phys. Rev., 109:603 (1958).
5. É. V. Shpol'skii, Atomic Physics [in Russian], Gostekhizdat, 1951, Vol. 1.
6. I. A. Lesk, N. Holonyak, et al., IRE Wescon Conv. Record, Pt. 3, p. 9 (1959).
7. C. Zener, Proc. Roy. Soc. (London), 145:523 (1934).
8. K. B. McAfee et al., Phys. Rev., 83:650 (1951).
9. H. J. Hartman, M. Michelitsch, and W. Steinhäuser, Arch. d. Electr. Übertragung, 15(3):125 (1961).
10. W. Franz and L. Tewordt, Halbleiterprobleme, Braunschweig, 1956, Vol. III.
11. L. V. Keldysh, ZhÉTF, 33:994 (1957).
12. L. V. Keldysh, ZhÉTF, 34:962 (1958).
13. R. N. Hall, J. Bardeen, and F. J. Blatt, Phys. Rev., 95:559 (1959).
14. C. W. Bates, Phys. Rev., 121:1070 (1961).
15. I. I. Ivanchik, FTT, 3:103 (1961).
16. P. J. Price and J. H. Radcliffe, IBM J. Res. Develop., 3:364 (1959).
17. E. O. Kane, J. Appl. Phys., 32:83 (1961).
18. V. L. Bonch-Bruevich, Radiotekhnika i élektronika, 5:2033 (1960).
19. F. A. Trumbore and A. A. Tartaglia, J. Appl. Phys., 29:1511 (1958).
20. L. Esaki, Solid State Physics in Electronics and Telecommunications, Academic Press, London, 1960, Vol. 1.
21. N. Holonyak and I. A. Lesk, Proc. IRE, 48:1405 (1960).
22. R. A. Smith, Semiconductors, Cambridge University Press, 1961.
23. H. S. Sommers, Proc. IRE, 47:1201 (1959).
24. R. N. Hall, Proc. IRE, 40:1512 (1952).
25. R. F. Rutz, IBM J. Res. Develop., 3:372 (1959).
26. R. N. Rubinshtein and V. I. Fistul', Zavodskaya laboratoriya, 27:1242 (1961).

27. M. Michelitsch, Naturwiss., 47:274 (1960).

28. V. I. Fistul' and I. D. Abezgauz, Proc. Conf. on Impact Ionization and Tunnel Effect in Semiconductors [in Russian], Izd. AN AzSSR, Baku, 1962, p. 151.

29. H. S. Sommers et al., IRE Wescon Conv. Record, Pt. 3, p. 3 (1959).

30. G. Dermit, H. Lockword, and W. Hauer, Proc. IRE, 49:519 (1961).

31. R. C. Sims, E. R. Beck, et al., Proc. IRE, 49:136 (1961).

32. V. I. Fistul' and N. Z. Shvarts, UFN, 77:109 (1962).

33. H. S. Sommers, Solid State Conference, Cornell University, 1959.

34. N. A. Belova and A. N. Kovalev, Radiotekhnika i élektronika, 6:160 (1961).

35. Y. Furukawa, J. Phys. Soc. Japan, 15:730 (1960).

36. W. Cady, Colloque International sur les Dispositifs à Semiconducteurs, Paris, February 20-25, 1961.

37. E. Spenke, Electronic Semiconductors, McGraw-Hill, New York, 1958.

38. R. N. Hall, IRE Transactions ED7, p. 1 (1960).

39. B. M. Vul, A. P. Shotov, and S. P. Grishechkina, FTT, 3:667 (1961).

40. N. Holonyak, I. A. Lesk, et al., Phys. Rev. Letters, 3(4):167 (1959).

41. T. Yajima and L. Esaki, J. Phys. Soc. Japan, 13:1281 (1958).

42. S. L. Miller et al., Phys. Rev. Letters, 4:60 (1960).

43. R. A. Logan and A. G. Chynoweth, Bull. Amer. Phys. Soc., 5:160 (1960).

44. A. G. Chynoeth et al., .. Phys. Rev., 121:684 (1961).

45. E. O. Kane, Proc. Intern. Conf. on Semiconductors, Prague, 1960.

46. W. Franz, Handbuch der Physik, Berlin, 1956, Vol. 17, p. 1955.

47. N. Holonyak, J. Appl. Phys., 32:130 (1961).

48. Proc. Conf. on Tunnel Diodes, USA, February, 1961.

49. V. I. Fistul', FTT. 6:3738 (1964).

50. V. I. Fistul' and A. M. Agaev, FTT, 7:3691 (1965).

51. E. Spenke, Electronic Semiconductors, McGraw-Hill, New York, 1958.

52. V. S. Davidson et al., Electronic Design, February, 3 (1960).

53. A. G. Chynoweth, Phys. Rev., 120(5):1620 (1960).

54. Fukui, Dénki tsusin gakkai dzassi (in Japanese), 43:1351 (1961).

55. Electronic News, 199:22 (1960).

56. V. I. Fistul', Doctoral Dissertation [in Russian], Moscow, 1965.

57. Nachrichten Tech. Z. (NTZ), 13:3-7 (1960).

58. Electronic News, 4(175):13 (1959); 4(176):41 (1959).

59. L. Esaki, Paper presented at Twelfth General Assembly, London, 1960.

60. A. A. Andronov, A. A. Vitt, and S. É. Khaĭkin, Theory of Oscillations [in Russian], Fizmatgiz, 1959.

61. H. W. Bode, Network Analysis and Feedback Amplifier Design, 1945.

62. M. A. Aizerman, Lectures on the Theory of Automatic Control [in Russian], Fizmatgiz, 1958.

63. A. K. Baum, I. Ya. Bilinskii, and P. P. Treis, Tunnel Diodes in Industrial Electronic Circuits [in Russian], "Energiya," 1965.

64. Collection: Tunnel Diodes (ed. V. I. Fistul') [Russian translation], IL, 1961.

65. J. V. Morgan and E. O. Kane, Phys. Rev. Letters, 3:466 (1959).

66. E. O. Kane, Proc. Intern. Conf. on Semiconductors, Prague, 1960.

67. A. G. Chynoweth et al., Phys. Rev. Letters, 5:548 (1960).

68. A. R. Calawa et al., Phys. Rev. Letters, 5:55 (1960).

69. M. J. Nathan and S. L. Miller, Bull. Am. Phys. Soc., Ser. II, 5:265 (1960).

70. A. M. Agaev and V. I. Fistul', FTT, 7:3042 (1965).

71. V. L. Bonch-Bruevich and P. S. Serebrennikov, Radiotekhnika i élektronika, 6:2041 (1961).

72. V. L. Bonch-Bruevich and P. S. Serebrennikov, Radiotekhnika i élektronika, 8:1002 (1963).

73. S. T. Eng, IRE Trans. MTT5, p. 419 (1961).

74. M. M. Kaufman, Solid State Design, 3(2):23 (1962).

75. J. Przybylski and G. N. Roberts, J. Brit. IRE, 22:479 (1961).

76. E. Gottlieb and J. Giorgis, J. Electronics, 36(27):26 (1963).

77. W. C. Folmer, ·Proc. IRE, 49:1939 (1961).

78. L. A. Logunov, N. K. Rudneva, and E. B. Chernyak, Radiotekhnika i élektronika, 9:1258 (1964).

79. L. A. Logunov, Dissertation for Candidate's Degree [in Russian], Moscow State University, Moscow, 1964.

80. I. A. Lesk and H. A. Jensen, Solid State Electronics, 1(3):183 (1960).

81. A. Tewes, Radio und Fernsehen, 11:314 (1962).

82. O. N. Krokhin and Yu. M. Popov, Supplement to B. A. Lengyel's "Lasers" [in Russian], "Mir," 1964.

83. M. I. Nathan, W. P. Dumke, et al., Appl. Phys. Letters, 1:62 (1962).

84. R. N. Hall, G. E. Fenner, et al., Collection: Lasers [Russian translation], IL, 1963, p. 176.

85. T. M. Quist, R. N. Rediker, et al., Appl. Phys. Letters, 1:91 (1962).

86. J. R. Haynes, M. Lax, and W. Flood, Proc. Intern. Conf. on Semicond. Physics, 1961, p. 423.

87. N. G. Basov, O. N. Krokhin, and Yu. M. Popov, ZhÉTF, 39:1486 (1960).

88. N. G. Basov, O. N. Krokhin, and Yu. M. Popov, ZhÉTF, 40:1203 (1961).

89. O. N. Krokhin, FTT, 4:822 (1962).

90. J. I. Pankove and M. Massoulie, Bull. Am. Phys. Soc., 7:88 (1962).

91. D. N. Nasledov, A. A. Rogachev, S. M. Ryvkin, and B. V. Tsarenkov, FTT, 4:1062 (1962).

92. N. G. Basov, O. N. Krokhin, and Yu. M. Popov, ZhÉTF, 40:1879 (1961).

93. K. Weiser and R. S. Levitt, Bull. Am. Phys. Soc., 8:29 (1963).

94. I. Melngailis, Bull. Am. Phys. Soc., 8:202 (1963).

95. B. M. Vul, A. P. Shotov, and B. S. Bagaev, FTT, 4:3676 (1962).

96. T. Deutsch, R. C. Ellis, and D. M. Warschauer, Phys. Status Solidi, 3:1001(1963).

97. D. F. Nelson, M. Gershenson, et al., Appl. Phys. Letters, 2:182 (1963).

98. N. Holonyak and S. F. Bevacqua, Appl. Phys. Letters, 1:82 (1962).

99. I. Melngailis et al., TIIRÉ [Russian translation], 51:1157 (1962).

100. K. Weiser and R. S. Levitt, Appl. Phys. Letters, 2:178 (1963).

101. I. Melngailis, Appl. Phys. Letters, 2:176 (1963).

102. G. Burns and M. I. Nathan, Proc. IEE, 52:770 (1964).

103. L. M. Vallese, Semicond. Prod., 6:25 (1963).

104. R. H. Rediker, Solid State Design, 4:19 (1963).

105. B. Lax, Science, 141:1247 (1963).

106. L. Libov, S. Meskin, D. Nasledov, V. Sedov, and B. Tsarenkov, PTÉ, No. 4, p. 14 (1965).

107. A. F. Ioffe, Semiconductor Thermoelements [in Russian], Izd. AN SSSR, 1956.

108. H. J. Goldsmid, Thermoelectric Refrigeration, Plenum Press, New York, 1964.

109. V. S. Vavilov, Effects of Radiation on Semiconductors, Consultants Bureau, New York, 1965.

110. V. P. Zhuze, Collection: Semiconductors in Science and Technology [in Russian], Izd. AN SSSR, 1957, Vol. I.

111. V. I. Fistul', Doctoral Dissertation [in Russian], Moscow, 1965.

112. C.S. Smith, Phys. Rev., 94:42 (1954).

113. A. Sagar, Phys. Rev., 117:101 (1960).

114. C. Herring and E. Vogt, Collection: Problems in Physics of Semiconductors [Russian translation], IL, 1957.

115. C. Herring, Bell System Tech. J., 34:237 (1955).

116. V. S. Shadrin, Dissertation for Candidate's Degree [in Russian], Novosibirsk, 1964.

117. A. F. Gorodetskii, V. S. Shadrin, M. I. Iglitsyn, and V. I. Fistul', Author's Certificate No. 164698 (1964).

118. M. Dean, Semiconductor Strain Gauges (ed. by A. F. Gorodetskii) [Russian translation], "Energiya," 1965.

Index